MW00683062

Multivendor Networking

Linking PCs, Minis, and Mainframes over LANs and WANs

Related Titles

ISBN	AUTHOR	TITLE
0-07-0005119-4	Baker	*Network Security: How to Plan for It and Achieve It*
0-07-005089-9	Baker	*Networking the Enterprise: How to Build Client/Server Systems That Work*
0-07-004128-8	Bates	*Disaster Recovery Planning: Networks, Telecommunications, and Data Communications*
0-07-004674-3	Bates	*Wireless Networked Communications: Concepts, Technology, and Implementation*
0-07-005076-7	Berson	*Client/Server Architecture*
0-07-005203-4	Berson/Anderson	*SYBASE and Client/Server Computing*
0-07-005570-X	Black	*Network Management Standards: SNMP, CMOT, and OSI, 2nd Ed.*
0-07-005560-2	Black	*TCP/IP and Related Protocols, 2nd Ed.*
0-07-005592-0	Black	*The V Series Recommendations: Standards for Data Communications Over the Telephone Network, 2nd Ed.*
0-07-005593-9	Black	*The X Series Recommendations: Standards for Data Communications, 2nd Ed.*
0-07-015839-8	Davis/McGuffin	*Wireless Local Area Networks: Technology, Issues, and Strategies*
0-07-020359-8	Feit	*SNMP: A Guide to Network Management*
0-07-020346-6	Feit	*TCP/IP: Architecture, Protocols, and Implementation*
0-07-021625-8	Fortier	*Handbook of LAN Technology, 2nd Ed.*
0-07-024842-7	Grinberg	*Computer/Telecom Integration: The SCAI Solution*
0-07-033754-3	Hebrawi	*OSI Upper Layer Standards and Practices*
0-07-032385-2	Jain/Agrawala	*Open Systems Interconnection: Its Architecture and Protocols, rev. ed.*
0-07-035968-7	Kumar	*Broadband Communications: A Professional's Guide to ATM, Frame Relay, SMDS, SONET, and B-ISDN*
0-07-060362-6	McDyson/Spohn	*ATM: Theory and Applications*
0-07-042586-8	Minoli	*1st, 2nd, and Next Generation LANs*
0-07-042591-4	Minoli/Vitella	*ATM and Cell Relay Service for Corporate Environments*
0-07-046461-8	Naugle	*Network Protocol Handbook*
0-07-911889-5	Nemzow	*Enterprise Network Performance Optimization*
0-07-046321-2	Nemzow	*The Token-Ring Management Guide*
0-07-049663-3	Peterson	*TCP/IP Networking: A Guide to the IBM Environment*
0-07-057063-9	Schatt	*Linking LANs, 2nd Ed.*
0-07-911880-1	Signor/Creamer/Stegman	*The ODBC Solution: Open Database Connectivity in Distributed Environments*
0-07-057442-1	Simonds	*McGraw-Hill LAN Communications Handbook*
0-07-060360-X	Spohn	*Data Network Design*
0-07-067375-6	Vaughn	*Client/Server System Design and Implementation*

Multivendor Networking

Linking PCs, Minis, and Mainframes over LANs and WANs

Dr. Andres Fortino, P.E.
Jerry Golick

McGraw-Hill
New York San Francisco Washington, D.C. Auckland Bogotá
Caracas Lisbon London Madrid Mexico City Milan
Montreal New Delhi San Juan Singapore
Sydney Tokyo Toronto

McGraw-Hill

A Division of The ***McGraw-Hill*** *Companies*

©1996 by The McGraw-Hill Companies, Inc.

Printed in the United States of America. All rights reserved. The publisher takes no responsibility for the use of any materials or methods described in this book, nor for the products thereof.

hc 1 2 3 4 5 6 7 8 9 DOC/DOC 9 0 0 9 8 7 6 5

Product or brand names used in this book may be trade names or trademarks. Where we believe that there may be proprietary claims to such trade names or trademarks, the name has been used with an initial capital or it has been capitalized in the style used by the name claimant. Regardless of the capitalization used, all such names have been used in an editorial manner without any intent to convey endorsement of or other affiliation with the name claimant. Neither the author nor the publisher intends to express any judgment as to the validity or legal status of any such proprietary claims.

Library of Congress Cataloging-in-Publication Data

Fortino, Andres G.

Multivendor networking : linking PCs, minis, and mainframes over LANS and WANS / by Andres Fortino, Jerry Golick.

p. cm.

Includes index.

ISBN 0-07-912190-X (hc)

1. Computer networks. I. Golick, Jerry. II. Title.

TK5105.5.F67 1995

004.6—dc20 95-32898

CIP

McGraw-Hill books are available at special quantity discounts to use as premiums and sales promotions, or for use in corporate training programs. For more information, please write to the Director of Special Sales, McGraw-Hill, 11 West 19th Street, New York, NY 10011. Or contact your local bookstore.

Acquisitions editor: Jennifer Holt DiGiovanna

Editorial team: Marc Damashek, Editor
David M. McCandless, Managing Editor
Joanne Slike, Executive Editor
Joann Woy, Indexer

Production team: Katherine G. Brown, Director
Jan Fisher, Coding
Brenda M. Plasterer, Coding
Rhonda E. Baker, Desktop Operator
Linda L. King, Proofreading
Toya B. Warner, Computer Artist
Brenda S. Wilhide, Computer Artist

Design team: Jaclyn J. Boone, Designer
Katherine Lukaszewicz, Associate Designer

912190X
WK2

To my beloved wife, Kathleen.

AGF

To Little One, for all the unusual reasons.

JBG

Contents

Acknowledgments

This book is the product of a dozen years of association with many colleagues who have supported us in perfecting our knowledge of multivendor networks. Our gratitude goes out to all of them.

First to my friend and co-author, Jerry, who has traveled down this road with me from the beginning. Then to all those wonderful folks at Learning Tree International who put out the best educational product on the planet. Thank you, John Moriarty, Rick Adamson, Richard Beaumont, Karen Snyder, Beverly Voight, Stu Ackerman (gnarly cartoons, guy!), Bruce Wadman, Lori Sheridan, Mike Lopez, and John Ruthke. A special note of thanks to Eric Garen and David Collins for creating Learning Tree and giving us the privilege to work in their company. We wish to thank our Learning Tree students who over the years helped us shape the seminar and this book. A special note of thanks goes to our colleagues Dr. Karanjit Siyan, John McDermott, Peter Curran, Ken Avellino, Arthur Messenger, John Page and Brad Waller, as well as many other instructors who corrected these pages, informed us, taught our seminar and in general enjoyed this wonderful technical field with us.

Our gratitude rightly extends to our manuscript editor Marc Damashek, who polished our stilted prose with great care, and to our acquisitions editor, Jennifer DiGiovanna, who had great faith in us.

Lastly, I wish to acknowledge the selfless work of my wife, Kathleen, who worked as my editor and cheerleader. It was truly a joint project, and the credit belongs to her as much as to the author. Thank you for your patience and understanding, dear.

AGF

What can I possibly add to Andres' acknowledgment? Everyone at Learning Tree International has been very supportive of the entire project. In addition to all of the people Andres mentions above, I would like to add a special thanks to Judith Montagano for finding me the extra time I needed to do this work.

"Thank you" is too small a phrase for my good friend Andres Fortino. He has been a driving force in this project, and it would not have happened without him. I have always admired his competency and professionalism.

Finally to Jeffrey, Julie, Julie (yes, there are two of them!), and Robert. They gave up time with me so that I could have the time to get this job done. Thanks, kids.

JBG

Introduction

Remember when networks were simple? In those bright early days of cyberspace it was easy to build networks. Large or small networks were always implemented with a single vendor's product line. Even when plug-compatible equipment was used, integration of components was simple and straightforward. During the era of proprietary specifications, the vendor assumed responsibility for making the whole mess work together. While we may have forgotten all those late night calls to come into the office and get the system back on line, the fact is that we knew how to do it. Data communications was still a subject that could be understood by a single individual.

Well those good old days are now past. Today's data networks have grown up fast and are still growing. As with any rapid growth, we have discovered that there are some "growing pains" that are a significant part of modern network life. Today's networks are complex, support all sorts of applications and data types, link thousands of user workstations, and are being called upon to handle more traffic on a daily basis. No longer is a single vendor available who can provide for this sophisticated environment. Networks are now comprised of technologies that come from many different vendors. Even in a single LAN, the variety of products from different vendors can be amazing. While all this sounds exciting, it can be

terrifying to the individual who has been told to make all this stuff work together. That brings us to the reason that we wrote this book.

Multivendor networks are today's reality. Individuals are now being mandated to integrate these heterogeneous systems into a single, seamless enterprise system. We call this system the well-integrated network. The well-integrated network is an environment that can support products and protocols from a variety of vendors. It does this while preserving the local autonomy of individual departments or workgroups that still wish to control their local networks. This book will show you how this can be done.

This book is intended for those individuals who have been given the task of finding a way to integrate the data and applications that currently exist on seemingly incompatible platforms and networks. You may be a LAN administrator, an application designer, a network manager, a consultant, or even a local user who has become interested in this area and is looking for more information. Regardless of your background, you will learn not only the basics of building a well-integrated network but also some of the more advanced techniques that can be used to ensure that you have a solid foundation for future growth, regardless of what new technologies become available in the next few years. We will also give you insight into why things work as they do. We believe that it is important to know the background of a technology if it is to be correctly used, rather than misused.

In Chapter 1 we introduce you to the field of multivendor networking and help you determine the scope of your integration project. We also cover the terminology of network integration and provide the definitions that will be used throughout the book. We also present a simplified model of data communications that will facilitate our exploration of networking technology. Most importantly, we introduce the two principal concepts of integration: interoperability and internetworking.

Chapter 2 explores issues in the selection of standards. Unlike other books that detail the specifications, we examine the various pros and cons in the selection of a protocol as a primary integration specification. We examine the major open and proprietary protocols.

We conclude Chapter 2 by examining a number of methodologies that can be used to implement a well-integrated network, such as protocol reduction, standard backbone, and others.

Chapter 3 covers the topic of internetworking. This is a somewhat brief examination. Part of our reason was that there is already a vast amount of reference material available, and there was little more that we could contribute. Instead we examine the major issues involved in the selection of internetworking components, such as bridges, routers, and gateways.

If we were somewhat light in our treatment of internetworking, it was only to save space for Chapter 4, which focuses on the difficult issues of interoperability. To our knowledge, it represents the first attempt on anyone's part to try to formally define and list the techniques that are required to enable heterogeneous systems to share data and process information. We explore subjects such as distributed file systems, e-mail, file transfer, database access, and others. The intent is not to explain the technology, but rather to describe how it can be linked, in a seamless manner, across a wide range of vendor offerings.

Chapter 5 examines the user's desktop in detail. This device will be a primary integration component and requires special attention. Also in this chapter we provide some case studies in integration and provide sample solutions to a series of commonly faced problems.

In Chapter 6 we discuss emerging concepts in the area of multivendor integration, such as client/server and resizing. Different client/server frameworks are shown. Some of the myths of both sizing and client/server are exploded. We also examine a very exciting technology known as the Distributed Computing Environment (DCE).

Chapter 7 discusses the important issue of network management in a multivendor network. These issues are important because distributed network management will require a fundamental change in thinking on the part of network managers. A brief overview of the Simple Network Management Protocol is included. We conclude the chapter by examining both security and support issues.

Chapter 8 provides some predictions for the future. While we don't claim to have a crystal ball, there are certainly a number of trends that should be tracked carefully. We will identify them so you can track them on your own.

Most of the chapters include a "concept review" so you can test your newly-acquired knowledge. We recommend that you read the book chapter by chapter. Concepts are developed in Chapters 1 and 2 that are used in all the other chapters. This book was intended to be read from beginning to end, rather than to act as a reference book.

This book should also have a CD-ROM included. If the CD-ROM has useful publicly available data files and utilities that will be of interest to anyone involved in integration. Just to whet your appetite, it contains the complete set of Internet RFCs (Request For Comments) that was current at the time we went to press. Now you don't have to spend hours searching for the RFC you need: it will be right at your fingertips!

Both of us have been involved in the design and implementation of mutivendor networks. We have also been active as trainers, teaching others how to build well-integrated networks. We decided it was time to share our experience with as many others as we could reach. This book seemed like the best way to do it.

We hope you have as much fun reading it as we had writing it. We believe that it will provide valuable assistance in building your own well-integrated network.

Enjoy!

Andres Fortino
Jerry Golick
April, 1995

1

The motivation for integration

CHAPTER 1

WHY integrate? What is the motivating force driving many of today's organizations to connect a wide range of hardware and software into a single seamless system? What is the potential payoff, and how is it being measured? How will integration reduce operating costs, generate new revenues, or increase productivity? What sort of commitment is required, and by whom? Does integration lead to client/server, open systems, downsizing, objects, and all the latest "bleeding edge" technology? How can we separate the reality of what is possible from the hype of future fantasy?

The questions continue. What are the deliverables for an integration project? How do we set realistic milestones and estimates? How do we measure the effectiveness and efficiency of the integrated system? What are the metrics to be used and what are the industry standards that we can compare our results against?

What about the individual(s) tasked with the integration project? How does one select integrators? Should they be end users, network managers, LAN administrators, MIS staff, outside consultants, or senior management? Who has overall accountability? What is the set of skills required to successfully complete an integration project? After the system has been deployed, how will it be managed? What are the lines of authority and accountability?

Finally, given the dynamic, volatile nature of today's computer industry, how do we assure ourselves that our system will be able to meet future requirements and adapt itself to new technologies as they become available? What are the critical technological decisions that must be made? What trends should we track, and which organizations should we partner with as we build these advanced systems? How can we balance the need of local departments to maintain autonomy over their own technology against the overall organization's requirement to ensure integration of data and process?

These questions and others face any organization contemplating the creation of an enterprise system. The answers are varied. Much depends on the scope of the integration scenario. Is the integration taking place at a local level, or is an enterprise-wide task envisioned? Local integration may be a fairly straightforward task that will require only a few tools or technologies to succeed. On the other hand, the task of integrating the entire organization can be daunting, and fraught with the potential for failure. Given the risks involved, why are so many organizations actively contemplating the transition to fully integrated systems?

Benefits of a well-integrated network

Purchasing flexibility

Purchasing flexibility will be enhanced with respect to new technology. Organizations will be able to deploy the technology that is correct for a particular problem, since the well-integrated network must tolerate heterogeneous hardware and software. This ability to choose technology based on requirements rather than proprietary specifications will also put organizations in a better position when new purchases are being considered. We expect that vendors will respond by offering products that are more competitive.

Hardware leverage

Leverage of existing hardware will be improved. The introduction of new technology should not require the termination of the existing infrastructure. The well-integrated network takes advantage of existing resources through the use of bridges, routers, gateways and other internetworking technologies. Large scale computers (such as minis and mainframes) will also be integrated through a variety of techniques, including server emulation software, "front ending," POSIX compliance, data warehouse, and peer-to-peer transaction processing. These large hardware platforms still represent the best choice for many activities. Organizations may discover that the data handling capability (disk space, channels, access time, etc.) of minis and mainframes make them ideal for many of the large object and relational databases being planned. In addition, the well-integrated network ensures that new hardware purchases can be used to their full potential for as long as possible, thereby increasing the likelihood of a

positive return on investment (ROI). This will be done through the use of well-published specifications and guidelines. While the dynamic, volatile nature of user desktop hardware poses a special set of challenges, the well-integrated network offers a variety of solutions, such as multiple stacks, mixed stacks, fat/thin clients, fat/thin servers, etc. By the way, if this terminology is new to you, don't panic. We will be discussing it all as we go through the book.

Software leverage

A well-integrated network will "surround" existing software (such as the so-called "legacy" applications) in a seamless manner so that they can be used as long as they remain productive. When the time comes to replace these application systems, the migration will also be easier. Since many of the presentation functions will be moved to the user's desktop, it will be possible to replace the back-end process without major impact on how the application is displayed. This modular approach to integration is a prime benefit. The well-integrated network does not require tremendous amounts of new technology. It provides for the gradual introduction of this technology in a phased approach.

New business opportunities

New business opportunities will become available through the rapid exchange of information. These may include cost reduction schemes such as telecommuting and shared office space. Additional cost savings will occur through the removal of duplicate systems and data. Data will not have to be reentered. The ability to share processes and data will lead to lower administrative and support costs. Beyond cost savings, organizations will discover that having a well-integrated network will provide other business opportunities. While the phrase "knowledge is power" may be greatly overused, there is nowhere that it is more applicable than in integrated systems. As information flow increases, individuals and groups will be better able to predict and react to change. New organizational structure will be possible, such as flat structures and autonomous workgroups. The flexibility of a well-integrated network to support new business strategies and quickly

adapt to change will provide important leverage for those organizations that wish to remain competitive.

Industry segment integration

Integration does not stop within a single organization. Many industry segments have already realized the benefits of well-integrated networks. The EFT (electronic funds transfer) network links most financial institutions. We might speculate that without such a network, it would prove impossible to have a global financial community. Currently, many organizations are being linked via EDI (electronic data interchange). These systems provide the ability to move "forms," such as invoices, shipping orders, and statements between suppliers and purchasers. Better cash flow management, inventory control, and distribution are only some of the benefits.

Global integration

We may also think of integration at a global level. The implications of having a data communication network that may someday rival the phone system in magnitude are not yet clearly understood. On the other hand, early experiments such as the Internet have demonstrated that there is sufficient interest to make them appealing. The combination of computers, networks, and data/service providers is powerful and synergistic. While the size of the Internet changes from moment to moment, it is probably a safe bet to set the total number of active users well into the tens of millions. While we may not be sure of the Internet's objective (if it has one), the fact that so many individuals are contributing their time and effort to this global system is a justification in itself. The Internet works because it conforms to the rules of a well-integrated network.

Challenges in integration

Many of the benefits just stated represent a potential rather than an accomplished fact. Regardless of the size of the system to be

deployed, the integrator will be faced with a number of challenges that must be overcome. The ability of the integrator to meet these challenges successfully will largely determine the nature and degree of the benefits. Organizations are currently faced with a wide variety of computer technology which, on the surface, appears to resist any attempt at integration. (See Fig. 1-1.)

Figure 1-1

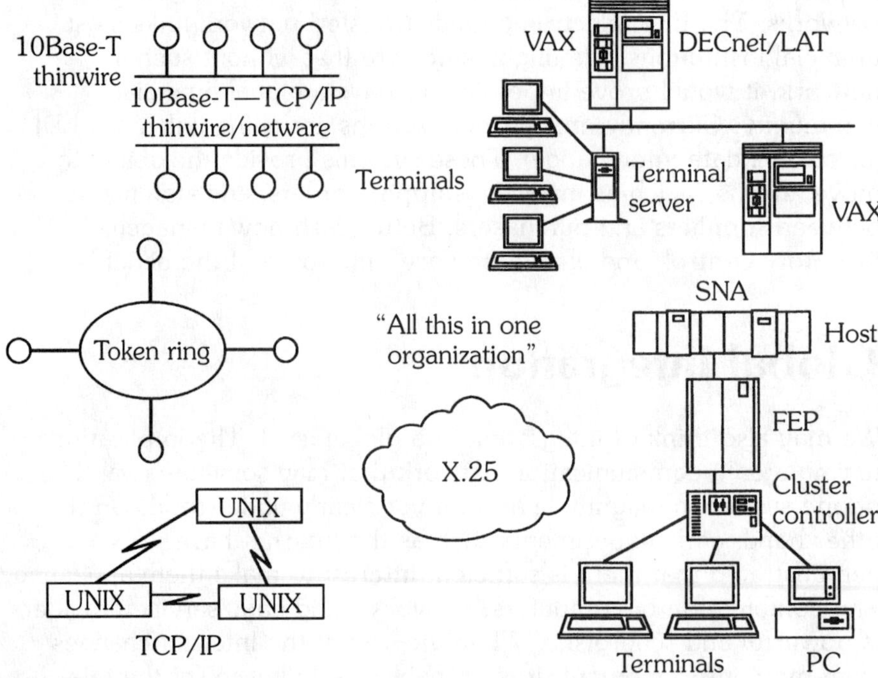

The challenge of integration. Learning Tree International

These challenges are formidable. They include hardware selection for the user desktop, server platforms, and the network. Software must be purchased or developed to implement business applications, utilities, and system processes. The support and administration of a multivendor environment may prove difficult to implement effectively. A number of factors will modify the complexity of these challenges.

Centralized vs. decentralized purchasing

Some organizations favor decentralized over centralized purchasing for office automation technology. This complicates integration, since it is more difficult to ensure that all components support common specifications. On the other hand, by staying strictly with a central purchasing model, users may not feel committed to the technology. This intangible "buy-in" is a powerful motivator, and should not be underestimated. Local autonomy over technology acquisition often leads to the level of dedication required to implement local application systems successfully. In general, central purchasing tends to favor global optimization, while decentralized purchasing favors local optimization. Traditionally, "big-ticket" items have been centrally purchased, while smaller items have been purchased locally. This is no longer true. Which costs more, a piece of mainframe software for a one-time license fee of $50,000, or one thousand copies of PC software at $50/copy? Managing this balance between local and global issues is one of the key challenges of the integrator. While guidelines and specifications may help, legitimate questions will be asked as to their source.

Shifting technology

There is no sign that we are slowing the development and adoption of new technology. Often this technology may represent a major shift in specifications or applications. For example, the introduction of video conferencing is likely to require new hardware, software, networks, and protocols. End-user demand for new services and functions continues to increase. While the early adoption of leading-edge technology may have advantages, it often presents additional challenges to the integrator. New technology is often vendor specific and proprietary. This increases the difficulty of designing integrated networks. How can we plan for an environment whose components change on a daily basis?

Accounting

Today's desktop technology and LANs represent one of the largest undocumented costs in many organizations. Financial officers have begun to ask for some form of accounting with respect to the usage of these systems. How will this be accomplished? In a well-integrated system, we must anticipate that accounting will be required for capital expenses, usage, and support. These systems have to work with a wide range of heterogeneous technology. To further complicate the issue, we are still unsure of how resource utilization should be billed. There are a variety of options, such as flat rate, usage-based, class of service, quality of service, etc. Selecting among these options and finding automated accounting packages that satisfy the requirements are difficult tasks. A further complication will likely be political, since many local departments view such systems as unnecessary "red tape."

Reliability, availability, serviceability (RAS)

As mission-critical applications are deployed on integrated networks, designers will have to ensure that they meet the same stringent RAS requirements as today's terminal-host based systems. Designing for RAS in heterogeneous systems is challenging. Which metrics should be used? Classic metrics, such as response time, throughput, MTTR (mean time to repair), etc., may be difficult to define. Even simple tasks like monitoring network utilization are challenging, since the ratio of overhead to useful data is often difficult to determine when techniques such as encapsulation are being used.

Measurement tools that support heterogeneous systems may not be available. In a multivendor environment, there is a tendency for vendors to point fingers prior to addressing a problem. The rapidly declining price of hardware technology presents challenges with respect to serviceability. Should a discard or replace strategy be followed? Software—in particular those components deployed on the user desktop—is a source of concern. As the user's workstation plays an increasingly crucial role as the primary interface to mission-critical

applications, its stability must be ensured. Since the bulk of the software portfolio on these machines is "shrink-wrap" in nature and subject to constant change, how can this be accomplished? Once again politics may enter the equation, since users may feel a degree of autonomy over their desktop environment (remember, we do call them Personal Computers!), and may resist hardware or software that has been specified by external sources.

In any event, the integrator must keep in mind that the user desktop can rarely be controlled, since it can be modified at any time by the insertion of a disk or a new hardware card. By definition, every desktop machine is a custom environment. How can RAS be maintained in this diversity? How can such an environment be managed and administered?

Security

How can security be provided in a heterogeneous environment? Today's networks are optimized for ease of access. This results in a reduction of security. As we just mentioned, the nature of the user's desktop implies that security enforcement is difficult. Each workstation represents a potential entry point for viruses, worms, and other nasty software. Authentication and validation is a concern, since passwords can be easily recorded as they traverse shared media systems like LANs. Sensitive data can be passively tapped. It is anticipated that in an integrated network, users will require access to multiple machines (i.e., servers) during their normal working day. Will multiple passwords be required? In addition, new software might contain security loopholes that can be exploited.

Role of the integrator

Table 1-1 indicates that the focus of the integrator varies, depending on who is selected to have overall accountability. In general, this is related to job function and mandate. End users wish to focus on those areas that impact their ability to access applications and data in a seamless manner, with a minimum amount of training. If local departments are funding the integration effort, it is likely that cost

Table 1-1 **The focus of different integrators.** Learning Tree International

Tasks and opportunities	Users	Developers	Network managers	Consultants
	— Productivity — Ease of use — Consistent interfaces	— Application integration — Distributed database management system (DBMS) — E-mail — Groupware — Portable systems	— Bridging — Routing — Enterprise network — Network-management systems — Virtual terminals — Multimedia	— System integration — Outsourcing — Facility management — Global networks

will also be a key focus. On the other hand, application developers are more interested in guaranteeing that the integrated environment provide portability, scalability, and flexibility of new code. Leverage of existing systems is also a key focus to the developer. A third point of view is represented by the network manager, who is often given the task of integration and support. Here the emphasis is likely to be on internetworking technology, such as bridges, routers, and gateways. The network manager will also wish to guarantee that the integrated system have sufficient capacity (bandwidth) to support the various applications deployed.

The introduction of new technology, such as multimedia, is also a prime concern. Note that outside organizations such as consultants are often given responsibility for integration. Their focus has a great deal to do with previous experience. The tendency is to use technologies and techniques that have worked well in similar situations. Outsourcing of facilities might also be contemplated. Regardless of who is given overall authority for the project, building well-integrated networks is a multidisciplinary effort requiring input from all sources. Need we be redundant in mentioning the political implications?

A further word about the integrator is required.

Selection of an integrator poses its own challenges. Developing in-house expertise promotes well-designed custom solutions that are well-suited to the needs of the organization. In addition, in-house staff are able to provide long-term commitment to the support and maintenance of the system. These benefits are offset by the higher cost of training and the time required to develop expertise. Using outside consultants mitigates these last two effects, since they can apply their expertise directly to the problem. However, external integrators might lack the in-depth understanding of the organization's culture and goals. The solution might not be as well-tailored to current and future needs. Either way, it's bound to take time to implement a well-integrated network.

Authority for integration

Regardless of the background of the integrator, the organization has to define the limits of their authority and accountability. It is unreasonable to ask anyone to be accountable for a project without giving that person the authority to manage. The question then becomes how far this authority extends. Does it include the network interface card (NIC) on the user's desktop? What about the operating system that controls the NIC? Shall we include application software that uses the network? In a number of the previous challenges, we indicated that internal politics and integration may overlap. Nowhere is this more true than in the area of authority and accountability. Often the integrator is informed that the technology has already been selected, and it is his or her responsibility to make it all work together! This is frustrating at best, and impossible at worse. If you are selected as the integrator, make sure that you have the authority to get the job done. (See Fig. 1-2.)

Selection of standards

Perhaps the largest single problem facing today's integrator is the selection of integration standards. While the computer and networking communities have been urging their clients to standardize

Figure 1-2

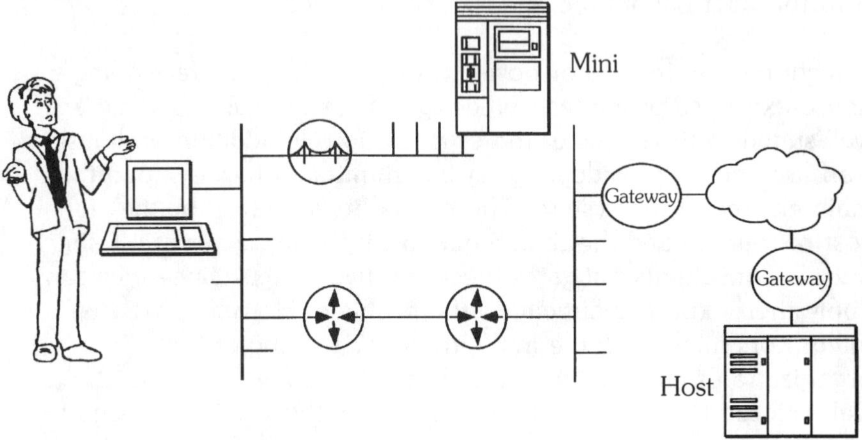

Before you can integrate, you have to know the scope of responsibility.
Learning Tree International

on open specifications, it seems (with the possible exception of networking standards) that just the opposite has happened. It is the rare environment that does not have two or more "standards" governing data communications, operating systems, e-mail, file transfer, databases, etc. The fact that suppliers of these technologies have not been able to agree on a single standard may be partly to blame. Sophisticated marketing by vendors to promote their own proprietary standards has also contributed to the problem.

Most organizations now have heterogeneous technological environments. The technology, regardless of the standard, will probably find its way into the organization, one way or the other. In many cases, organizations select products based on proprietary standards because they already have the status of a *de facto* standard. Microsoft's Windows, Novell's NetWare, and IBM's SNA are good examples. Furthermore, in many cases, organizations reject *de jure* standards outright. Open Systems Interconnect (OSI) is probably the best example. Regardless of the decision to choose *de facto* or *de jure* standards, the selection of these specifications is a primary task of the integrator. We cover this important topic in more detail in Chapter 2.

The bottom line on integration

Building a well-integrated network can be an enormous undertaking. The full set of products and specifications that potentially must be made to work together would fill an encyclopedia, and would require daily updating. Attaining a full and complete understanding of how to perform network integration will never be possible, due to the dynamic nature of the industry. Does this mean that our integration efforts are doomed to failure? You may have noticed that the number of challenges exceeds the number of potential benefits. Does this mean that building a well-integrated network is risky? Yes it is, but don't let that stop you. The benefits are there, and their potential magnitude justifies the risk.

However, the reasons for integrating may be much simpler than evaluating the potential ROI. The reality of today's organization is heterogeneous. The investment has already been made. These technologies are now deployed. Over time, as applications are developed that use them for the implementation of business-critical functions, they become entrenched. It is increasingly difficult to encourage users to migrate to other technologies solely for the benefit of integration. With increased decentralization, user departments are selecting their own technology, hiring technical staff, and performing local implementations. At the same time, centralized MIS departments are trying to coordinate the deployment of corporate backbones, based on multiple protocols, that will provide "any-to-any" connectivity. Integration is no longer a "wish list" item. It has become a critical requirement to preserve the *existing investment*. The only issue is how long to wait before getting started.

Assistance may be some time in arriving. At a global level, public carriers, governments, and industry groups are struggling with the problem of providing a world-wide data communications network. If governments cannot agree on the technology to be used to build these systems, how can organizations be asked to commit to specifications? Once again, we may not have the luxury of waiting for everyone to make up their minds. Users are demanding access to

data, new technologies are being developed and deployed, and the industry shows no indication of slowing down or consolidating. Just the opposite—multimedia, virtual reality, data visualization, and other leading-edge technologies indicate that we will be living in a multivendor world for the foreseeable future. Our task as integrators must be to define a strategy that will accommodate this change while preserving the existing investment. Integration is no longer an option—it has become an imperative.

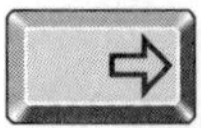

Defining the terms

The problem of terminology

Before we can discuss building a well-integrated network, we have to agree on the terminology. This can be challenging. Many vendors have already defined terms and named products after their definitions. One vendor's *gateway* is another's *bridge*. *File transfer* by one definition might mean *distributed file system* by another. As another example, the term *server* used to refer to a distributed file server running a Network Operating System (NOS), such as NetWare or LAN Manager. Today, it may refer to any process that responds to client requests. In addition, integration technology creates new terminology, such as *internetworking* and *interoperability*. The problem does not stop with the definition of terms. The industry is filled with competing standards and models. Vendors of technology create new terms to help them better market their wares. Purchasers are faced with a bewildering set of options. (See. Fig. 1-3.)

Of course, we have the ever-present TLA (three letter abbreviation) syndrome. The computer industry seems to be in love with these acronyms. Often they are created just for the apparent "cuteness" of the acronym itself. For example, many are aware that the acronym GUI (pronounced *gooey*) stands for Graphical User Interface. But were you aware that the next generation of person-machine interface is likely to be called a "Self-Teaching Intuitive Computer Interface," or STICI? That's right: after the GUI comes the STICI (i.e., sticky)! This ongoing love affair with acronyms shows no signs of abating.

Figure 1-3

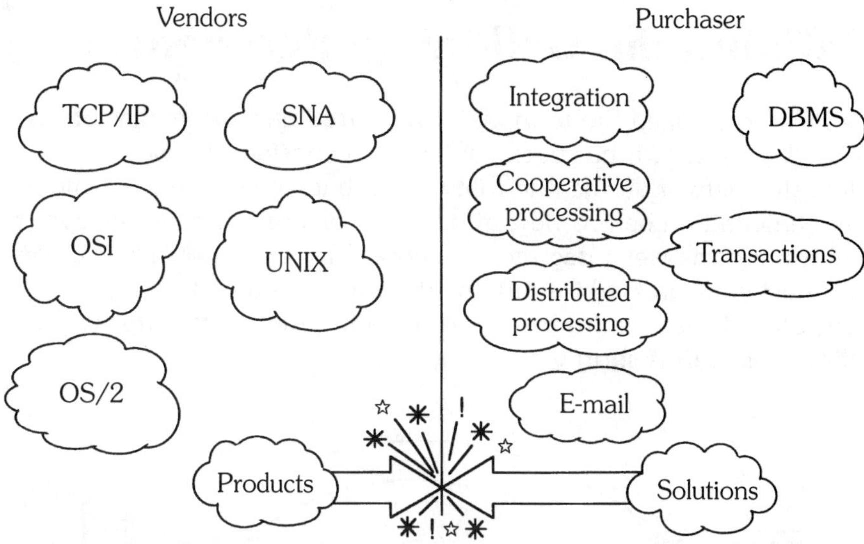

TCP/IP = Transmission Control Protocol/Internet Protocol

SNA = Systems Network Architecture

OSI = Open Systems Interconnection

Terminology vs. solutions. Learning Tree International

Glossaries and indexes can be created using nothing more than TLAs. We should know—look at the end of this book! So before we can examine the issues and solutions offered by the well-integrated network, we require some agreement on terms.

Please note that we do not require that you adopt these definitions as your own. We will be using them throughout the book to achieve consistency. How you choose to define the same tasks or items outside of this book is very much your own affair. Wherever possible, we have tried to remain consistent with generally accepted industry definitions. We recognize that definitions are not static, and therefore we recommend the use of the *duck* test when considering terms and definitions, namely, "If it walks like a duck, swims like a duck, and quacks like a duck, it's a duck, no matter what it's called." Definitions are less important than concepts.

Defining the well-integrated network

We have mentioned the term *well-integrated* quite a number of times. Up to now, it has been sufficient to consider the well-integrated network to be one that works, but we will require a more precise definition before we can proceed. Figure 1-4 provides our definition of the well-integrated network. There are two major areas that must be addressed for a network to be considered well-integrated. These are *internetworking* and *interoperability*. We will define these terms shortly.

Figure 1-4

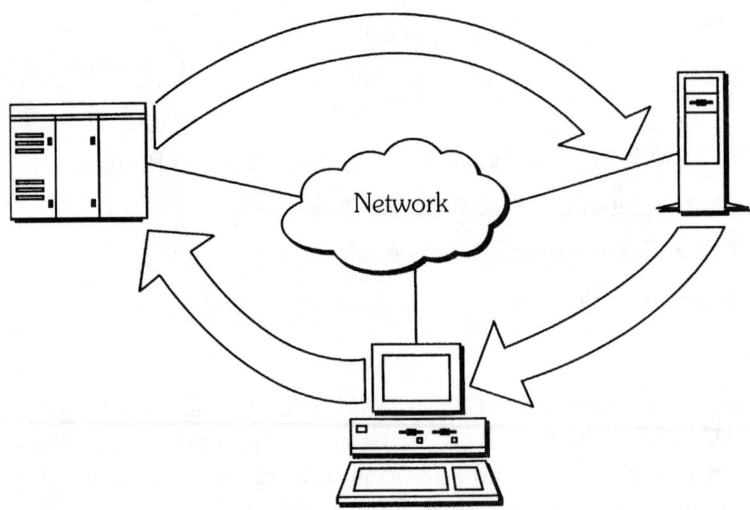

The well-integrated network. Learning Tree International

In our vision of a well-integrated network, users can access data and processes in a seamless manner. For example, users can select any printer in the organization for output of a document without regard to the location or type of printer. Data can be retrieved from any database, regardless of the particular SQL (structured query language) required for high-performance access. E-mail can be easily sent among proprietary vendor implementations. The well-integrated network achieves this capability through a variety of techniques to be described.

The well-integrated network also offers ease of support and administration by balancing the issue of local autonomy and backbone integration. It meets the organization's RAS requirements. It allows the organization to leverage its existing investment. It provides a solid foundation for the introduction of new technology and services as they become available.

Without waxing philosophical, we can say that a well-integrated network is in many ways the result of a strategic approach to networking. It requires vision, and a commitment to that vision. The vision, as you will see, is of a standard virtual backbone that links every corner of the enterprise. Local departments can connect at well-defined entry points. Local autonomy and control over data and technology are preserved. The backbone acts as a "glue" that binds local departments. However, the backbone does much more, since it also provides access to a variety of applications and services that are used by the enterprise as a whole. The well-integrated network recognizes that certain applications, and data, are best provided from a central location. As we shall see, it can provide these services without impacting local autonomy.

In the end, it will be your choice as to how you define a well-integrated network. Remember the duck test. It doesn't matter what you call it. It's the vision and commitment that are important. The vision we may be able to give you. The commitment you will have to discover on your own.

Terminology of integration

While we will be offering many new terms throughout this book, there a number that will be used globally. We now provide these definitions. Each definition is followed by a narrative that spells out the term in more detail.

Internetworking The physical and logical connection of networking components.

Internetworking comprises the set of activities required to physically and logically join networking components. In essence, the mandate of

internetworking is to describe how the components will be linked. Figure 1-5 highlights some of these components. Other internetworking components include repeaters, network protocols, and network management systems. As a general rule, internetworking should be invisible to the users of the system. Internetworking is one of the first challenges faced by the integrator. We discuss internetworking in Chapter 3.

Figure 1-5

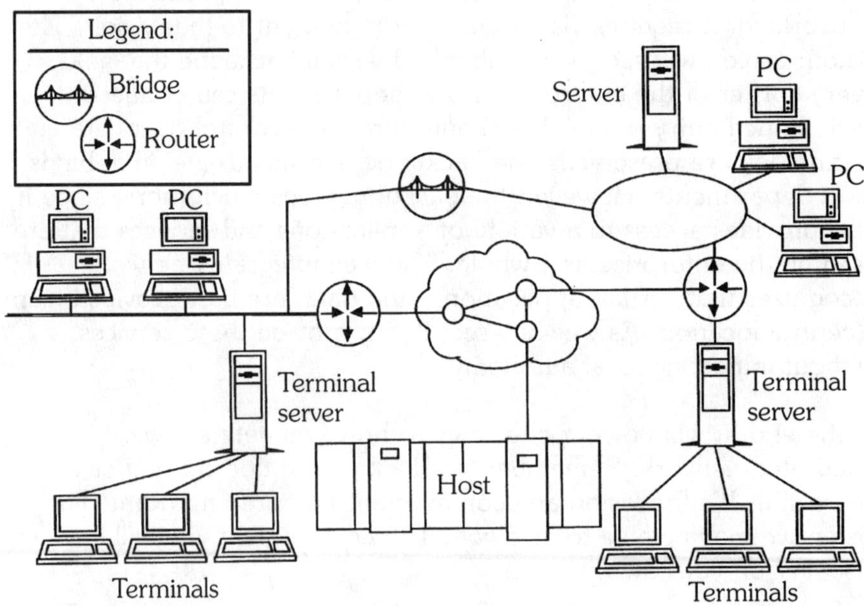

The components of internetworking. Learning Tree International

Internetworking can be costly, but a variety of products exist to assist the integrator. To counterbalance the expense, it is possible to purchase a solution with off-the-shelf components, thereby making solutions easier and faster to implement. Interoperability is a different story.

Interoperability The ability of heterogeneous software applications to interchange data and process information.

Internetworking is not enough to build a well-integrated network! Interoperability is the study of high-level integration. Interoperability

describes how data, and related processes, will interact. While internetworking addresses the question of how we integrate, interoperability tells us what we should be able to do with the completed system. This task is in many ways much more complex than internetworking. Conflicting definitions, proprietary protocols, a dynamic marketplace, plus RAS concerns, all increase the difficulty of achieving a high degree of interoperability. Most solutions involve compromises. Even when adopting standards, different vendor implementations may result in incompatibility problems. Furthermore, interoperability may be required from highly dissimilar systems. For example, while it may be possible to integrate all the e-mail systems with one another, how can they then be integrated with database retrieval?

It is not possible to cover all the possible permutations and combinations of interoperability. There are too many products, protocols, and applications to cover in a reasonable amount of space. Moreover, in many cases, interoperability may be required among custom-built applications. These areas are beyond the scope of this book. We have therefore selected what we believe are the major challenges of interoperability facing today's organizations. There are:

- Remote terminal access
- Remote program execution
- File transfer
- Distributed file systems
- PC network services
- E-mail
- Database access

We expect that there will be a greater need in the future to achieve interoperability among multimedia, groupware, and virtual reality systems. For the present, these applications are still in their infancy, and it is difficult to provide guidance due the immaturity of both the market and the specifications. Wherever possible, we will try to

provide guidance on these new technologies, but no attempt will be made to address them in detail.

Model A conceptual view of a real-world situation that defines the interaction among components.

A model is a simplified way of examining a complex process or environment. In the field of integration, the term *model* generally refers to a communication model. In the next section, we provide you with one such model. However, the term *model* can be used to refer to any abstraction. As we will see in Chapter 2, we can use a model to describe both internetworking and interoperability in a well-integrated network.

Open system A computing/communicating environment that is vendor-independent and commonly available.

Many definitions exist for *open system*. We feel that this one is best suited to our requirements because it is easy to apply to any technology. For a specification or product to be considered "open," it must exhibit two main properties. First, it must be independent of any vendor. This generally means that the specification has been developed by a standards-setting body with no vested interest in products that implement the standard. The IEEE standards for LANs are good examples. While vendors may participate in these standards-setting associations, they cannot control them. The second attribute of an open system is that it must be commonly available. An open system is not of great interest if only one or two vendors have implementations. Open systems stand for flexibility of choice. If it is not commonly available, then there is no choice. An example of a commonly available open system might be the set of products based on the TCP/IP protocols. They are available from many different vendors. In general, an open system should provide interoperability among its components, and permit portability of applications, data, and users among scalable platforms.

Some would argue that open systems should also be certifiable. This means that they can be tested against published specifications. While we agree in principle, this is sometimes difficult to achieve in practice. Many specifications are available, but it is often the responsibility of the purchaser to perform any certification testing. Often, if the product works, then that is deemed successful certification testing.

As we mentioned earlier, there is some debate regarding the practicality of using open systems rather than proprietary ones to build well-integrated networks. While we would like to avoid this contentious discussion as much as possible, we examine it in more detail in Chapter 2.

Subnet The smallest addressable unit in an internet, generally a LAN characterized by a single subnet protocol.

Internet An aggregate of subnets, not to be confused with *the* Internet.

Protocol A defined interaction among processes or entities, generally detailed in a specification.

We group these three definitions together because they are somewhat interrelated. Subnets are the basic building blocks of our well-integrated networks. They are generally administered as a single unit, and tend to support a particular business function. For many organizations, subnets will be implemented as one or more LANs. In addition, a subnet will generally implement a single subnet protocol. Normally this will be Ethernet (IEEE 802.3), Token Ring (IEEE 802.5), or Fiber Distributed Data Interchange (FDDI). The LANs can be joined by bridges, but not routers. This is because bridges implement virtual LANs, while routers implement internets. We will have more to say on this in Chapter 3.

Note that while we have stated that subnets are generally implemented as LANs, this is not always the case. A WAN may also be a subnet if it is defined as a single addressable entity. The Internet (note the capital letter used to differentiate it from all other internets) is comprised of many subnets. Some are WANs. However, in almost every case the subnet is administered as a single entity. Very large

subnets may be broken into smaller networks, but these are also referred to as subnets.

An internet is a collection of subnets. Our well-integrated network will be an internet. Beyond this definition, an internet is generally characterized as having a common network protocol. The Internet uses TCP/IP. Novell-based networks use SPX/IPX. We refer to these protocols as *end-to-end service* protocols, and will explain them in greater detail shortly.

As an interesting side note, an internet can be considered a subnet of a larger internet. For example, many organizations support their own internets that comprise many subnets. However, when these organizations access the Internet, they are considered a single subnet. The addressing technique required to do this is sometimes referred to as *subnetting*, and is commonly used by organizations that implement TCP/IP-based backbones. (See Fig. 1-6.)

A protocol refers to any specification that defines how services can be offered, requested, or negotiated. In essence, a protocol is the specification of a model. Protocols are used extensively in both the data communication and data processing industry to define all types of interactions. The development of protocols is a key role played by

Figure 1-6

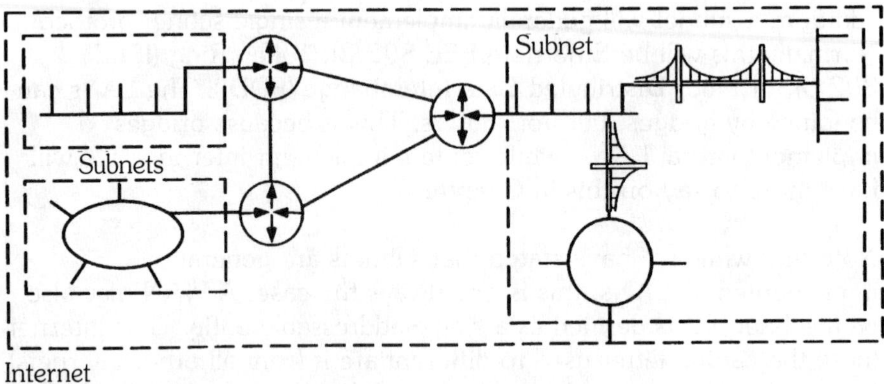

Internets and subnets. Learning Tree International

vendors and standards-setting bodies. Protocols may be proprietary, which implies that they are controlled by a single agency and must be licensed (or might be kept a trade secret). Other protocols are defined to be open, which means that they are commonly available. Open protocols might not be free of charge, since standards-setting bodies might charge for the specification, its licensing, or certification. Once again we defer our more detailed discussion of protocols to Chapter 2.

 Related terminology

While not strictly a part of integration terminology, there are a number of related terms that we will use frequently throughout this book. Furthermore, since these terms are also subject to multiple definitions, we will provide our definitions now so that we can resolve any conflicts with other viewpoints.

Client/server system	A cooperative processing environment that provides a single system image to the user.
Client	A general-purpose request process.
Server	A specialized request-satisfaction process.
Cooperative processing	Interaction of two or more processes to complete a task.
Single system image	A common view into a series of heterogeneous processes; a single syntax/interface to multiple systems.

Client/server is considered to be the foundation of a well-integrated network. It is best described as teamwork, among a number of processes, operating on different hardware platforms, to accomplish a particular task. The client, which normally resides on the user's desktop, requests services provided by servers that execute on remote hardware platforms. The client/server model of interaction is very

powerful, since it implies that existing resources can be leveraged. In addition, client/server also maps well to the object-oriented approach to programming.

Some degree of confusion and disagreement exists when defining clients and servers. For many, both the client and server can refer to hardware, application software, or operating system software. Often, the user's desktop machine is referred to as a client, and the hardware that implements the distributed file system is called the server. We admit to being guilty of the same loose definitions! At different points in the book, we may use the terms *client* and *server* to refer to any or all of these interpretations. We apologize, but industry acceptance of these terms as referencing any number of things has become somewhat *de facto*. Where possible, we will try to avoid discussing hardware in terms of client and server, but it will not always be possible. We could make up some new terms, but this would probably only add to the confusion.

Single system image is another powerful concept. It refers to the idea of the user becoming independent of the location of resources or services. The entire network becomes hidden and the user can access remote services and resources using a common syntax. This will often be the same syntax that is used to access local resources. The intent is to minimize training and provide an interface that can be customized to meet the needs of the individual user. A variety of techniques can be used to accomplish this task. We will explore this in more detail in Chapter 6.

Distributed processing Support processes that operate across an internet.

Client/server is a special case of distributed processing. However, beyond the functions offered by client/server, users and programs will require access to a set of services that will be distributed throughout the well-integrated network. These may include, directories, security servers, and communication gateways, for example. While client/server generally focuses on satisfying user requests, distributed services generally provide a set of services that support both the clients and the servers. These services may be implemented at both the local and the global level.

Internets, subnets, and models

A simplified communication model

We have described an internet as a collection of subnets. As mentioned, a subnet may be physically implemented as a single LAN or a collection of bridged LANs. Subnets are available in a variety of topologies including star, bus, and ring. These topologies may be used to describe both the physical implementation and the logical perception of the subnet.

The description of subnets and internets will be facilitated by the use of a reference model. We have indicated that many vendors and standards groups provide their own reference model. For the sake of simplicity, we have developed our own reference model to be used when building the well-integrated network. It provides a common reference to compare with other models, as well as facilitating the discussion of integration.

Figure 1-7 introduces our three-layer reference model. It consists of a Subnet layer, an End-to-End Services layer, and an Application layer. Each layer provides a service to the layer above, and requests services from the layer below. Two processes can be considered to be peers of one another if they both implement the same layer functionality. These peer layers will communicate as required to perform various tasks.

One reason for choosing a three-layer model has to do with how communication technology has been implemented. Most LAN protocols specify both layers 1 and 2 of the OSI reference model. Furthermore, it is generally difficult to "mix and match" technology within these layers. For example, an Ethernet card cannot be placed on the same physical wire as a Token-Ring card. On the other hand, it is possible to mix and match technology among the layers. For example, it is possible to implement TCP/IP (an end-to-end service) over both Token Ring and Ethernet. While purists may rightly claim that there are quite a few exceptions to this general rule, we feel justified in using this model because it does make our discussion of

Figure 1-7

Three-layer model

3	Application services X.400, X.500, SMTP	How applications interface with the network
2	End-to-end services TCP/IP, NetBIOS, SPX/IPX	How to move data through the network — sometimes called the "transport sublayer"
1	Subnet IEEE 802.2, HDLC, SDLC	How to put data on the media

HDLC = High-level Data Link Control
IEEE = Institute of Electrical and Electronics Engineers
NetBIOS = Network Basic Input/Output System
SDLC = Synchronous Data Link Control
SMTP = Simple Mail Transfer Protocol
SPX/IPX = Sequence Packet Exchange/Internetwork Packet Exchange

A simplified communication model. Learning Tree International

integration easier to digest without compromising the power of a layered approach to networking.

The Subnet layer

The Subnet layer is roughly equivalent to OSI layers 1 and 2 (physical and data link). It describes how data will be placed on the media, and how point-to-point communications will occur. Point-to-point communication is always required, even in LANs, since most communication takes place between a single source and destination. It is important to recognize that the subnet layer is not concerned with the final destination of any piece of data. It is only responsible for moving the data to the next location in what might be a chain of devices between the end points of a communication. Figure 1-8 shows some of the more common subnet protocols and how they might be implemented in an integrated network.

Figure 1-8

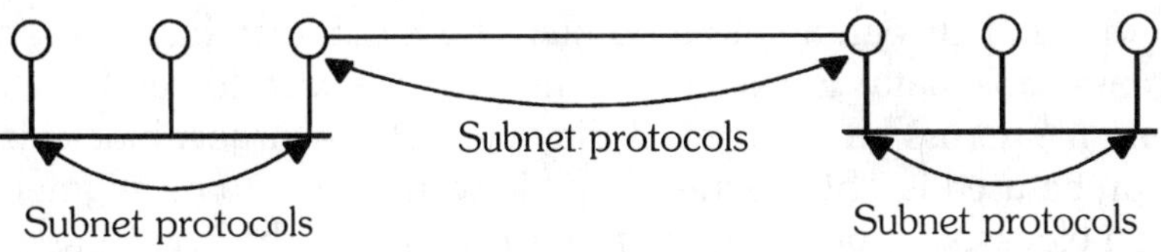

Multiple subnet protocols in an internet. Learning Tree International

One of the main advantages of the well-integrated network is its ability to support a wide variety of subnet protocols. This is important because the adoption of a subnet protocol is often considered to be a local issue, and therefore should be considered the responsibility of local implementers. This implies that a number of subnet protocols will be present in any internet. Since LAN and WAN subnet protocols are quite different, at least two protocols will have to be supported.

As might be expected, the primary use of subnet protocols is to implement subnets! For example, a subnet might be based on Token Ring or Ethernet technology. It is a characteristic of the subnet layer that all communicating devices within the subnet share a common protocol. While a subnet may have Ethernet or Token Ring, it will not have both at the same time.

The End-to-End Services layer

Sitting above the subnet layer is the End-to-End Services (ETES) layer. This layer is equivalent to OSI layers 3 and 4 (Network and Transport). It is sometimes referred to as the *Transport Sublayer*. The motivation for an ETES derives from the nature of internets. Since an internet comprises a network of subnets, a mechanism must exist to provide for efficient transport of data packets among the subnets, i.e., moving data through a series of incompatible point-to-point subnet links. End-to-end communication requires a protocol that can be transported across these subnets. For example, we require that TCP/IP (an ETES protocol) operate over Ethernet, Token Ring, and other subnet protocols.

The ETES layer is also responsible for the end-to-end error detection and error correction of the entire data transfer. In contrast, the subnet protocol is only concerned with the error-free transmission of a single

block of data between any two points. The ETES may have to segment and reassemble data. For example, suppose we wish to send a 15-kilobyte file across an internet. It may be that the largest block size that can be used is 1500 bytes. The file would have to be segmented into ten blocks, shipped, and then reassembled at the receiving station. TCP/IP is an example of a typical ETES. (See Fig. 1-9.)

Figure 1-9

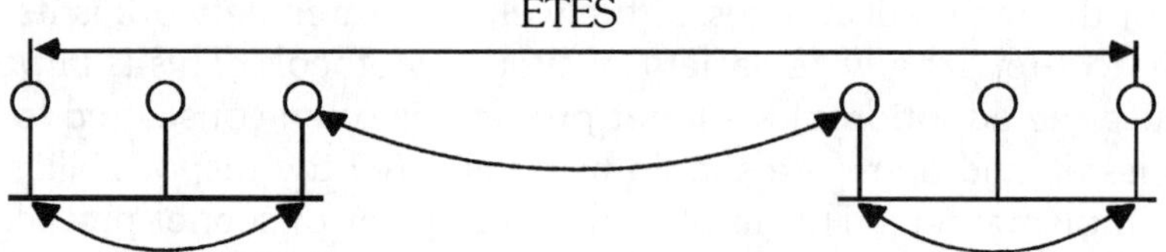

End-to-End Service protocols. Learning Tree International

Selection of an ETES protocol is one of the primary tasks of the integrator, and must be considered a strategic decision. The ETES protocol will be deployed on the backbone and used as the common "glue" that provides a major part of the internetworking infrastructure. Furthermore, once selected and deployed, it will prove difficult to change to another ETES protocol, because many of the distributed services, applications, and gateways that are purchased and deployed must be compatible with it. The well-integrated network will try to minimize the number of ETES protocols in order to avoid the use of gateways wherever possible.

The Application Services layer

The Application Services layer provides a set of services and APIs (Application Programming Interfaces) for programs that wish to access the network. It provides a wide range of features and functions, including file transfer, e-mail, directories, data access, peer-to-peer, etc. Other services, such as session control, recovery, and security, may also be included. Implementations of Application Layers differ considerably, depending on the standard in use. While OSI uses three layers (Session, Presentation, Application), TCP/IP provides all of this through a single layer (Application). For our purposes, a single layer will be sufficient. (See Fig. 1-10.)

Figure 1-10

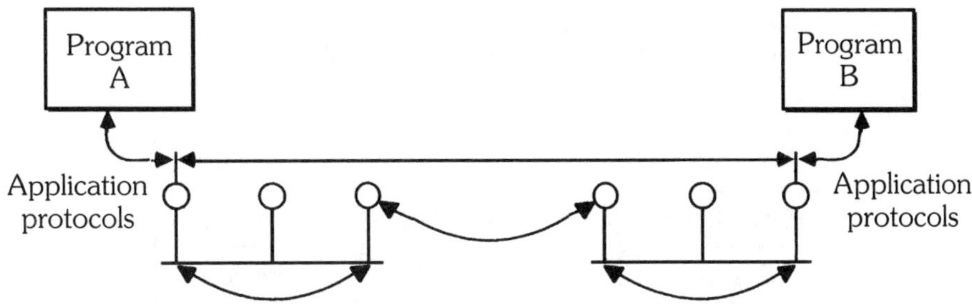

Application Services layer provides an entry point to the network and offers distributed services. Learning Tree International

In many cases, the selection of an ETES protocol will dictate the selection of the Application Services protocols. In most cases, these are rather tightly coupled. For example, if you select TCP/IP as your ETES protocol, it is likely that you will use the TCP/IP application services such as FTP, TELNET, SMTP, etc. While techniques exist to allow mix-and-match between these layers (middleware, mixed stacks, and gateways, for example), performance or integrity implications may need to be considered.

A final look at the integration model

Some have described internetworking as covering OSI layers 1–3, while interoperability covers layers 5–7, as well as applications 1–4. In addition, we extend interoperability above layer 7 to include databases and other utility services that must be made to work together. We tend to agree with this overall allotment. By separating the subjects of internetworking and interoperability along these lines, we can select areas of focus. Much has been written on the subject of internetworking. R. Perlman's *Interconnections* and M. Miller's *Internetworking* both provide excellent sources of information. (See Fig. 1-11.)

Information on interoperability is more difficult to obtain. While vendors will provide information on how their technology may interoperate with other offerings, there has been a lack of material

Figure 1-11

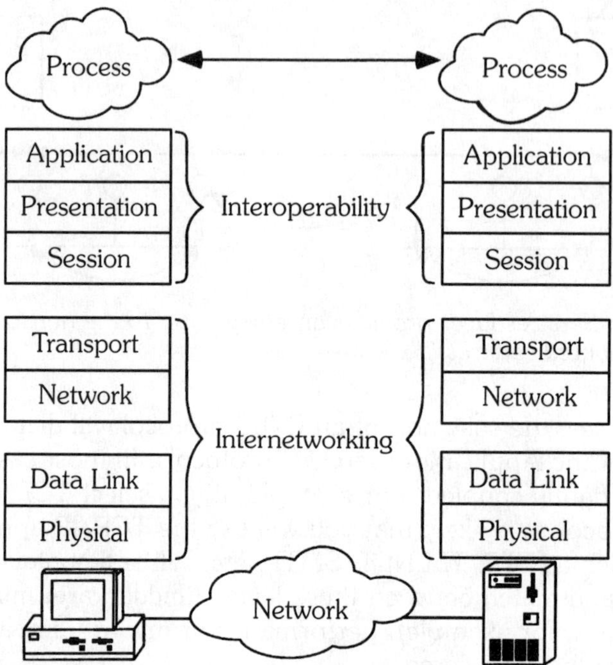

Internetworking, interoperability and the OSI reference model. Learning Tree International

devoted to a general examination of the field. This is one of the reasons that we have devoted so much attention to the subject of interoperability in our book. We have shown that the well-integrated network must provide both internetworking and interoperability. The integrator must pay attention to both areas if there is to be any chance of building a well-integrated system.

This leads us to a final look at our integration model (Figure 1-12). Integration can only be achieved by careful, informed selection of standards and technologies that address these two issues. We have also shown that in many cases, there may be a close coupling between the technology of internetworking and interoperability. Understanding the ramifications of choosing a particular protocol is also a prime requirement for the integrator.

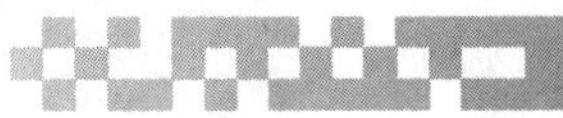

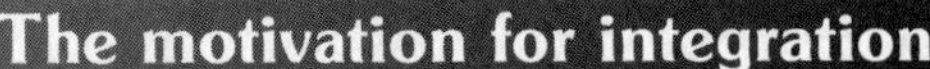

Figure 1-12

Internetworking	+	Interoperability	=	Integration

— Media

— Components

— Communication software

— Network operating system

— Network-management tools

— OSI layers 1–4

— E-mail

— Database

— Client/server

— Application gateways

— Groupware

— Electronic data interchange

— Distributed file systems

—OSI layers 5–7

Internetworking = *How* we get it connected
Interoperability = *What* we do with it when it's connected

The integration model. Learning Tree International

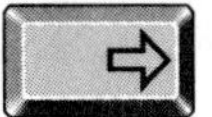

Summary

Organizations find themselves facing a series of challenges when attempting wide-scale network integration. While the potential benefits are attractive, the complexity of the task makes implementation hazardous and prone to failure. However, the reality is that today's organization is heterogeneous; integration must be attempted so that existing technology can be leveraged. This must be done while balancing the issues of local autonomy and integration. A well-integrated network will achieve both of these objectives, and still remain flexible enough to accommodate new technology as it becomes available.

In examining the challenges, we have seen that there are issues involving the authority of the integrator, the selection of integrator,

terminology, RAS, accounting, etc. Each must be addressed. Often the problems will be political, rather than technological. Resolution will require commitment at a strategic level in the organization. Since the well-integrated network will be used by almost everyone in the organization, it is important that this commitment be as widespread as possible. The implication is that integration must be approached as a planned, phased activity that requires significant investment.

Integration brings with it a new set of terminology to be learned. Lack of precise terminology means that suppliers, purchasers, and users will be unable to communicate effectively. This is especially true in the case of terms such as *internetworking*, *interoperability*, and *client/server*. While we have attempted to provide some definitions, they are far from being universally accepted. It is therefore important that integrators assure themselves that all participants in the integration project are using the same set of definitions. In many cases, this may prove to be a major undertaking.

Much has been written on the topic of internetworking, and many excellent books are available. Dealing with interoperability is more challenging, both because of the lack of published material and the fact that interoperability can only be achieved through custom-fit solutions. This is why we have chosen interoperability to be the major focus of this book.

Before we can address the subject of internetworking or interoperability, we require a better understanding of the communication models that are available, and how they compete with or complement each other. This will be the topic of the next chapter. But first, here is a short optional quiz that you might like to try.

Concepts review

1. What are some benefits of a well-integrated network?

   ______________________, ______________________

2. What are some current problems in the definition of terminology in multivendor networks?

   ______________, ______________, ______________

3. What characterizes the vendor's view of multivendor networks?

   ____

   What characterizes the customer's view of multivendor networks?

   ____

4. What are the two components of integration?

   __________ + ______________ = integration

5. Define:

   Internetworking______________________________

   Interoperability______________________________

   Subnet______________________________

   Internet______________________________

   Protocol______________________________

   Model______________________________

6. Name some subnet protocols:______________________________

7. Name the seven layers of the OSI model:
   1.______________________________
   2.______________________________
   3.______________________________
   4.______________________________
   5.______________________________
   6.______________________________
   7.______________________________

8. Name some end-to-end service protocols:

   _____________________, _____________,______________

9. Name some components of internetworking:

   _____________________________, _____________________________,

   ____

10. Name the major components of interoperability:

   ____

   ____

   ____

   ____

   ____

   ____

   ____

   ____

Standards for integration

CHAPTER 2

GIVEN that a heterogeneous mix of protocols exists within the organization and that it is difficult to standardize on just one, we must find a way to integrate products built to these different specifications. While detailed knowledge may not be required, some understanding of these protocols is required in order to discuss the issues and select from the available integration options.

In the previous chapter, we indicated that the selection of an ETES protocol would be a strategic decision. This may not always be the case. In the event that the protocols are to be isolated from each other, no integration strategy is required. If this proves not to be the case, then a strategy will be required when approaching the integration of each protocol. Having an overall vision of how this can be accomplished will be beneficial. We discuss a number of approaches including standard and multiprotocol backbones, dual stacking, and others. The selection of this protocol integration strategy is a key task of the integrator.

Standards do not remain static. Over time, existing specifications are modified, and new ones are proposed. Currently there are a number of efforts, such as the Open Software Foundation's Distributed Computing Environment (OSF/DCE), the Object Management Group's Common Object Requester/Broker Architecture (OMG/CORBA), and others, that are attempting to develop a level playing field for integration. We now provide an overview of this area, and examine standardization trends that might have an impact on the integrator in the short term.

An integration blueprint

Think of enterprise integration in the same way a builder thinks about construction. A builder has a wide diversity of parts (wood, nails, windows, etc.), that must somehow be combined into a single structure. The builder is faced with a wide range of choices when selecting material. Should the frame be built of wood or steel? What type of piping should be used? Where should electrical cable run, and at what current-carrying capacity? What type of ventilation system is required? In other words, the builder must select and integrate heterogeneous building supplies into a single entity called a building. The integrator faces largely the same task.

In the same way that a builder requires a blueprint prior to starting a construction project, an integrator requires a blueprint before attempting to build an enterprise system. A blueprint provides two types of information. First, it must specify the components to be used in the project, and second, it must describe the relationship between the components so that the builder can see how they are joined.

We have identified a number of key areas that are required to perform successful integration. These have been classified as internetworking and interoperability components. While not specifically stated, we recognize that application programs are a third major component that must be addressed when building a well-integrated network. Within each category there are a variety of alternatives. Furthermore, these products may or may not be capable of working with one another. A frequently asked question is "How do all the standards and products relate to one another?". Clearly some form of enterprise blueprint is required. What form should it take?

 IBM's Open Network Blueprint

In the late 1980's, IBM introduced their "Network Blueprint." It was an attempt to describe the relationship between IBM and non-IBM technology. An early example of the blueprint is shown in Figure 2-1. It is a layered approach to integration. At the base is a *Subnetworking* Layer (OSI/RM 1-2) that describes the physical connections of the network. It is equivalent to the *Subnet* Layer that we described in Chapter 1. Figure 2-2 shows some of the technologies specified at this layer. Four broad categories are shown: *LAN*, *WAN*, *Channel*, and *Emerging*.

The next layer up, the *Transport network*, basically map to OSI/RM 3-6. It provides the same functionality as the *End-to-End Services* layer described in Chapter 1. It is important that this layer be independent of the Subnetworking Layer. For example, we may wish to run TCP/IP over LAN technology like Token-Ring or Ethernet, and at the same time also use TCP/IP over a variety of WAN systems. We do not wish to have a separate version of TCP/IP for each subnetwork specification. Having a Transport Network level makes sense, since it enables us to consider the merits of a particular

Figure 2-1

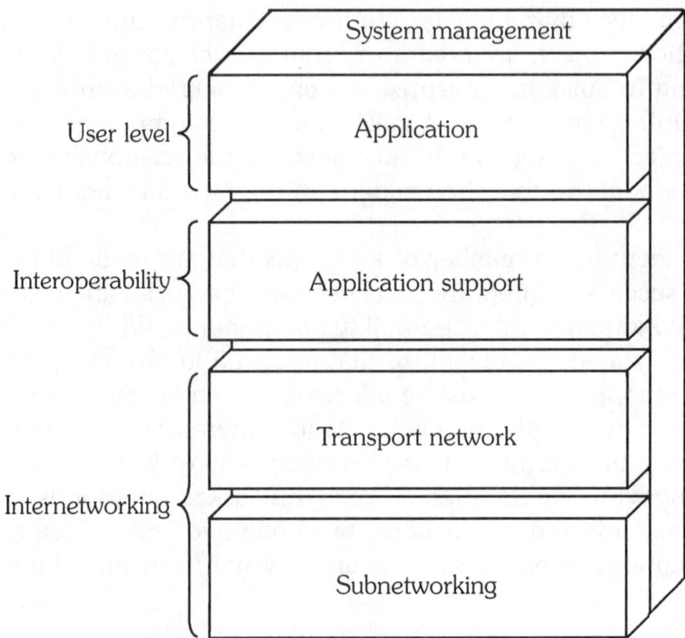

IBM's Open Network Blueprint. Learning Tree International and IBM

Figure 2-2

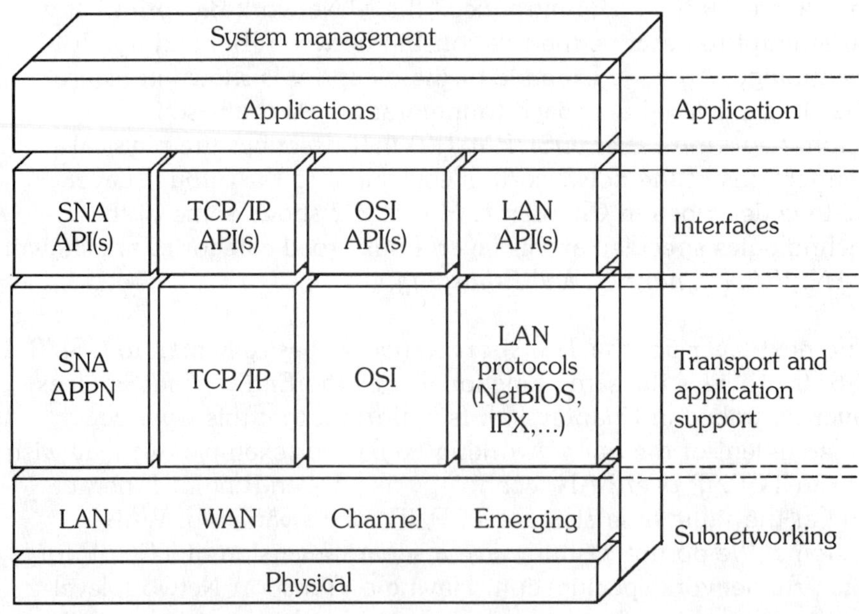

Blueprint with network protocols. Learning Tree International and IBM

protocol without becoming tied to the applications or subnetwork protocols. This is an important element in trying to balance local autonomy and integration. Figure 2-2 shows some of the specifications available at this level.

Above the Transport Network layer is an *Application Support* layer. While this layer provides OSI reference model layer 7 functions, it also provides tools, utilities, and services that may be deployed. There are two broad categories of specifications and/or products at this level. The first describes the forms of communication among peer devices in the network. The second category is for standard applications and distributed services. Figure 2-3 indicates that this total set of products and services is referred to as *Application Support*. Note that the interface points may be middleware offerings (which we discuss shortly) in order to allow for independence and decoupling among the various specifications.

Figure 2-3

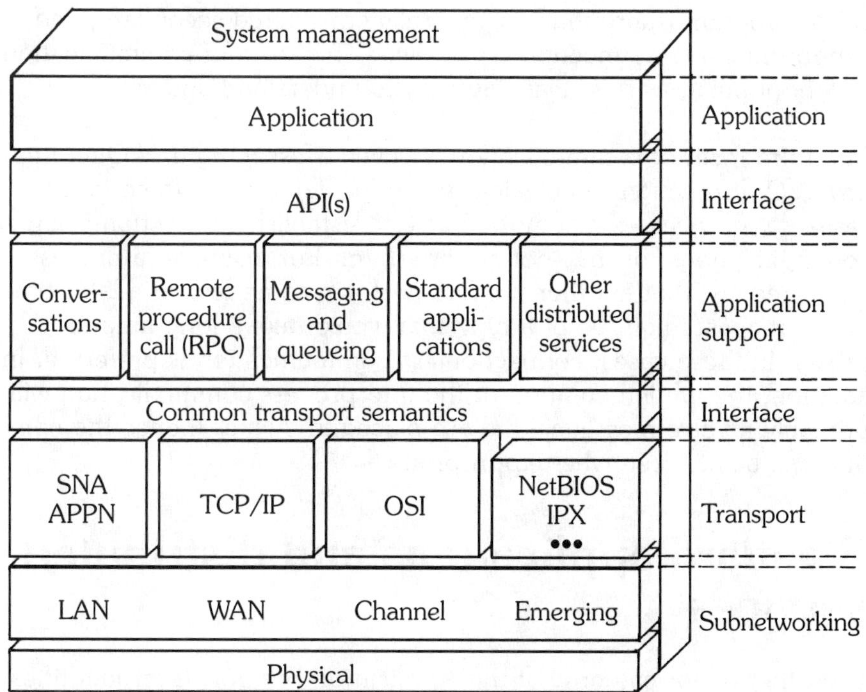

Application support and the blueprint. Interface layers represent potential implementations of middleware. Learning Tree International and IBM

Peer process communication

Peer processes can communicate in one of three modes. The first is called *conversational*. It is used when the nature of the communication requires an ongoing dialog, as in a telephone call. Multiple sends and receives may occur during a single session. A conversational communication is connection-oriented, and relies on the network to provide most of the required error detection and correction. It also assumes that the communicating processes will require bidirectional communications to accomplish their task. Transaction processing generally requires this form of communication.

The second form of communication is the *Remote Procedure Call* or *RPC*. The RPC is equivalent to a subroutine call, except that it is distributed across the network. It differs from the conversational mode of communication, since there is only one send and receive. The called program (subroutine) is generally considered secondary, and cannot initiate a connection on its own. This form of communication is very popular, as it is relatively easy to understand and use.

The third form of communication is called *Messaging and Queuing* or *M&Q*. It is roughly equivalent to the postal system. In certain cases, two processes may not be able to establish connection-oriented communications, or may not require them. For example, a process might request that another process perform some task at a later time, or the request might be of a type that would take a long time to satisfy. In these cases, connectionless communication is preferred. In addition, the implementation of the interprocess communication will probably involve messages that are placed in queues (hence the name) that can be serviced when appropriate.

Standard applications and distributed services

The other major category at the Application Support layer identifies standard applications and distributed services. Standard applications

are services offered by the network that reside primarily at the source and/or destination. File transfer is a good example. Most of the code that implements a file transfer system is resident at both ends of a communication session. In addition, file transfer is a type of service that is typically required by most systems. Therefore, instead of developing custom code to implement this function, we implement it as a standard application. This means that the same code is resident on each machine in the network. Other examples of standard applications include e-mail, databases, object technology, etc.

Let's contrast standard applications with distributed services. Like standard applications, distributed services offer a set of functions to the communicating processes. The difference lies in where the process is running. One example of a distributed service is a directory. Generally a directory is implemented on a machine separate from the source or destination. It provides locator services that are distributed through some domain of an internet. A variety of distributed services may be offered, such as security, file systems, administration, etc. As the complexity of the network increases, a set of distributed services that can run over a wide variety of Transport Networks becomes of particular importance. One example of a comprehensive set of integrated distributed services is the Open Software Foundation's *Distributed Computing Environment* or *DCE*. Proprietary solutions from many vendors also exist.

The application layer

The top layer of the blueprint is for the applications that run over the network and take advantage of the services offered. This might include business applications, end-user utilities like word processing, groupware, etc. The set of applications is unique to each environment, and is therefore beyond the scope of this book. However, it is important to recognize that many of these applications can be purchased as "shrink wrap" commodities. Frequently they are built to work with a particular transport network such as NetBIOS or TCP/IP. If middleware is available, these purchasing decisions can be made easily. If not, gateways are required to perform conversions or encapsulations.

System management

The final layer is called the *System Management* layer. It might be best described as a "metalayer," as it spans all the layers of the blueprint. System Management provides services and utilities to monitor the network, diagnose and correct problems, and carry out other functions that may be required. It must span all the blueprint layers because these functions must be implemented throughout the system. It would be beneficial if these services could be integrated. In Chapter 7, we will discuss one of the more common system management implementations, namely the *Simple Network Management Protocol* (*SNMP*).

Middleware

Finally, you will notice that there are interface layers shown between the Transport/Application Support and Application Support/ Application. These are sometimes referred to as *middleware*. Middleware is important because it provides a mechanism to achieve portability and flexibility in our networks. Middleware shelters a process that runs at a particular layer from having to know the details of processes running at other layers. Given the heterogeneous nature of most environments, middleware makes good sense. Of course, there is generally a performance tradeoff involved with middleware, since conversion and translation must take place.

Blueprint evolution

In 1994, IBM provided an expansion to the Network Blueprint simply called *The Open Blueprint*. It modified the original blueprint in a number of ways. Application Support was renamed *Distributed System Services*. A new category, *Object Management Services*, was added. The Standard Applications are now listed as part of the *Applications and Application Enabling Services* layer. Also note that the "upper" middleware layer has been removed, since the layer below is now supposed to provide these functions. (See Fig. 2-4.)

Figure 2-4

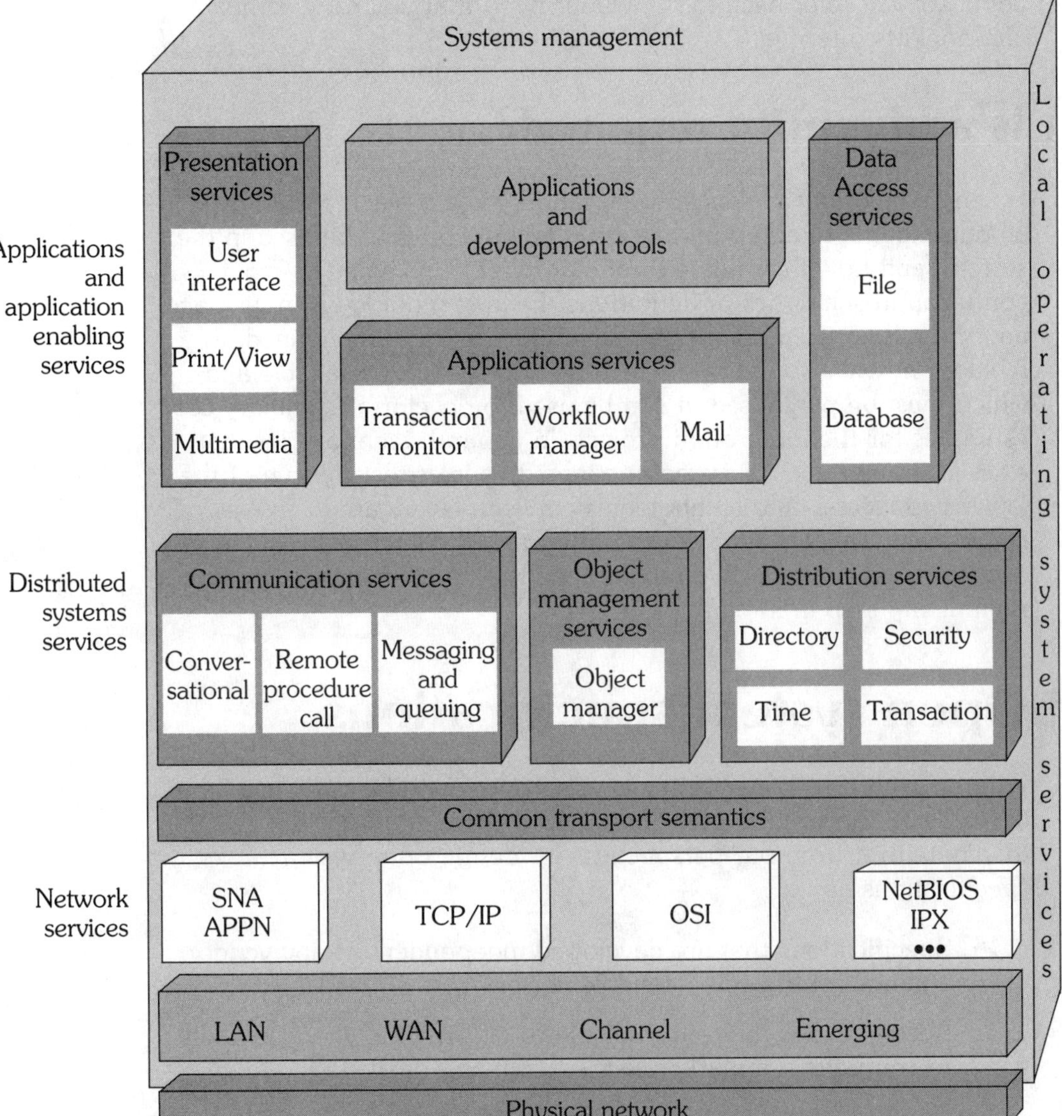

The Open Blueprint. Open Blueprint Technical Overview, ©1994 International Business Machines

Regardless of which version is used, the blueprint provides assistance by indicating the technology required for implementation; it is also a convenient model with which to compare specifications, standards, and products. Over the course of this chapter and the rest of the

book, we will refer back to the blueprint so that you can see how components interrelate.

Is a blueprint required?

Having a blueprint makes an integration project much easier. It allows all participants to agree on the exact technologies to be used in the system, and how they relate to one another. This helps to avoid confusion in semantics or definitions. Each technology is in its place, and the relationship among the technologies is clearly defined. Since a blueprint will also indicate which technologies are available and which must be purchased, it can be used to develop preliminary cost estimates for the integration project. In general, technologies that exist at the same layer are competitive (this is especially true of the lower two layers), and technologies between layers are complementary. This enables the integrator to determine which decisions are strategic. We recommend the use of blueprinting as a preliminary step in integration projects.

Open system specifications

What are open systems? No clear answer exists—the response depends on the vendor, user, or standards-setting group that you talk to. Therefore, from our perspective, we define open system specifications as:

- Specifications that are developed independent of any vendor, and not controlled by a single vendor.
- Specifications that are available to all vendors who would like to compete in a marketplace.
- Specifications that can be used to test vendor products in order to ensure conformance and interoperability.
- Specifications that are implemented in more than one vendor's product line.

While this set of rules goes beyond the simpler definition set forth in Chapter 1, it is required, since it provides more criteria for

determining what is, and is not, open. Many vendors make their specifications freely available. This does not make them open. In other cases, a vendor may develop a specification, but then turn control over to a standards-setting body. This has the effect of making a proprietary specification open.

 Open communication protocols

When it comes to network integration, there are two primary open standards. We will examine both the *Internet Protocol Suite* (commonly known as *TCP/IP*) and the ISO's *Open Systems Interconnect* (OSI). As we will see, both offer a comprehensive set of standards. However, their present role in our blueprint is open to some debate. See the soapbox later in this chapter for our views.

Before proceeding, it should be noted that many organizations still prefer to use proprietary specifications. Integration is easier to achieve. Less "finger pointing" is likely to occur. Products offered by a vendor that support the vendor's own specifications tend to perform better. Furthermore, the vendor can take advantage of leading-edge technologies and adapt them quickly. The general feeling is that vendors provide better support for their own specifications. IBM's SNA and NetBIOS, Novell's SPX/IPX, Digital's DECnet still account for the majority of network protocols in use today within commercial enterprises. It is unlikely that the open protocols will dominate until organizations are convinced that there are compelling reasons to make the switch. This might happen if governments were to commit to open protocols.

 The Internet Protocol Suite: TCP/IP

In the early 1970's, the United States Government, through its armed forces, created a network to promote cooperation among researchers at universities, research installations, and certain military installations. This early network, called the Defense Advanced Research Projects Network, or DARPANET, was one of the first attempts to build a heterogeneous network.

The variety of software and hardware to be supported posed certain problems. It would have been impossible to develop products for each target system, not just because of the diversity of products, but because many of the specifications that were implemented were proprietary and secret. Instead, it was felt that a series of simple specifications could be developed that would enable each research installation to develop its own implementation. Since many of the researchers used computers as a primary tool, it was possible to leverage the expertise of all the potential users in the creation of products that would be custom-developed for their particular computer system. With this decision, the foundation for the Internet and TCP/IP was created.

Many books have been written on the subjects of the Internet and TCP/IP; we have no interest in rehashing these works. Nor do we have the time, in this book, to do justice to either subject. We will therefore consider those issues concerning TCP/IP and the Internet that impact multivendor integration options.

Today, the Internet is arguably the world's largest network. Consisting of thousands of networks and millions of users, the Internet continues to grow at an estimated 15% per month. The protocol that binds the Internet together is generally referred to as the *Transport Control Protocol/Internet Protocol*, or *TCP/IP*. This is a bit of a misnomer, since TCP/IP actually refers to only two of the standards that are available in the Internet Protocol suite, but we will follow convention.

In addition to being used to bind the Internet, TCP/IP has emerged as the dominant open multivendor protocol suite. With a broad base of support, a wide choice of vendors, and an available pool of skilled staff, the TCP/IP protocol offers the only choice for organizations wishing to use an open specification to perform multivendor integration. It offers a full set of services, including file transfer, remote terminal, directories, and mail services. The specifications are freely available via the Internet. Organizations and individuals can easily provide extensions to the protocol. An excellent example is Sun Microsystem's development of both the Network File Service (NFS) and Remote Procedure Call (RPC). Developed to meet internal Sun networking challenges, the company chose to release the specifications. This means that many vendors' products now include implementations of these standards. (See Fig. 2-5.)

Figure 2-5

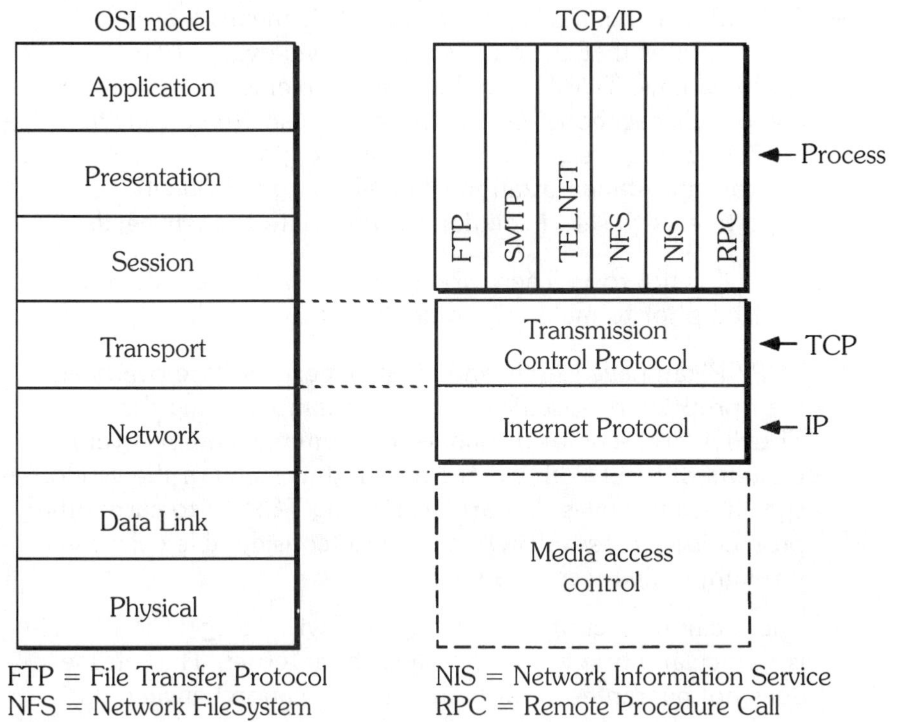

TCP/IP and its relation to the OSI reference model. Learning Tree International

Should you select TCP/IP as your primary ETES protocol? Here are some potential benefits:

- TCP/IP has been widely adopted. It is an open specification that shows every sign of being with us for the foreseeable future. This means that investment in TCP/IP is probably secure. Since it is an open specification, you do not have to commit to a single vendor's product line. In addition, the size of the Internet guarantees that the protocol will continue to be enhanced in order to take advantage of new technology as it becomes available.
- TCP/IP facilitates access to the Internet. While the battle for the commercialization of the Internet is ongoing (see our soapbox), it is clear that access to the Internet is advantageous for inter/intracompany communications.

- TCP/IP is a "neutral" specification. This means that in organizations that currently support a wide variety of specifications, TCP/IP can be easily implemented on the company's backbone. Gateways can be used to convert to TCP/IP as required. This will satisfy the global requirements for support and standardization while allowing individual departments to select solutions appropriate to their needs.

Of course, all is not rosy. There are a number of concerns that must be considered prior to making a commitment to TCP/IP.

- TCP/IP can have significantly higher performance overhead than proprietary specifications. For example, using the TELNET protocol for terminal–host communication is very wasteful of bandwidth when compared to many of the vendor-specific alternatives. In particular, using TCP/IP to carry other protocols such as NetBIOS can cause considerable delays in throughput and response time.
- There can be security risks associated with using TCP/IP. This is particularly true when accessing the Internet. The Internet does not guarantee security in communication between the source and the destination.
- Scalability of the TCP/IP specifications is an ongoing concern. At present, the number of network addresses available on the Internet is rapidly diminishing. A new addressing scheme has recently been adopted. Migrating the entire Internet community and all other TCP/IP-based networks to this new addressing structure will require considerable effort. For the moment, a number of short-term fixes are available. However, as the new addressing scheme becomes more commonly available, organizations that have adopted TCP/IP will face a major migration task.
- At the risk of offending some, it should be mentioned that TCP/IP is an aging protocol—it is over twenty years old. This means that it cannot always easily take advantage of current research in multivendor integration.
- There is a high administrative cost associated with TCP/IP. Many activities such as name and address assignment are not easily automated, and must be done manually. Furthermore,

interoperability testing is generally the responsibility of local organizations, as there is no certification agency currently available or endorsed. Organizations should be prepared to invest in additional staffing resources for their TCP/IP-based networks.

While organizations should be aware of the possible disadvantages of using TCP/IP, the potential benefits clearly outweigh them. In summary, TCP/IP provides an excellent choice for a near-term integration specification. The size of the Internet guarantees that the specification will have longevity. However, as we will see later, proprietary specifications are still viable and must also be considered.

Open Systems Interconnect

Consider the global telephone network. In magnitude, it clearly overshadows even the largest data communication networks currently available or being contemplated. What are the mechanisms that enable us to connect to any one of the approximately four billion telephones in the world? Clearly, the complexity of implementing the multivendor integration of the global telephone system is many orders of magnitude greater than building a well-integrated data communication network in a particular organization. Furthermore, since most governments wish to control their internal phone systems via publicly owned utilities, some agency is required to act as a coordinator to promote international telephone connections.

That agency is the International Telecommunications Union (ITU). One of the ITU mandates is to develop specifications that enable different companies' phone systems to communicate. ITU has broad international support for its specifications. Participants in ITU are governments or their appointed representatives.

The ITU has been active in the area of data communications for many years. International acceptance of protocols such as X.25 (packet switching) can be directly attributed to ITU. One of the major subcommittees of ITU is the Technical Standardization Sector (ITU-T). It is largely responsible for the development of the standards that are used in data communications. In the 1970's, it was determined that a global, comprehensive set of specifications was required to link data communication devices. They requested that the International

Organization for Standardization (ISO) develop these specifications. Working with other international standards groups and major vendors, the ISO developed a suite of specifications that are commonly referred to as Open Systems Interconnect or OSI.

The original intent of OSI was to act as a gateway between proprietary vendor specifications. In other words, one should be able to use OSI to link DECnet to SNA. Rather than intending the specifications to be implemented within private organizations, it was viewed as a common global backbone to link organizations. This is similar to the way the phone system works today. Each organization can choose its own PBX systems, but they are all linked via common specifications.

OSI's history has been rocky at best. Many of the specifications have gained rapid acceptance. For example LAN protocols such as Token-Ring and Ethernet have been widely adopted. In addition, wide area network specifications such as X.25 and others have been broadly implemented. At the top end of the OSI specifications, protocols such as X.400 (for electronic mail) and X.500 (directory service) will probably be quickly accepted. However, adoption of the OSI protocol, as a single entity, still appears far off; the chief reason for this may have to do with the Internet. (See Fig. 2-6.)

Should you be considering OSI as an open integration protocol? Here are some of the potential benefits:

- OSI provides a comprehensive set of specifications to meet organizations' requirements for reliability, security, functionality, and scalability. It was designed with integration as a key goal.
- OSI still enjoys widespread support by most of the world's governments and public carriers. While this support is still not coordinated, it is anticipated that, over time, a global OSI foundation will be put in place.
- OSI provides a vendor-neutral set of specifications. This is a benefit that it shares with TCP/IP. Commitment to OSI does not require selection of any particular vendor's products.
- OSI was designed for growth. It is unlikely that any organization will ever reach the capacity limits provided by the specifications.

Figure 2-6

<table>
<tr><th>OSI layer</th><th colspan="6">Example ISO protocols</th></tr>
<tr><td>Application</td><td colspan="2">ISO 9040/9041
VT</td><td>ISO 8831/8832
JTM</td><td colspan="2">ISO 8571/8572
FTAM</td><td>ISO 9595/9596
CMIP</td></tr>
<tr><td>Presentation</td><td colspan="6">ISO 8823/CCITT X.226
connection-oriented presentation protocol</td></tr>
<tr><td>Session</td><td colspan="6">ISO 8327/CCITT X.225
connection-oriented presentation protocol</td></tr>
<tr><td>Transport</td><td colspan="6">ISO 8073/CCITT X.224
connection-oriented presentation protocol</td></tr>
<tr><td>Network</td><td colspan="3">ISO 8473
connectionless network service</td><td colspan="3">ISO 8208/CCITT X.25
packet level protocol</td></tr>
<tr><td rowspan="2">Data Link</td><td colspan="4">ISO 8802-2</td><td rowspan="2">ISO 7776
CCITT X.25
LAP/LAPB</td><td rowspan="2">ISO 7809
HDLC</td></tr>
<tr><td>ISO 9314-2
FDDI</td><td>ISO 8802-3
CSMA/CD
bus</td><td>ISO 8802-4
Token
Ring</td><td>ISO 8802-5
Token
Ring</td></tr>
<tr><td>Physical</td><td colspan="6">Options from EIA, CCITT, IEEE, etc.</td></tr>
</table>

OSI reference model with some sample specifications. Learning Tree International

As always, here are some of our concerns:

- ➢ There is a definite lack of products, skilled staff, and vendor support. This makes implementation difficult and challenging.
- ➢ TCP/IP is the prevalent technology. This means that as an OSI implementer, an organization will probably be isolated from the mainstream of data communication networks.
- ➢ Early OSI implementations might have considerable performance overhead problems. It will take some time to "fine tune" the specifications so that acceptable performance levels can be achieved.
- ➢ OSI is very complex. Not only is the terminology different, but the specifications, by trying to be all things to all people, are difficult to implement and support. This means that there are also training costs that must be considered.

The net result is that there is a definite lack of motivation to migrate to OSI in the short term. At some later date, when the

world's governments put their full support behind OSI, there will be time to consider migration. On the other hand, use of X.400 and X.500 to provide a common mail and directory system should be considered if possible.

Andres and Jerry's soapbox: The open networking wars

Want to make enemies real fast? Just suggest that the Internet should migrate to OSI specifications. Even better, make the suggestion on the Internet. Then prepare to go into hiding for a long, long time. Few issues have generated more technical controversy than the OSI vs. Internet debate.

Originally the Internet was seen as the logical place to launch the OSI specifications. It was, by definition, open, available, and required updating. Also, since the Internet does not guarantee security or performance, it would have provided an excellent test bed for the new OSI technology. And, as near as we can make out, that was the plan. But someone forgot to take into account the philosophical difference between TCP/IP people and OSI people.

To understand this difference it is important to understand that TCP/IP specifications are developed by consensus. Anyone can comment on a new proposed standard. In addition, many of the Internet's most popular services were developed by the users. There is a strong sense of ownership, a sense of pride, and a chaotic anarchy that governs the Internet experience. While overall decisions may be made via the Internet Activity Board, the Internet community, as a whole, can and does strongly influence the direction of the specifications and the network.

Now, consider the reaction of these cyber-pioneers to the seemingly draconian decision on the part of the U. S. Government to deploy OSI on the Internet—to convert the Internet to a set of specifications that had been developed by companies and standards-setting bodies. The perception was that the Internet was being taken out of the hands of users, and being put into the hands of the government. The reaction was, in hindsight, predictable: absolute rejection of OSI. Of course, the fact that many of the members of the IAB had been significant contributors to the development of the TCP/IP specifications meant that this rejection would find support at all levels of the Internet.

Sadly, too many people still see the issue in terms of either OSI or TCP/IP. In our opinion, this is the major obstacle to OSI acceptance. In fact, the world probably requires at least two major global networks. The Internet should be used as a common meeting ground to link individuals and organizations that wish a forum for discussion on a wide variety of topics. It provides a low-cost alternative for information sharing. The overall purpose of the Internet has never been clear. It should be left alone to develop as it may.

Commercial organizations and governments have different networking requirements. Reliability, security, and availability must take priority. To accomplish this, a second network is required. This is where OSI should play a key role. It was developed with these requirements in mind. It should be implemented by governments and public carriers. While the cost of implementing and operating such a network is considerably higher than the Internet, commercial organizations should be willing to support this effort.

What will happen? Who knows? That's what makes this industry so exciting.

Portable Operating System Interface (POSIX)

A major interoperability issue when considering the portability of programs between different operating systems is the fact that each system offers its own set of unique calls to access system services. It would be preferable if a single set of calls were available on each operating system. This is the motivation for POSIX. The IEEE has been developing these specifications since 1981; they are covered under the general specification called 1003.1. The IEEE has now been joined in these efforts by the ISO. The task is to create a single set of Application Programming Interfaces (APIs) that will be implemented on a wide variety of proprietary platforms. The *X* was added in recognition of the Unix roots of POSIX. In 1987, the European X/Open Group endorsed POSIX as the standard set of calls for UNIX.

Today POSIX, while having been adopted by many of the major standards-setting groups, is still struggling for commercial

acceptance. Part of the problem is that many still consider that the POSIX calls are too watered-down to be of value. By trying to be all things to all groups, POSIX may end up being too general to be useful. Nonetheless, POSIX calls have been adopted by many major operating systems, such as IBM's MVS, Digital's VMS, and Microsoft's Windows NT. Sadly, there are few major applications that have been written to these specifications.

The IEEE and ISO now control the major direction of the POSIX specifications. In September 1992, IEEE approved the POSIX 1003.2 specifications, which included the base shell programming language needed for a shell-script language. It also defined POSIX 1003.2a, which defines a series of utilities that should be offered to POSIX-compliant applications.

However, performance considerations, end-user applications, and lack of completeness will probably delay the widespread adoption of POSIX. In addition, there is the question of who will promote POSIX. Once again, it is likely that vendors will promote their own specifications for their own products, rather than POSIX. Since customers seem to be willing to accept this state of affairs, we anticipate no major POSIX adoption in the near future.

Open Software Foundation's Distributed Computing Environment (OSF/DCE)

What happens when 300 hardware and software vendors agree on a standard? No one knows for sure, but let us call it *OSF/DCE*. While there is some debate over the reasons why OSF was formed, the results are rapidly becoming clear. OSF/DCE is emerging as key middleware for multivendor integration. By providing a level playing field for technology, OSF/DCE can be viewed as the best bet for fully integrated networks.

How does OSF/DCE work? We will discuss this subject at more length in a later chapter. For the moment, we consider *why* OSF/DCE works.

OSF is a consortium of key players in the computer industry. Founding members included IBM, HP, Siemens, Digital, and others. The mandate given OSF was to select the best technology from each member company, integrate it, and then license it back to the members. For example, the OSF/DCE directory services, called *Cell Directory Service*, was provided by Siemens. Siemens provided the code, which was then distributed to each member company. This means that regardless of who sells it to you, every version of CDS is the Siemens code (as modified by OSF). The implication is that there is an excellent chance that DCE components provided by different manufacturers will interoperate. This enables vendors to compete based on the functionality and quality of their products, rather than on specifications. Furthermore, it enables purchasers to have much greater freedom of choice in selecting vendors.

As with other new standards, there are certainly concerns. Beyond the relative immaturity of the products, there is a lack of software development tools, qualified staff, and broad-based exposure. In addition, foundation applications such as databases, e-mail, and others are still missing. Even more disturbing is the lack of participation by Microsoft and Novell in the OSF. Both companies have provided, at best, lukewarm support for DCE. This is a problem for organizations that have a major commitment to products from these companies. Even though IBM offers a native DCE client for MS Windows, it would have been preferable if Microsoft had chosen to license the DCE technology.

Nevertheless, we consider OSF/DCE to be a significant technology. While it may still be too early to adopt on a global organization level, it is not too early to investigate the technology and learn its capabilities. That is why we have chosen to devote a chapter to OSF later in this book.

 Object Management Group

Is object technology the ultimate middleware? Many people think so. The OMG now has over 300 members. As a nonprofit, international consortium, OMG is defining the standards for distributed objects. The key word is distributed. While using objects on a single platform

is being explored by a number of vendors (for example, Microsoft's Compound Object Model and IBM's System Object Model), making objects work across a network is a very different problem.

There is a logical link between object technology and the well-integrated network. Interoperability of different vendors' software can be achieved in a number of different ways (which we will explore in Chapter 4). In most cases, this boils down to a choice between two basic approaches. Either you convert, or you use a common standard. Object technology provides a third, and arguably better, alternative.

At their most simplistic level, objects can be considered containers of both process and data. Rather than having to know the internal workings of a process, all a calling process needs to know is

- how to find the object
- how to invoke the object
- how to pass messages to the object

This defines some of the OMG mandate. The OMG's Object Management Architecture Guide defines four major components required to implement distributed object technology. These are:

- The Object Request Broker: the mechanism that enables objects to transparently receive/transmit requests with other objects. It is commonly referred to as *CORBA*, or the *Common Object Request Broker Architecture*. CORBA 2.0 will define a usable interoperability specification on how this will be accomplished.
- Application Objects: objects specific to an end-user application. These typically communicate with other common objects to provide end-user services.
- Common Facilities: fundamental end-user services. Typical examples might be e-mail, database access, and compound documents. These are of course optional, and will likely be one of the last areas for detailed specifications.
- Object Services: required services needed by every system that wishes to participate in CORBA. Services include object naming, event notification, security, etc.

What's the current status of all this technology? Well, companies such as Sun, HP, and IBM are all using it. IBM has released a version of CORBA called DSOM (Distributed System Object Model). Sun's offering is called DOE (Distributed Objects Everywhere). Component Integration Labs (a consortium of Xerox, WordPerfect, Apple, IBM, and others) will shortly begin releasing their OpenDoc products, which are compound documents that can operate across multivendor platforms. Imagine creating a document on a Sun workstation, modifying it on a Macintosh, and displaying it on an IBM PC running OS/2. Sound impossible? Let's really stretch your imagination. Not only will this be done, but this compound document will be able to call on services from the remote machines when displaying the document! OpenDoc technology may represent one of the first "killer" apps based, in part, on the OMG specifications.

If technologies such as CORBA and OpenDoc live up to their promise, the era of true multivendor interoperability will be upon us sooner than many of us thought possible. With OpenDoc developer kits now shipping, and with the release of the second revision of the CORBA specifications that define interoperability between different vendors' object brokers, the possibility of seeing real-world applications in the very near future is high. For the moment, it's one part wishware, one part vaporware, and some parts realware. But it sure sounds exciting, doesn't it?

One last word. What about Microsoft's Object Linking & Embedding (OLE) and its underlying Compound Object Model (COM). Is it better than the OpenDoc/CORBA offerings? Well, rather than get embroiled in that dangerous debate, we will avoid it through a convenient loophole. Since this discussion is about *open* specifications, no proprietary specification need be included. Second, since at the current time OLE/COM can only operate on a single machine, it does not provide a vehicle for multivendor integration. All this is subject to change. Microsoft is planning to provide distributed OLE/COM with the release of the Cairo operating system. Digital has said that they will provide a gateway between COM and CORBA. A wait-and-see attitude must be adopted.

SQL Access Group: SAG

Founded in 1988, SAG has a mandate to provide an interoperable standard to allow database clients to talk to any database server, and to provide a Call Level Interface (CLI) that defines a common API set for multivendor databases. Products such as Microsoft's *Open DataBase Connectivity* (ODBC) are based on this standard, with extensions. SAG presently has 44 database vendors participating.

Unfortunately, vendors tend to implement the CLI specifications with extensions that once again obligate you to use a particular vendor's products. In addition, the specifications are still evolving. There is still a great deal of politics in the area of database access, which waters down the influence of groups like SAG. A wait-and-see attitude is recommended.

Proprietary specifications

We have seen how open specifications can be very appealing. They provide a framework in which to implement true peer-to-peer networks, and a way to achieve interoperability among components. While the size of the Internet guarantees that TCP/IP will be in strong demand for many years to come, most network communications are still based on proprietary specifications. In fact, we might even say that a prime task of the integrator is to provide a gateway between these proprietary specifications. Later on in this chapter, we take the opportunity to examine how the gateway can be implemented. Chapter 4 deals with gateways between open and proprietary specifications. But first we provide a brief overview of these proprietary specifications.

Adopting a standard based on a proprietary protocol has a number of advantages. Vendors will probably support their own specifications much better than those of others, and the corresponding protocols are therefore easier to integrate. Products built to these specifications are generally mature and stable, since they have had many iterations to get them right. In addition, using proprietary specifications allows

a degree of very tight integration difficult to achieve using open protocols. This tends to make the proprietary specifications a better choice from a price/performance point of view. In general, upward compatibility will more likely be offered within a single vendor's proprietary product line than across the broad range that an open specification will impact.

We should also point out that sticking with a single vendor does not guarantee upward compatibility. For example, consider the Atari computer line. Rarely was a new model compatible with its predecessor. After a while, developers and end users became frustrated rewriting or repurchasing the same software. So proprietary is no guarantee of safety.

In fact, we might say that upward compatibility is one of the single most important requirements that must be addressed when considering how to integrate a wide range of proprietary protocols. Without a solid approach to upward compatibility, the current investment in software (which will always be substantial) cannot be preserved. This means that organizations will constantly have to retrain their end users and invest heavily in new, incompatible software. While this issue may not be critical if the software is low-cost shrinkwrap, it becomes very important if the end user has any software that has been custom built. Having to rewrite software applications is always a drain on an organization's resources.

It should also be mentioned that support and administration costs using proprietary specifications will probably be lower than using their open cousins. In particular, the network management products specific to a single vendor that are available to manage resources are much better than the general-purpose network integration consoles using SNMP. The Simple Network Management Protocol provides one of the only viable approaches to integration based on an open system specification. That does not make it better than the proprietary systems. A great deal depends on both the budget and the resources available to implement a network control center. Later on, we will have more to say about network management systems (NMS) and the role they play in modern network systems.

As we have seen, there are potential benefits in using a proprietary protocol, and there are concerns. We have saved the most serious

concern for last. Selecting a proprietary protocol increases the potential for "vendor lock-in." Once you choose a proprietary protocol, it is very difficult to change at a later date. The vendor knows this and, in some cases, might take advantage of your position.

IBM's SNA

The most widely implemented proprietary enterprise-wide networking system in use today is IBM's SNA. It provides a stable set of interoperable products, and also has broad-based industry support from many third-party developers. Originally developed to integrate large IBM networks consisting of mainframe computers and thousands of terminals, SNA has now been adapted to support peer-to-peer communications. This work goes under the name *Advanced Peer-to-Peer Networking* or APPN.

One of the prime benefits of APPN is that it allows for much of the network configuration to be handled automatically. This enables the network to be more responsive to dynamic conditions that may occur during normal operations. Unlike TCP/IP, the IBM networking protocol is connection-oriented. In general, this will provide a more reliable, secure network, with the tradeoff involving the implementation expense. In addition, APPN supports a variety of protocols, including TCP/IP and NetBIOS, through encapsulation. This means that APPN can be used as an integration technology for those wishing to build multiprotocol backbones.

For those organizations with a strong commitment to IBM or SNA products, APPN appears to be a good choice. In addition, it may be of interest to those who have examined TCP/IP and find that it is a bit "light" for their requirements. Organizations with a heavy transaction-processing load will also want to examine APPN.

Recently IBM announced a mixed-stack approach to interoperability called *Multiprotocol Transport Network* (MPTN). Implemented in a series of products called the Anynet family, this strategy allows a variety of peer-to-peer protocols to operate over a variety of End-to-End Services. This means that it should be possible to run CPI-C over TCP/IP, or run RPCs over APPN. However, this product is still

in its infancy, and not enough is known to allow us to recommend it at this stage.

Anynet products are available for many platforms, including MS Windows, OS/2, AIX, and MVS. You can therefore test the products to determine whether this approach will satisfy your requirements. More products of this nature will become available over the next few years as vendors try to integrate their products with other vendors' offerings.

NetBIOS

In 1984, IBM and Sytek introduced a protocol for use on local area networks called *NetBIOS*. This protocol has become the de facto standard for local area networks, and is supported by all major software and hardware vendors. NetBIOS provides both connection-oriented and connectionless End-to-End Services. By providing a simple, intuitive interface, NetBIOS can be used by programmers to implement new applications easily. The fact that it is widely available assures software developers a large potential market.

Given the fact that NetBIOS represents a de facto standard and that it is implemented on virtually all LANs, it might be thought that it would be the ideal choice for an integration standard. This might have been the case if, back in 1984, IBM and Sytek had been thinking in terms of enterprise integration rather than local area networks. However, at that time, IBM still considered SNA to be the protocol that would integrate organizations. NetBIOS was seen, at best, as a peripheral protocol.

This attitude impacted the design of NetBIOS. It lacks many of the advanced features required by an enterprise protocol. Chief among these is the fact that NetBIOS cannot be routed. This means that it is impossible to make optimal route selection or dynamically modify paths when handling NetBIOS traffic. As a side issue, the fact that NetBIOS cannot be routed also means that it becomes impossible to implement network layer addressing. This means that discovering the path from the source to the destination using NetBIOS is cumbersome and difficult.

Other issues that confront the users of NetBIOS include its lack of standard naming conventions (in particular, support for the Internet naming scheme), authentication, and security. Finally, programs that have been written to conform to the NetBIOS interface are tightly coupled to the underlying network infrastructure, and portability is affected.

In summary, NetBIOS is the *de facto* standard for most LAN communication. Its lack of scalability, portability, and flexibility make it an unwise choice for an enterprise system.

Novell's SPX/IPX

Novell's SPX/IPX is based on the Xerox Network System (XNS). It has become a major LAN-based protocol for use in both small LAN systems and larger enterprise systems. It has been widely implemented and has a wide degree of third-party support. Like NetBIOS, it has become a de facto standard.

SPX/IPX provides high performance and a wide variety of distributed services. These include security, directories, file services, and print services. Novell's network product line implements many of these functions. Additional functionality is provided by third parties.

There is some concern over Novell's use of SPX/IPX for very large systems. Some of these have to do with the fact that SPX/IPX was never intended to work over wide area networks. Novell has been addressing these concerns by continually improving the protocol. Perhaps of more concern is the somewhat political issue of relying on a proprietary vendor's protocol for enterprise integration. While we have indicated that IBM's SNA may be a suitable choice for some environments, Novell's offering does not have the maturity or sophistication of the IBM protocol. In addition, Novell's long term commitment to the protocol cannot be guaranteed. A legitimate question might be "Will Novell continue as a full-service LAN provider, or will they focus on server technology?" Their recent acquisitions seem to indicate that they are more interested in becoming a competitor to Microsoft rather than IBM. This leaves the future of SPX/IPX in some doubt.

In any event, SPX/IPX is a major *de facto* standard in the LAN industry. For the short term, it still represents a good compromise between ease of use and sophistication. The integrator should consider continued support for SPX/IPX, but may not wish to use it as the primary protocol for integration. Of course, those organizations that have widely implemented Novell products will wish to use SPX/IPX as their primary protocol.

 Other proprietary specifications

Many other proprietary specifications exist. Apple's AppleTalk architecture provides a well-integrated approach for Macintosh computers. It operates over Token Ring, Ethernet, and LocalTalk. Many router vendors support AppleTalk on their products. However, AppleTalk does not make sense as an enterprise protocol. It is exclusive to the Macintosh environment. It does not offer the sophistication required to implement heterogeneous products. While it may be ideal for Macintosh environments, it should be limited to departments that want a simple, low-cost solution.

Another proprietary protocol is Digital's *Digital Network Architecture* (DNA). It is implemented in a product line known as DECnet. The vast majority of Digital's clients use DECnet. However, we have chosen not to cover this protocol because Digital is in the process of migrating their customers to an OSI-based version of DECnet. While they have run into some resistance in performing this migration, it appears clear that it is the direction that Digital has chosen. Later on, we will cover both internetworking and interoperability with DECnet products.

 Proprietary protocols and multivendor networks

As we mentioned, many proprietary protocols exist. A number of potential benefits will probably result from selecting them as your primary integration protocol. However, the protocol must be sophisticated and powerful enough to accommodate the needs of a heterogeneous environment. It must also provide for flexibility,

scalability, and ongoing support. In our opinion, with the exception of IBM's APPN (and perhaps DECnet), none are acceptable for large network integration.

Protocol integration strategies

We conclude this chapter with an overview of building the well-integrated network. In general, we use a series of methodologies in building our enterprise system. Chief among these are protocol reduction and the establishment of a standard backbone. Since this is not always possible, we examine other alternatives, such as mixed and dual stacks, as well as multiprotocol backbone approaches.

Protocol reduction

Protocol reduction adheres to the dictum "less is more." Whenever possible, organizations should attempt to limit the total number of protocols that must be integrated and supported. By selecting a few well-known protocols, the integrator's task is simplified. Of course, we always need to ask which protocols should be kept and which should be eliminated. The decision must be based on the current demographics of protocol utilization, plus the long-term commitment that an organization has to a particular vendor or protocol. In general, protocols such as TCP/IP, SNA, OSI, and in some cases NetBIOS, all are good to keep. (See Fig. 2-7.)

Standard backbone

It is not always possible to perform protocol reduction. In many cases, local departments may wish to preserve their autonomy. In addition serious investment may have already been made at a local level in products that require a particular protocol. Somehow, the need to preserve these "compartments" of protocol must be balanced against the organization's global requirement for protocol reduction. One way that this can be accomplished is through the use of a standard backbone.

Figure 2-8 shows a simple standard backbone implementation. A single protocol such as TCP/IP or OSI is used on the backbone.

Figure 2-7

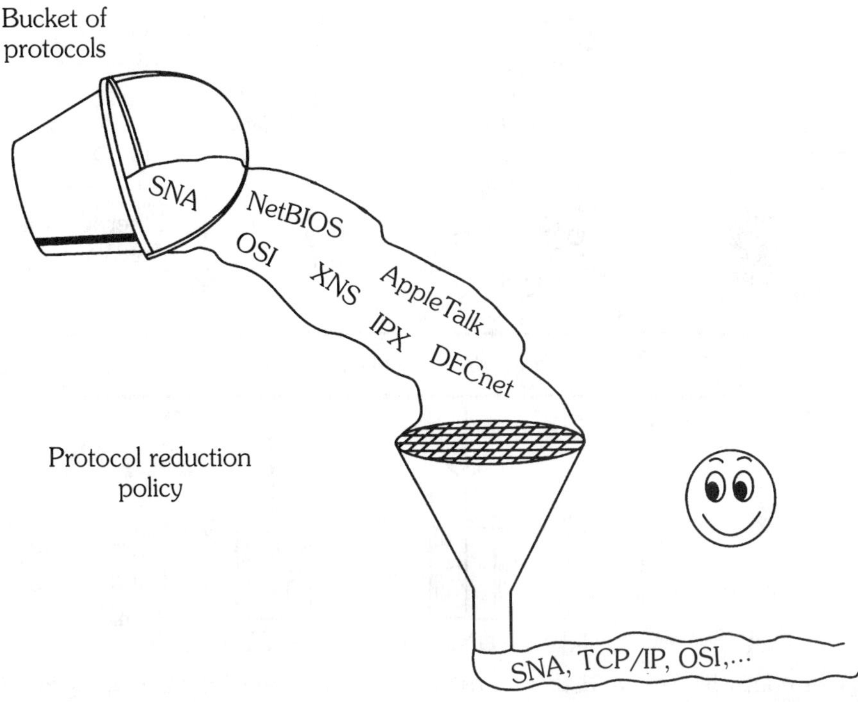

A protocol reduction strategy is one of the preferred methods of simplifying the task of integration. Learning Tree International

Subnetworks are attached to the backbone via a router or a gateway. The function of the router/gateway is to convert between the local protocol and the standard backbone. This allows for local autonomy and it facilitates support and administration on the backbone.

Of course the standard backbone approach has a number of limitations and concerns. Any time a conversion takes place in a network, there is a possibility of errors being introduced through the transformation process. Also, it may not be possible to find a gateway that will perform the required conversion. In any event, conversion will have some impact on performance.

Beyond straight conversion, there exists the possibility of protocol encapsulation. In this approach, the local protocol is "wrapped" in the standard protocol and then shipped across the backbone. This approach assumes that the receiving station will understand the

Figure 2-8

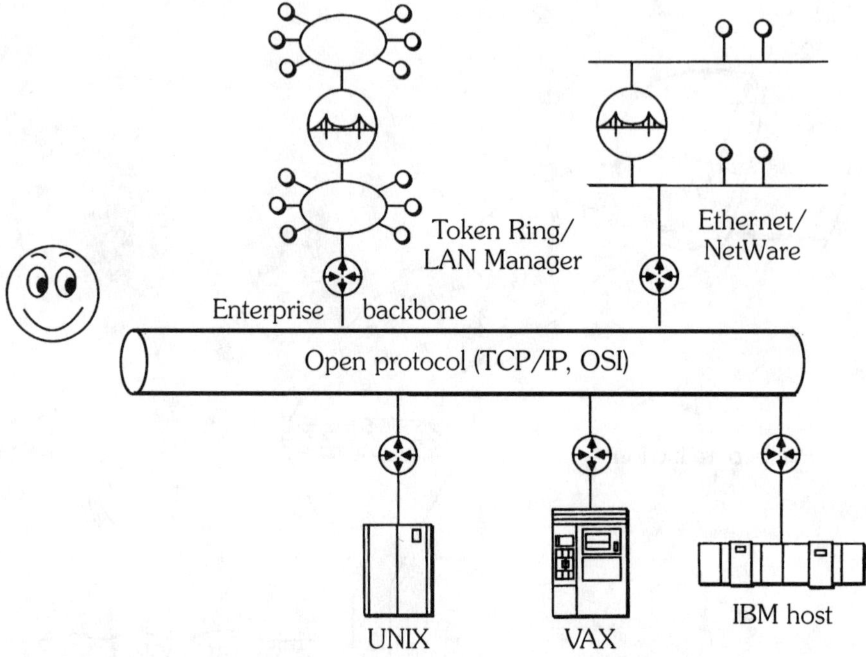

A standard backbone facilitates the implementation of the well-integrated network. Learning Tree International

protocol that was used by the transmitter! Except in special cases, encapsulation should be avoided. It is inefficient, has high overhead, and does nothing to provide for interoperable systems at an enterprise level.

Since it may not always be possible to implement a standard backbone what alternatives exist?

Single-protocol stacks

Figure 2-9 shows one possible alternative. By using a single protocol, complete integration can be achieved. Selecting a major protocol such as TCP/IP or SNA is probably a good choice, as they are widely available on most vendors' hardware and software. The benefits are obvious: ease of administration and support, full interoperability, and ease of internetworking. Unfortunately, the disadvantages far outweigh the benefits of this solution. They include the lack of

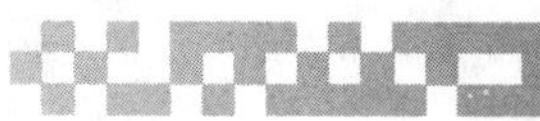

Figure 2-9

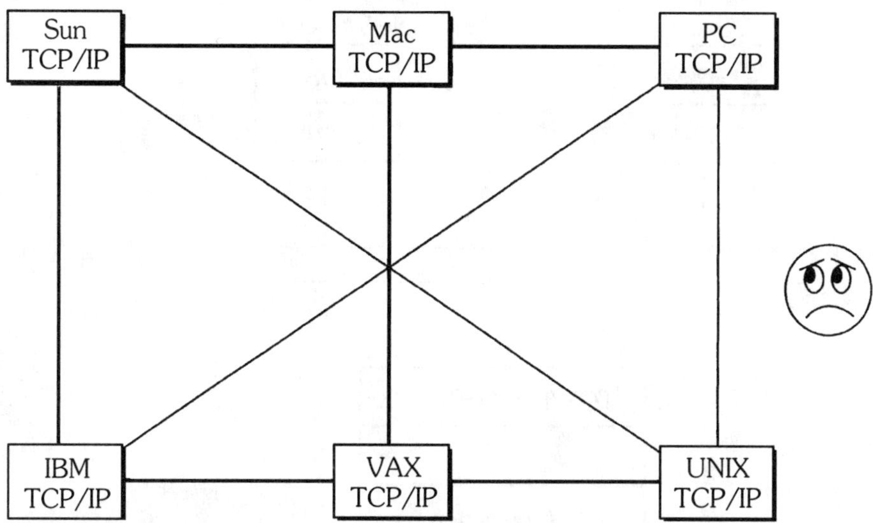

A single-protocol strategy lacks the flexibility required to make it attractive.
Learning Tree International

autonomy available to local departments, the inability to perform local optimization, and the fact that some software packages will inevitably not work with the selected protocol.

By removing choices in protocol selection and working with a lowest common denominator, the long-term viability of the enterprise system is compromised. While this approach may eventually have more potential, it cannot be recommended at present. Of course if you do have full authority over all aspects of your internet, then by all means use this approach.

Multiple-protocol stacks

Figure 2-10 shows an approach used by many organizations to solve the problem of access to applications using different protocols. Sometimes referred to as *Dual Stacking* or *Multiprotocol Stacks*, this approach relies on having a transmitter or receiver with multiple "personalities." In the figure, we show how a single workstation can access multiple servers. The workstation is configured to support two protocols and can alternate as required.

Figure 2-10

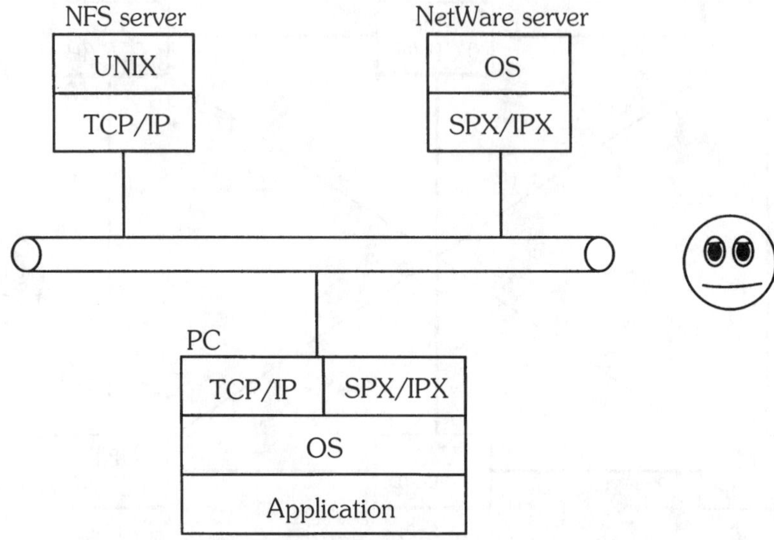

Multiple or dual stacking provides a quick fix for integration with potential longer-term costs in administration. Learning Tree International

As an interesting side note, implementing dual stacking on the workstation raises the issue of operating system selection. If a single-tasking operating system such as DOS or Windows 3.1 is selected, only one communication can take place at a time, with the user or program switching between the two servers. On the other hand, using a multitasking operating system such as OS/2, Windows NT, or UNIX enables the program to communicate with both servers concurrently. As the need to communicate with multiple servers increases, the need for a multitasking operating system on the user's desktop will also increase. We will discuss this at greater length when we talk about client/server technology.

In any event, multiple-protocol stacks are commonly used. While they may result in higher client costs and more difficult support and administration, they have the benefit of getting the job done. We will offer quite a few examples of this approach in Chapters 4 and 5.

Another way to implement dual stacks is to implement the protocols at the server. This approach may be preferable, since it minimizes the number of copies of the protocol that must be purchased, installed, and supported. It also allows network administrators better control over access to their systems. Most modern server operating systems such as NetWare, IBM's LAN Server, and Microsoft's Windows NT Server offer this capability. Whether implemented on the user desktop or on the server, dual stacking provides a viable short-term solution to integration. Since it does little to promote interoperability, it may not be suitable as a long-term solution.

Mixed stacks

Currently a very popular word in the computer industry is "middleware." Middleware refers to using a mixed stack approach to integration. Figure 2-11 shows a number of ways in which this can be implemented. A special program is inserted between two dissimilar protocols to perform emulation on both sides. This enables programs that have been written with a particular interface, say NetBIOS, to operate with other protocols.

Figure 2-11

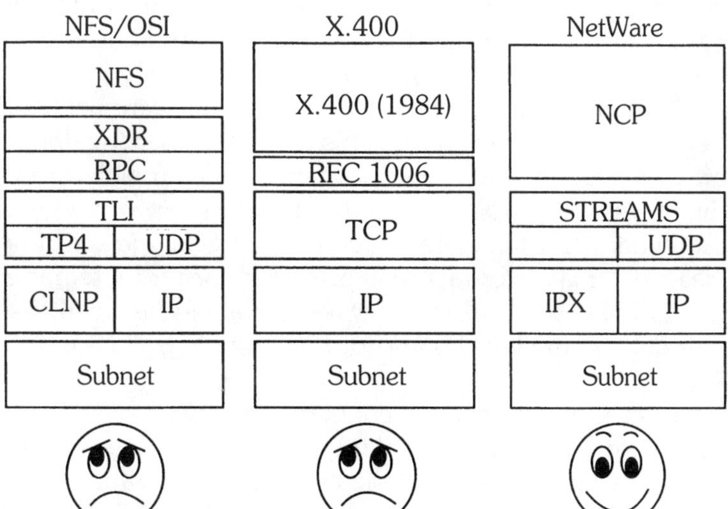

TLI = Transport-Level Interface
RFC-1006 = provision of TPO service over TCP
XDR = External Data Representation

Mixed stacks or middleware may provide the best long-term solution to integration once the technology matures. Learning Tree International

We talked about this briefly when we referred to IBM's Anynet products earlier in this chapter. As you recall, we stated that it might be too early to adopt this form of integration. The products are still immature, and performance is not yet comparable with single-stack approaches. However, because it offers the potential of providing a strong level of decoupling between the application and the underlying ETES protocol, it may have a significant role to play as this technology matures.

OMG CORBA is a good example. An available mixed-stack approach that we favor is offered by Novell in a product called STREAMS. It allows Novell file server protocols to be transported over both SPX/IPX and UDP/IP. However, this is only a single-case solution, and should not be considered a general-purpose integration strategy. The mixed stack has a great deal of merit, but the technology must still evolve.

Protocol conversion

To conclude this section on protocol integration strategies, we present the most popular form of integration, known as *protocol conversion*, or sometimes simply referred to as *gateways*. Ever since the first two incompatible computer systems were built, there has been a need for gateways. In this approach, a standalone hardware or software process translates between the incompatible protocols. The advantage of this approach is that it allows for a "many-many" connection. A single gateway can support a large set of users on both sides. As a matter of fact, our preferred form of integration, the standard backbone, uses gateways to convert between the local protocol and the backbone protocol. (See Fig. 2-12.)

However, there are serious concerns with gateways, from both a performance and integrity point of view. Gateways represent a single point of failure. They are difficult to implement and maintain. Any conversion requires time, and a gateway therefore always introduces a delay. So even though we may need gateways, that does not mean that we have to like them. Very few standards exist for gateways. In many cases, the approach taken to implementation is proprietary, even if we are linking open protocols. We will have an opportunity to look at gateways in greater depth in Chapters 3 and 4.

Figure 2-12

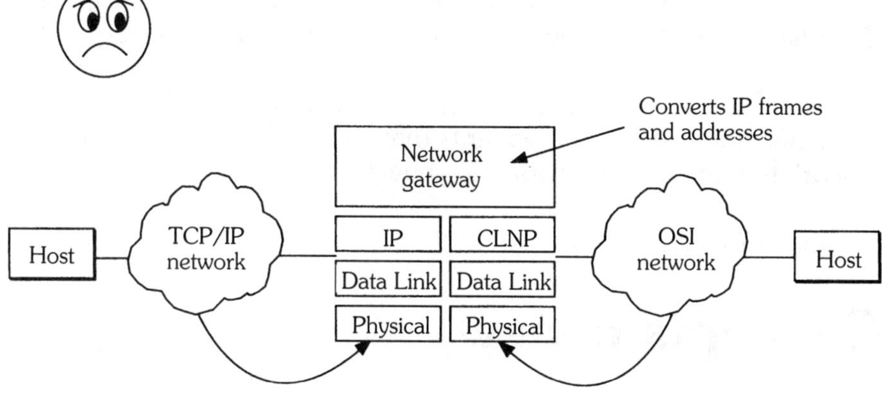

Protocol conversion via external gateways is the most commonly used form of integration, even though there are a number of concerns. Learning Tree International

Summary

In this chapter, we have provided a brief review of the major open and proprietary specifications. We have examined the role of TCP/IP and OSI. The conclusion that TCP/IP is the correct protocol for near-term integration seems obvious. It is not clear that all organizations will necessarily benefit by using open specifications; for those that will not, TCP/IP seems the current winner. The future role of OSI is still unclear. OSI itself will require a higher degree of commitment on the part of governments and major public carriers before it can succeed.

Other specifications also exist. IBM's APPN is a viable alternative for integration. In particular, organizations that are currently using IBM products may wish to consider the advanced functionality offered by this product line. Others, such as Novell's SPX/IPX, are useful for smaller integration scenarios. While the long-term choice appears to be open specifications, short-term decisions may require the use of proprietary specifications. Proprietary does not mean closed.

A variety of approaches exist to integrate protocols. We discussed the standard backbone, dual stacks, multiprotocol backbones, and others. Our preferred integration strategy involves both protocol reduction and a standard backbone. Any protocol reduction strategy must balance the

need for local autonomy with the need for administration and support of the backbone. The use of gateways is accepted as a necessary evil.

We now leave standards and specifications behind us and move on to the examination of the components that will enable us to glue our network together. Internetworking is next.

But first our usual concept review.

Concepts review

1. In today's multivendor networks, what is a key activity of an integrator?

   ____

2. Name three major network protocol suites that a network integrator needs to deal with in today's networks and the major promoters and developers:

   __________, ______________________
   __________, ______________________
   __________, ______________________

3. Name two popular LAN protocols that are used in today's PC networks, and name the major promoters and developers:

   __________, ______________________
   __________, ______________________

4. Name the major components of a successful integration blueprint:

   ____
   ____
   ____
   ____
   ____
   ____

5. Name three successful protocol integration strategies:

   ____
   ____
   ____

Internetworking

CHAPTER 3

ONE of the first areas that must be addressed when implementing the well-integrated network is internetworking. Before the area of interoperability can be addressed, the network must be glued together so that data can be efficiently and effectively routed between any two end points. Decisions about protocols and specifications to be used for internetworking must be made prior to this point. But internetworking is still viewed by many as the first "real" step in building a multivendor network.

In this chapter we examine the problem of internetworking, detail some of the components used in building an internet, and consider some of the challenges faced when choosing one technology over another. As we stated in Chapter 1, this is not intended to be a comprehensive examination of internetworking. A variety of excellent material is available that covers this subject in sufficient detail. However, we require internetworking as a prime element in achieving integration, and will therefore include the key pieces of information that the integrator will require. As previously mentioned, our primary focus will be on interoperability, which we will cover in Chapter 4. (See Fig. 3-1.)

Figure 3-1

Internetworking + Interoperability = Integration

— Media

— Components

— Communication software

— Network operating systems

— Network management tools

Internetworking is usually addressed prior to interoperability. Learning Tree International

 The internetworking problem

A heterogeneous mix of subnetwork technologies and standards currently exists in today's organization. Over the years, this technology has been acquired to meet a diverse set of information-processing requirements, including

- terminal/host application processing
- workstation and microcomputer resource sharing
- linking of geographically separated groups (WANs)
- high-speed local data transfer (LANs)
- local departmental systems

If the intent is to leave these areas isolated from one another, then no internetworking problem exists, and you can avoid reading this chapter! In most cases, the integrator will be assigned the task of linking these systems together to provide a foundation for the seamless sharing of data and process known as interoperability. The technologies used to provide the foundation for interoperability are jointly known as internetworking.

Internetworking is the selection and implementation of a variety of components that link subnetworks together to form an internet. These components include repeaters, bridges, routers, and gateways. Selection of the appropriate technology is a key task of the integrator. While we have yet to define these components, we can state a number of general guidelines that will assist us in making these choices:

- Repeaters implement physical LANs, where the technologies to be integrated are similar.
- Bridges are used to implement virtual LANs or subnets, where the subnetwork technologies are broadly similar.
- Bridges are preferred for connections where the subnetworks to be joined are physically adjacent to each other.
- Remote bridges should be avoided.

- Routers implement an internet by linking incompatible subnetwork technologies.
- Routers are the preferred solution for remote connections.
- Gateways are used to link highly incompatible subnetwork or ETES protocols, where no other solution is available.
- Gateways should be avoided if possible, due to possible performance and integrity problems.

As a final point, it should be mentioned that the integrator will generally find the task of internetworking to be easier than interoperability. Standards exist in many cases, and a great deal of reference material is available to assist the integrator in making the right choices. This does not mean that internetworking will be simple or straightforward, but it provides some assurance that a solution does exist to most internetworking problems. Selection of the appropriate solution can be challenging, and should be approached carefully, since many internetworking solutions require a strategic commitment to a particular technology or specification.

The components of internetworking

Figures 3-2 and 3-3 show the major components used in internetworking. These fall into four major categories: *repeaters*, *bridges*, *routers*, and *gateways*. The assignment of a component to one of these categories is generally based on which layers of the OSI reference model will be used by the component to achieve internetworking. Before defining these technologies, we can make the following statements:

- Repeaters operate using layer 1 technologies. They are concerned with media-to-media connections.
- Bridges use layer 2 to achieve connectivity. They allow more flexibility in the choice of physical media, since they are concerned with the point-to-point connections offered by the Data Link layer. The format of data blocks at layer 2, generally

called *frames*, will be the primary interest of bridges. By *format* we mean the control information, including source and destination address, that is described in the layer 2 specification. The data contained in these frames is of little interest to the bridge.

- Routers operate using control information described by the layer 3 (Network) specifications. What the bridge views as the data portion of a frame, the router sees as a packet containing additional control information plus data. This control information normally includes additional addressing details (network addressing) plus data on how the packet should be routed or relayed through the Internet.
- Gateways may operate on any or all of the seven layers. Since gateways are designed to handle specific conversion problems, it is difficult to provide a general definition. We might describe a gateway by a process of elimination. If it is not a repeater, bridge, or router, then it must be a gateway! We prefer to say that gateways are normally capable of protocol conversion. This implies that the gateway, regardless of the layer that it operates on, will always have incompatible technologies on either side. Its job is to translate data flow from one side into a format that is acceptable on the other. Gateways are sometimes referred to as "spoofers," since they pretend to be something that they are not. Formally this is known as emulation.

As we stated in the last chapter, there is a terminology problem when dealing with internetworking components. Various organizations may describe the same technology using different terms. For example, the Internet describes their routers as gateways. Novell called their router software a bridge. On the other hand, we often find the same term used to describe different technologies. The term relay can be applied to almost any internetworking component. This terminology problem does little to assist the integrator. To solve this dilemma, we recommend the use of the duck test that we described at the beginning of Chapter 2. The term used to name a component is less important than an understanding of its function. We have tried to use generally accepted industry definitions wherever possible. You may choose to accept or reject the use of these terms. To facilitate internetworking,

Figure 3-2

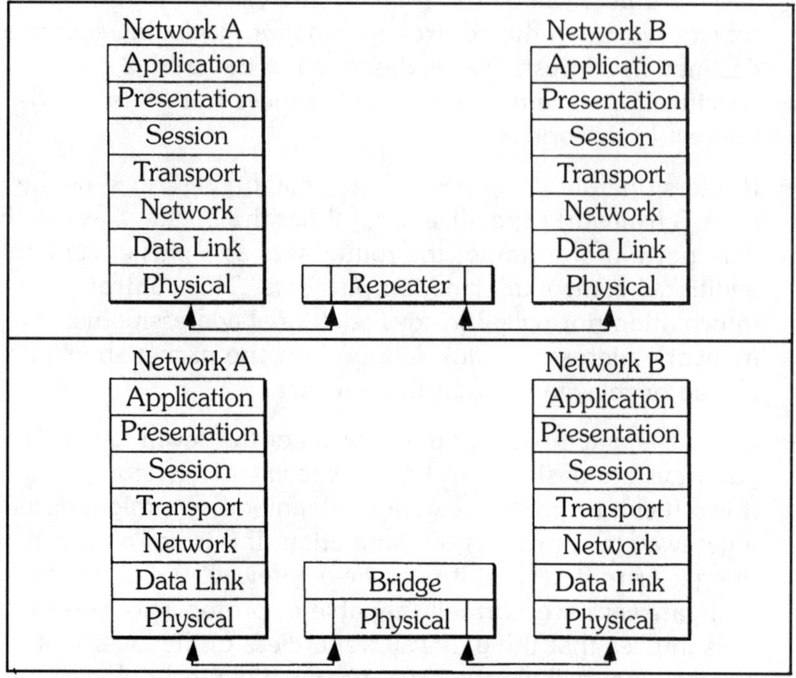

Components of internetworking part 1. Learning Tree International

we recommend that integrators have a set of definitions and terms that can be applied within their own internets. Make sure that vendors agree to these definitions before discussing their offerings!

Repeaters

Repeaters are often referred to as *physical relay devices*. They are used for a variety of purposes, including

- LAN extension
- recovering (cleaning) a signal that may have degenerated
- fault diagnostic point for cable testing
- implementation of a LAN (i.e., hubs)
- connection of different physical media types, such as twisted-pair wire and coax

Figure 3-3

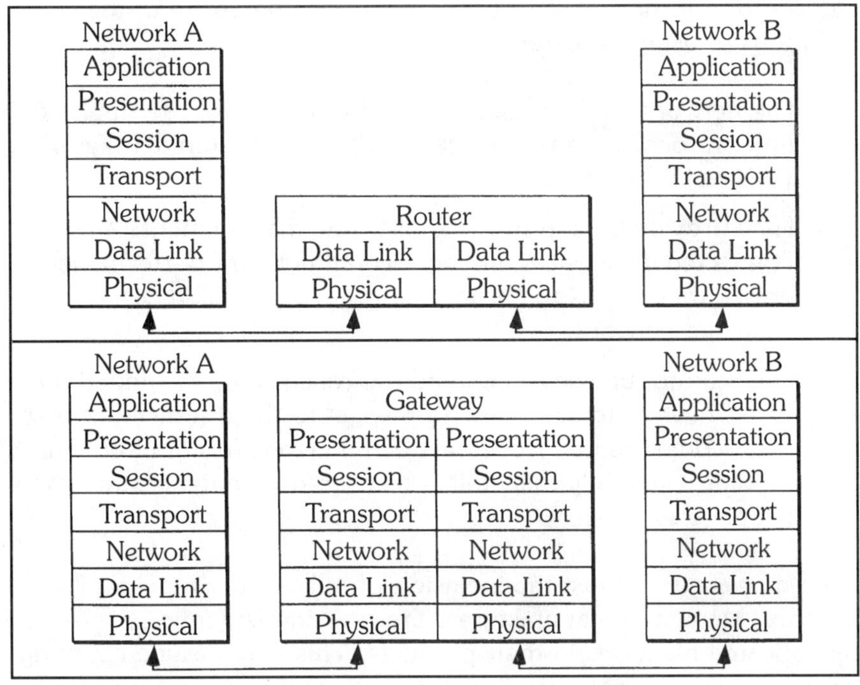

Components of internetworking, part 2. Learning Tree International

Repeaters are considered to be passive devices. Their sole function is to move a single bit from one side of the relay to the other. They are unaware of the relationships among the bits. A repeater cannot distinguish between good bits and bits that may be transmitted in error. This passive approach to internetworking implies that repeaters will be transparent to all systems that access the network. No networking components are ever aware of the repeaters' existence. This transparency can have its advantages, since it provides for a seamless connection between different physical media. A high level of transparency leads us to conclude that repeaters implement a physical network—generally a LAN. Since the repeater forwards all bits from one side to the other, anything that impacts the behavior of one segment impacts the behavior of all attached segments.

Repeaters are also low-cost and very easy to implement. Implemented in hardware, they provide a simple, easy-to-install

solution for internetworking. However, please be aware of the limitations of using repeaters:

- The networks to be joined must, at a minimum, use similar data-link protocols. A repeater cannot be used to join a Token-Ring LAN to an Ethernet.
- They provide little in the way of traffic management. All connected devices can impact overall network performance. A device that begins to "broadcast" will impact all devices on a repeated network.
- They cannot be used to extend the reach of a LAN indefinitely. LANs have a finite scale with respect to the distance between any two endpoints. Repeaters can increase this limit only up to a point. Generally this limit is measured in meters rather than kilometers.

Some may argue against the inclusion of hubs in the repeater category. Many hub manufacturers are now implementing sophisticated functions in their products. This may include counting frames, automatic recovery of lost circuits, and support for management protocols such as SNMP. Regardless of the "intelligence" embedded in the hub, it remains a repeater, since it cannot modify the data traffic. On the other hand, some vendors are now bundling their hubs with bridge or router software. Others have implemented advanced switching functions. With the inclusion of these capabilities, the hub has now become an internetworking device that can accomplish a variety of tasks. However, each task is distinct from the others. It would be better to call these integrated devices by some new name, and refer to their functions as *relay*, *bridge*, *routing*, etc. Industry marketing being what it is, you will probably be on your own to determine just what functions are being offered. Don't judge a product by its name!

Bridges

Repeaters are not sufficient to build subnetworks. As we have stated, they have a number of limitations that limit our ability to scale subnets to size and manage them effectively. In many cases, we may

wish to link LANs that are separated by large geographical distance. In other cases, we may have too many users to fit on a single LAN segment. As network traffic increases, the integrator requires some way to manage the load and keep it local to those users who will most likely use it, while segregating it from other users. This will free up bandwidth so that the overall traffic flow can be increased.

On the other hand, repeaters do offer transparency and ease of use. These are two very valuable attributes, especially in those cases where the LAN may be distant and no local support staff is available to work on problems. Is it possible to find a device that meets the joint requirements of traffic management, wide distance coverage, and transparency? Yes, it is. The device is called a *bridge*.

✻ Defining the bridge

Bridges can be defined as store-and-forward devices that operate using control information found at the data-link layer (OSI reference mode layer 2). The essential function of a bridge is described as "filter/forward." Since all data at layer 2 travels as frames (blocks), the bridge can choose either to forward the frame to an adjacent LAN or to discard (filter) it. We will shortly describe how this is done. In order to achieve transparency, bridges operate in a promiscuous mode. This means that they copy all frames that are sent by devices on all attached LANs. The filter/forward decision can be made without the knowledge of the transmitting or receiving device. This transparency is a key benefit of bridges. It is also defined as "plug and play" or "fire and forget." We prefer the term transparency.

Since bridges filter local traffic, they achieve a much higher degree of traffic management than that afforded by repeaters. A large file transfer or video conference between two local users on LAN A will not impact performance on LAN B, since all the local data is filtered. Furthermore, each LAN segment has its full bandwidth available for local use. This means that the aggregate bandwidth available in a bridged environment is equal to the sum of all bridged segments.

Bridges also offer fault management, because they can detect frames that are in error. Since they are programmed with the knowledge of what a frame should look like, they can identify the source of "malformed" frames. This implies that bridges can be used for fault

identification. As an added bonus, devices that exist on LAN segments other than the one having the problem are sheltered. They will be able to continue to operate as the faulty segment is being repaired. As a general rule, bridges will not forward frames that are in error.

Bridges are sometimes referred to as "selective transparent repeaters." We like this description because it gives a good idea of how the bridge operates, and its place in internetworking. The bridge is transparent to all networked devices (like the repeater), but it is selective in which frames it forwards. This is why we say that bridges implement virtual LANs. From the viewpoint of the networked devices, the bridged environment is a single, seamless LAN. The bridges know better!

❊ Physical addressing

In order to describe how a bridge provides its filter/forward capability, a brief review of physical addressing is required. Network interface cards for LANs always have a unique identifier. This identifier is normally created by the manufacturer and is called the physical address. The physical address is used by the Network Interface Card (NIC) to recognize incoming frames, and to provide a source identifier when transmitting. All LAN frames include both a source and destination physical address. Remember that the data link layer implements point-to-point communications, so that in a shared media environment such as a LAN these physical addresses are crucial. If they were not available, the determination of destination would have to made by software. Given the speed of LANs, such a solution would be impractical.

The decision to have the manufacturer implement the physical address was made to guarantee that all addresses would be unique. Normally the address is 48 bits long. The first two bits are used to identify broadcasts, and indicate if the address has been modified. The next 22 bits are generally described as the manufacturer's assigned code. They are unique to each manufacturer. The final 24 bits are called *manufacturer administered*, because they are assigned by the manufacturer at the time the card is created. As long as each manufacturer is given a unique code, and controls the assignment of the administered codes, uniqueness of the physical address is guaranteed. Often, such as in the case of Ethernet, the address may be "burnt" into the card so that it is not possible for users to modify it.

Many Token Ring cards allow this address to be overwritten. While this provides for additional flexibility in creating meaningful addresses, the integrator should be aware of the potential danger of having duplicate addresses on a LAN. Unfortunately, the uniqueness of the address will cause some problems as the total number of cards, and their location, increases. We discuss this issue in the *router* section of this chapter.

When a NIC wishes to transmit a frame, it will always place both the source and destination address in the frame header. It is this information that is used by the bridge to make its filter/forward decisions.

✻ Bridge operation

As mentioned earlier, the bridge copies all frames that are transmitted on all attached ports. Consider the simple network shown in Fig. 3-4. When device B transmits a frame to device A, it includes both the source and destination address in the frame header. The bridge copies the frame from the network and inspects the destination address. It then compares the address to all known addresses on the received port. In the simplest case the following algorithm is used:

```
IF (destination address exists on the received port) THEN filter
ELSE flood on all other ports
```

Figure 3-4

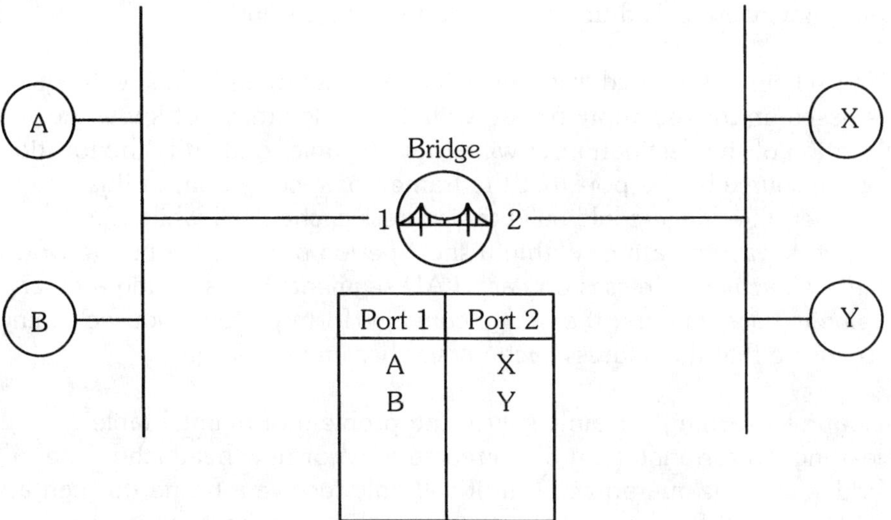

Operation of a simple bridge. Learning Tree International

Note that the term *flood* is used instead of forward. Since the destination address may not be known, and the bridge may have more than two ports, a copy of the frame must be placed on each. While this is somewhat inefficient, it can be tolerated, since most LAN traffic will remain local to the source segment and not be forwarded.

Given higher performance bridges, it might be possible to modify the previous algorithm as follows:

```
IF (destination address not found) THEN flood
IF (destination address is on the source LAN) THEN filter
IF (destination address is known) THEN forward to appropriate port
```

This provides for much better forwarding and reduces the redundant forwarding of frames. Most bridges tend to operate using a variant of this second algorithm.

The bridge uses address tables that are kept in a local cache to make filter/forward decisions. How is this cache created? In the early days of bridging, these tables had to be manually loaded. Although this worked well for small LANs, it quickly became apparent that it would not work well as the number of LAN segments, devices, and bridges began to grow. Updating the tables became a time-consuming operation that significantly impacted network performance. Bridge designers were called upon to come up with a solution.

The scheme developed was called "adaptive learning." This technique is based on the assumption that while the bridge may not know the location of the destination, it will always be able to identify the location of the source by the port that the frame was received on. In this manner, the bridge could build the address cache itself while the network was operating. Within a short period of time, the bridge would learn all active addresses on each LAN segment. By assigning a timer value to each address, the bridge could perform periodic housecleaning to ensure that the address cache never became too large.

Adaptive learning certainly solved the problem of manual table loading. Unfortunately, it also created a major new headache. The bridge must be guaranteed that it will only receive a frame through a single port. If the frame arrives at more than one port, which one should be assigned to that address in the bridge's table? In other

words, no loops could be allowed in the topology of the LAN segments. This posed a serious problem for network designers who relied on loops as a fundamental building block for networks. Loops are still used today to provide redundancy, load leveling, optimal path selection, class of service, etc.

A number of solutions exist to solve this problem. For Ethernet-based systems there is the Transparent Spanning Tree (TST) algorithm. In this approach, the bridges communicate amongst themselves and determine where loops exist in the topology. By assigning values to each path it is possible for the bridges to create a minimum spanning tree that guarantees that only one path exists between each source and destination. Any bridges that are redundant remain active and monitor the overall network status. A failure on the part of a bridge or a LAN will cause these redundant bridges to force a rebuild of the tree. This does not offer load leveling or optimal path selection, but it does offer some amount of redundancy in the event of failure. The cost of this redundancy is quite high, since the standby bridges are blocked and not performing any useful function until there is a failure. However, for LANs that require high availability this can be an important feature. (See Fig. 3-5.)

Figure 3-5

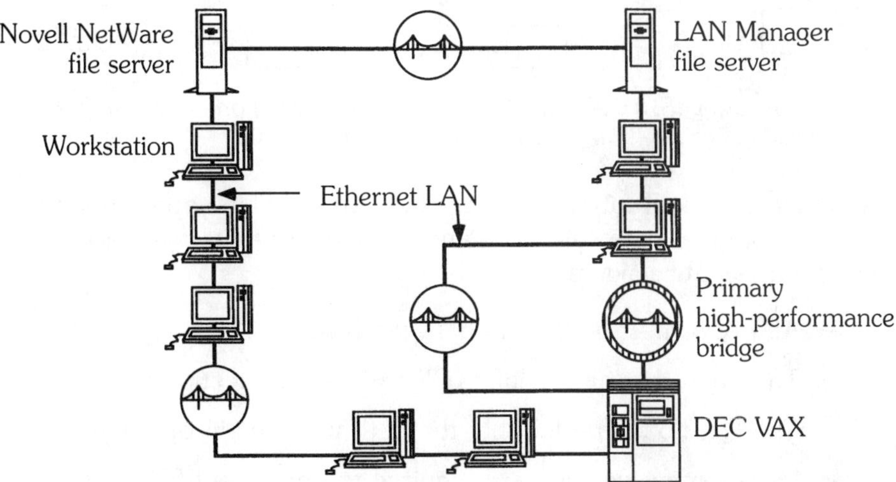

The Transparent Spanning Tree. Although it provides good redundancy, at least two of these bridges will not be active during normal operations.

For those organizations that have Token-Ring LANs, another alternative to the adaptive learning problem is available. It is called *source routing*, or SR. In this approach, the source attempts to discover all possible paths that exist to the destination, and then selects the optimal one based on minimum transit delay. This is done by broadcasting a frame across all potential paths and having the destination send each one back along the route that it traversed from the source. The first one back is deemed to have traveled the optimal path. Each frame contains a record of all the bridges and LANs that were traversed, so the source can use that information to specify the path to be taken when transmitting other frames. The bridges play a much more passive role, in that they only forward frames when they discover their identifier in the frame control header. (See Fig. 3-6.)

Figure 3-6

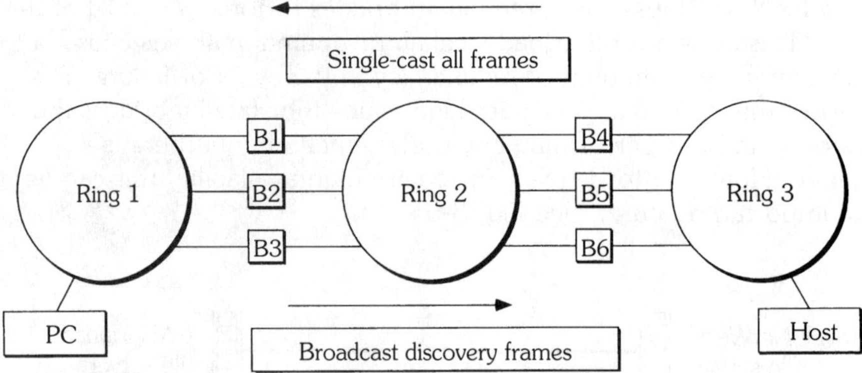

Source Routing (SR). Requires that all LANs be based on Token Ring. High overhead generally limits this to local connections only.

Although Source Routing does improve upon TST by offering load leveling, optimal path selection, and more continuous operation, it has some negative side effects:

- All LANs must be based on Token Ring.
- The broadcast nature limits SR to local connections.
- There is no guarantee that the path will remain optimal.
- More expensive NICs are required to implement SR.
- The elapsed time required to discover the route will negatively impact throughput for small file transfer operations.

As you can see, both SR and TST have their limitations. Neither solution is perfect, or even close to optimal. What other options are available? Bear with us for one more section.

✻ Remote bridges

What happens when the two LANs to be joined are not physically adjacent to each other? Is it still possible to use bridge technology. The answer is yes. It is called *remote bridging*, also referred to as *half bridging*. The concept is simple. Take a local bridge and cut it in half. Use a long wire (WAN) to attach the two halves together. You now understand how a remote bridge operates!

Figure 3-7 shows our previous example, now arranged as a half bridge setup. Note how each bridge is denoted by only half a circle. This is done to remind us that two remote bridges are required to implement this technique. Furthermore, the devices on each LAN segment are still unaware of the bridge (transparency). This means that they are unable to access any services that might exist on the WAN. It would never even occur to them to ask since, from their point of view, they are operating on a single virtual LAN. Of course the half bridges are aware of the WAN, and will try to use it as efficiently as possible, but we recognize that there will generally be a bandwidth mismatch problem between the high speed LAN and the lower speed WAN. This mismatch can be several orders of magnitude.

It might be easy to dismiss the bandwidth mismatch as inconvenient. After all, users might be willing to accept a slower transmission rate

Figure 3-7

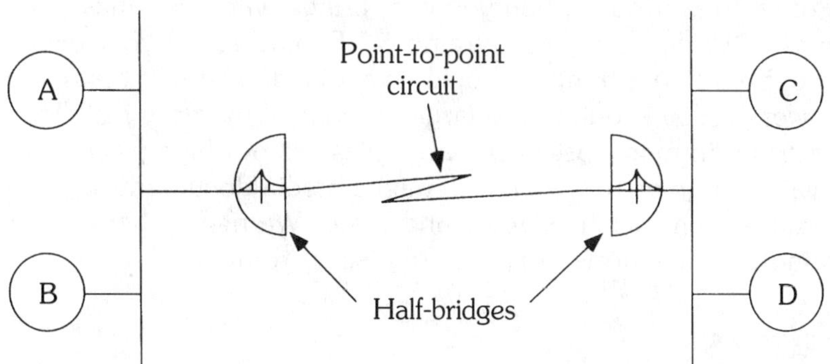

Remote bridging. Possible to implement, but the potential bandwidth mismatch can create significant problems.

when traversing long distances. Unfortunately, this mismatch problem can cause some severe transmission problems. To understand why, we must remember that the bridges are transparent. They are invisible to both the source and the destination, so they cannot implement any type of flow control. This means that as the buffers in the transmitting bridge begin to fill, it has no way to ask the source to stop transmission. The only thing it can do with the excess frames is discard them. This will cause the destination to request retransmission, since the data did not arrive correctly. In some cases this may cause the entire data set to be resent . . . which causes the buffers to overflow again . . . and so on, and so on. Get the picture? Increasing memory capacity at both bridges helps, but a sufficiently powerful source device will always be able to overload a half bridge. The other solution is to ensure that your WANs operate at LAN speeds. This is an effective option, though somewhat expensive.

Although bandwidth mismatch is the most significant problem, with remote bridges others do exist:

- The proprietary nature of the interface between the two remote bridges limits vendor selection.
- Valuable WAN bandwidth is lost to overhead such as broadcasts and interbridge communication when using either TST or SR.
- Timeouts may become more frequent as transmitting stations expect replies based on LAN speeds.

Peter Curan, the author of Learning Tree's Internetworking course, is fond of saying, "Route when you can, bridge when you must." We often say, "Bridge local, route remote." Regardless of your own philosophy, remote bridges should be avoided. The fact that many of them work today is due to the large amount of memory installed, and the fact that in most cases very large files are not being transmitted. This will change as we move into mixed-media documents and real-time data streams such as audio and video. We need a better solution for remote connections. It is time to discuss routers.

Routers

We have indicated that several problems exist when trying to use bridges to implement the well-integrated network. These include load-leveling, redundancy, optimal path selection, and the inability to design topologies using loops. Remote bridges suffer from a serious bandwidth mismatch problem, as well as a number of other related issues. If these were the only problems, we might be tempted to try to find a solution. Unfortunately, many other problems exist when bridges are used:

- It is difficult to implement security using bridges. The only information available is the physical source and destination address. Other details that might impact security are not known to the bridges.
- Accounting is difficult when trying to bill for WAN bandwidth utilization. Again, with only the physical address being known, we have no idea where the destination may be located. Furthermore, the bridge cannot differentiate between overhead frames (such as broadcasts) and frames containing user data.
- Classes of service cannot be offered. Since the bridge is transparent, the source will never request a particular class of service. This is frustrating when trying to tune a network for a variety of data streams, such as terminal/host, file transfer, and real-time video conferencing.
- Bridges have difficulty responding dynamically to changes in local or global conditions. Recovery is possible, but it will generally cause an interruption in service.
- Performance monitoring is difficult. If the ratio of overhead frames to user data cannot be measured, how can a monitoring system provide useful data?
- Bridges require that all LANs use the same frame format. All LANs must be Ethernet, Token Ring, or FDDI, but combinations do not work well. Each of these subnetwork protocols has its own unique frame format. Since the bridge does not modify the frame (it is transparent), it cannot deal with interoperability problems between these dissimilar LAN protocols. This limits the options available to the integrator, and severely impacts the well-integrated network's ability to provide both local autonomy and integration.

Even if it were possible to address all of these additional concerns, there is one remaining issue that bridges, by their nature, will never be able to solve. We call this the *structure problem*.

❊ The structure problem

Recall that bridges use a physical address to provide unique identifiers. These addresses are created by the manufacturer at the time the card is built. The format of the address enables us to identify who built it, but not its location in our internet. This is somewhat equivalent to being told to deliver a letter to a particular individual somewhere in the world, but not being told where that individual is located.

Another analogy that can be used is the world telephone system. Consider the situation if every time you purchased a phone, it came hard-coded with the phone number. This phone number would really represent a manufacturer code. If you bought four phones for your home, each one would have a unique number. Which phone number would you give to your friends? From the phone company's point of view, the situation would be much worse. As you started dialing a phone number, how would they know where to connect the call? It is hard to imagine having a single directory containing all possible phone numbers and their location. The size and frequency of updates would make such a directory impossible to manage. Even if it were possible, consider the extra time required to search for a number before the call could be completed. This is one of the reasons that the phone company assigns phone numbers rather than allowing them to be hard-coded into the handsets. Since this is done at a local level, how is the uniqueness of the phone number assured?

To address the problem of uniqueness, the governments of the world, through ITU, have agreed on a coding scheme. Each country is assigned a *country code* by ITU. Since a single agency is responsible for this work, the uniqueness of the codes is guaranteed. Within a country, the local public carriers can assign area codes. Within the area codes, the local business offices can assign exchange codes. This hierarchical system can be extended almost indefinitely by adding new country codes, area codes, and exchange codes as required. We see this happening in North America with some frequency as the total number of phones continues to grow.

Since the addressing structure is well understood by all participants, it now becomes possible to provide efficient routing of calls. In North America, a "1" as a prefix indicates that the call is to be placed outside the local area, and that the local switch should relay the remaining numbers to another switch responsible for long-distance calls. If no prefix is used, the local switch knows that it can satisfy the call locally, since the destination must be in the same area.

The addressing scheme does not require individuals to purchase a particular brand of telephone. All phones work to the same standard. Furthermore, organizations can choose to implement PBX (Private Branch eXchange) switches from any vendor, and take advantage of the proprietary services that are offered. However, even these switches must conform to global specifications if they are to be attached to the public telephone network.

This ability to create multiple levels of structured addresses is a prime requirement for building the well-integrated network. Without it, the network cannot be scaled, and it becomes impossible to effectively move data over wide area networks. Bridges and physical addressing cannot offer this capability.

We must also remember that some devices in our internet may not be attached to a LAN, which means that they do not have a physical address. Some devices may only have a WAN attachment. Consider a home or laptop computer, or perhaps a host computer, that is connected to an X.25 network. Figure 3-8 indicates that connections are required between both LAN- and WAN-attached devices. Bridges do not support this capability.

✳ Defining the router

As we mentioned at the beginning of this chapter, routers operate using information that is generally associated with layer 3 of the OSI/RM. This means that routers are part of the ETES protocol layer. Like bridges, routers are responsible for the forwarding of packets of information from a local source to a remote destination. The router may perform this operation directly, or may relay the packet to another router as required.

Unlike the bridge, the router does not operate in promiscuous mode. Since the bridge is transparent, it must copy all frames off the LAN and make filter/forward decisions on its own. The router is a well-advertised device. Frequently its location is hard-coded into the

Figure 3-8

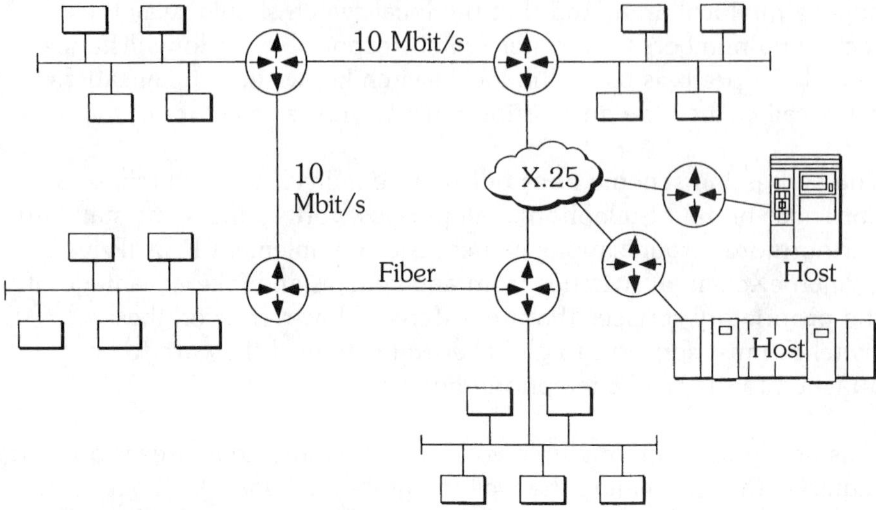

Routers support the connection of both LAN- and WAN-attached devices. They also allow the internet to be structured. Learning Tree International

devices on the subnet; the router might also choose to advertise its presence through an occasional broadcast. Devices can contact the router for assistance in forwarding packets for destinations that are off the subnet. Each device can make this decision, since it is always aware of its own network address and the address of the destination. If they are equal, there is no need to contact the router, since the destination is on the same subnet and can be reached directly. If the destination network identifier is different from that of the source, the packet is sent to the router. This is equivalent to dialing "1." It is the responsibility of the source device to contact the router.

By having the router visible, we also solve the problem of bandwidth mismatch between the LAN and the WAN. Since the source must contact the router directly, it is possible to implement a flow control scheme that enables the router to request that the source suspend transmission until bandwidth or buffer space is available. This is a primary advantage of using routers when connecting LANs via a WAN.

Routers can be implemented in both hardware and software. In many cases, manufacturers build special router "boxes" that only provide routing services. These boxes can be expanded to handle multiple

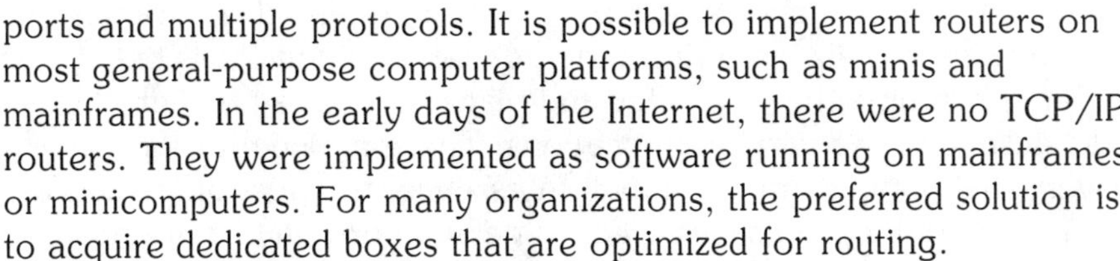

ports and multiple protocols. It is possible to implement routers on most general-purpose computer platforms, such as minis and mainframes. In the early days of the Internet, there were no TCP/IP routers. They were implemented as software running on mainframes or minicomputers. For many organizations, the preferred solution is to acquire dedicated boxes that are optimized for routing.

Another important problem that routers solve is their ability to link dissimilar subnetwork protocols. The layer 2 frame is viewed as nothing more than an envelope by the router. Recall that the original intent of layer 2 is point-to-point communications. The source sends the packet (inside the frame) to the router. The router examines the packet and then can use any available frame format as an envelope to transmit the packet to the next destination point. It can take advantage of whatever networking technology is available. Some find this concept of enveloping to be a bit confusing. Let us use an analogy to make the concept clearer.

Consider mailing a letter from New York to London. The letter writer puts the written pages in an envelope and writes the destination address and the return (source) address on the envelope *according to a well-defined protocol*. This means that the destination is placed in the center of the envelope and the return address is placed in the upper left-hand corner. A stamp must also be placed in the upper right-hand corner, since the mail system is based on a per usage charging arrangement. So, envelope in hand, the letter writer walks to the nearest mail box (router) and deposits the envelope (packet) in the box. This ends the responsibility of the source (letter writer) in terms of delivering the letter. The letter has now become the responsibility of the global postal network (internet).

At some point during the day, a mail carrier arrives and empties the mail box. The mail box is a very simple router, since it always transmits its packets to the same place, the local post office. The mail carrier places all envelopes in a large sack (frame) and brings it, along with other sacks, to the local post office. Note that the mail carrier does not really care where the letters are going. As a subnetwork, the only concern is the point-to-point connection between the mail box and the local post office. After delivering the sacks of mail, the mail carrier can go home and take a well-deserved rest. Those sacks are heavy!

At the post office, the sacks are opened and discarded. As frames, they have done their job. Within the post office, individuals or machines examine the envelopes and place them in a number of sacks destined for a variety of locations. These may be for local delivery, national, or international. In essence, the post office acts as a multiport router. The sacks used for delivery or relaying are not important, so long as they provide a solid point-to-point link. The addressing information on the letter is what makes the system work.

We could follow this analogy through the entire delivery process, but you should be getting the picture by now. The envelope will go to the airport, brought to a central post office in England, and then sent to a local post office, where another mail carrier will deliver it to the indicated address. The only person who must know the physical location of the address is the final mail carrier. Everyone else in the chain keeps forwarding it as required. Notice how a variety of subnetworks were used to deliver the mail. None of these systems, the planes, trucks, sacks, etc., were really compatible with one another. What allowed the system to work was the fact that interface points existed and routing decisions could be made based on the information contained on the envelope. These interface points are almost exactly equivalent to our concept of routers. The router does not really care how the packets are delivered. It only cares that the packets are in an agreed-upon format that all participants can use.

By the way, consider what would happen in our postal analogy if the envelope were not properly addressed or had incorrect postage. The postal system would probably discard the envelope. Since the source (return) address could not be read, the sender would be given no notification that there had been any problem, and would assume that the letter had been correctly sent. This is always a problem when dealing with connectionless networks. Almost all routed networks (with the exception of APPN and OSI) are connectionless.

As with the postal system, our routed network will require both a common method of addressing and an agreed-upon method (protocol) for relaying packets. Both are defined by the network protocol.

✻ Network protocols

Figure 3-9 lists the major routable protocols. SNA is shown as a separate protocol, since it is connection-oriented rather than

Figure 3-9

	TCP/IP	OSI	IPX/SPX	SNA/LU 6.2
7	FTP, TELNET, SMTP, SNMP	App. services	NCP	Application
6		Presentation		Presentation
5		Session		LU 6.2
4	TCP, UDP	COTS CLTS	SPX	Trans. control
3	IP	CONS CLNS	IPX	Data control
2	Subnets	Subnets	Subnets	Link control
1				Subnets

- To be routable, protocol must have a network layer

CLTS = ConnectionLess Transport Service

CONS = Connection-Oriented Network Services

COTS = Connection-Oriented Transport Service

LU = Logical Unit

SNMP = Simple Network Management Protocol

UDP = User Datagram Protocol

Major routable protocols. Learning Tree International

connectionless. Note that each protocol offers functionality at layer 3. This is where both network addressing and relaying functions are defined. NETBIOS is not indicated, since it does not provide any functionality at this layer. The selection of a network protocol is a key task of the integrator, since the preferred method of integration is via a standard backbone. Only one of these protocols can be selected. The reason that this is a crucial decision is that much of the internetworking equipment selected (routers, gateways, etc.) will be based on the standard protocol. The integrator would be well advised to carefully examine the various pros and cons of each protocol prior to selection. No single protocol is right or wrong. In Chapter 2 we gave you some of our views, but each case is unique.

✻ Network addressing

A key feature offered by any routable protocol is the ability to define network addresses that are independent of any physical addressing scheme. These addresses must be flexible enough for the network manager to define both simple and sophisticated network structures. The general form of network address is NETWORK_ID.HOST_ID. NETWORK_ID is roughly equivalent to an area code, while HOST_ID is comparable to the local phone number. As might be expected, there is no universal addressing scheme. Each protocol has its own format. These formats are not compatible between protocols, which is why gateways are required when integrating different ETES protocols.

Perhaps the best-known addressing scheme is the one used by TCP/IP. A 32-bit word is used to hold the complete address. It is generally expressed in dotted decimal notation of the form xxx.xxx.xxx.xxx, where each group represents a single byte that ranges in value between 0 and 255. Unlike the phone system, which provides four layers of addressing (country code, area, exchange, local), the TCP/IP protocol only provides two (network, host). Many have considered this a significant limitation. This has been corrected in the next version of IP, which will offer at least four layers. These additional layers are required to handle the growing complexity and size of the Internet. OSI also offers many layers of addressing. In fact, the OSI addressing scheme is quite flexible, since the address space is variable.

Regardless of the protocol chosen, all devices in a routed network will require the ability to map the network address onto the physical address. Remember that the network address is a convenient abstraction used in defining the network and relaying packets. Delivery must still be accomplished by the data link layer, which means that physical addresses will have to be used to handle the actual frame movement. In general, each participant in a routed network maintains a cache that maps known network addresses to physical addresses. Of course, this is only required for devices that share the same subnet. There is no need to know the physical address of remote devices, since they are never sent any frames directly.

The size of this cache depends on the number of devices on each subnet and the number of subnets that converge at a single point. For example, if a router has 10 ports, then it is attached to 10 different subnets. It must maintain a separate address table for each subnet so that frames can be sent as required. In a sufficiently large subnet, these tables can grow quite large. Techniques exist to reduce the total table size, but that is a topic best left for books devoted to internetworking of specific protocols.

Let us give a brief example of how network addressing, physical addressing, and routers all work together to move a block of data between two end stations on separate subnets. Figure 3-10 shows a small internet consisting of four subnets. The network id is listed for each. On network 137.10 there exists a host device. Its HOST_ID is 20.20. Its full IP address is shown. Since 137.10.20.20 is attached to a LAN, it also has a physical address, as shown. The address cache at 137.10.20.20 contains the physical address of the local router.

Figure 3-10

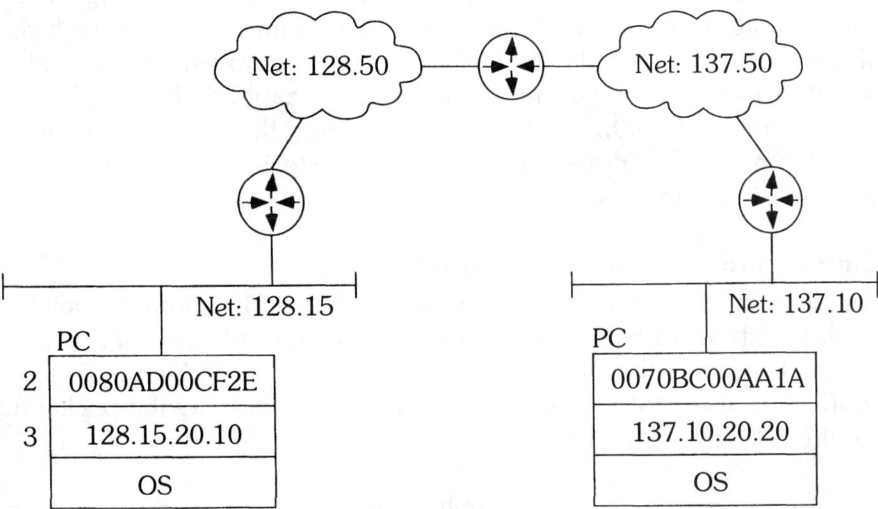

A simple routing example. Learning Tree International

Suppose that a process running at 137.10.20.20 requests that some data be sent to host 20.10 on network 128.15. The IP process on 137.10.20.20 is immediately aware that this destination is on a remote subnet, because the network id of the source and destination are different. In order to move the data, it is packaged in packets and sent to the physical address of the local router with the IP destination address being given as 128.15.20.10. Naturally, the physical destination address used in the frame is that of the local router.

When the frame arrives at the router it is opened to reveal the packet. The destination address is examined. The local router is only aware of two networks, 137.10 and 137.50. Obviously it will not be able to send the packet directly to the destination. Instead, it repackages the packet in a frame type suitable for 137.50 and sends it to the router that joins 137.50 and 128.50. Using the same process, this router forwards the frame to the router that joins 128.50 and 128.15. This last router has an address cache for 128.15.20.10 that it uses to send the packet to its final destination. When it is received, the destination discovers the original source by an examination of the source address contained within the packet header.

You may have found the use of network addresses cumbersome in this last example. So does everyone else. That is why most hosts are defined by a generic name rather than a number. Human beings find names easier to work with than numbers. This introduces a final layer of addressing for each device in the network. For example, it would be easier to define one of these machines as "zaphod@fool.org" instead of 137.10.20.20. The task of mapping these generic names onto the network addresses is the job of directories, which are discussed in Chapter 4.

✻ Single- and multiple-protocol routers

For a router to operate, it must know the networking protocol being used. If a standard backbone approach is being used, then only one protocol is required on the router. As long as all incoming packets conform to this protocol, the router will be able to relay the packet as required. (See Fig. 3-11.)

Many backbones must support multiple protocols concurrently. One's first assumption might be that a series of single-protocol routers are

Figure 3-11

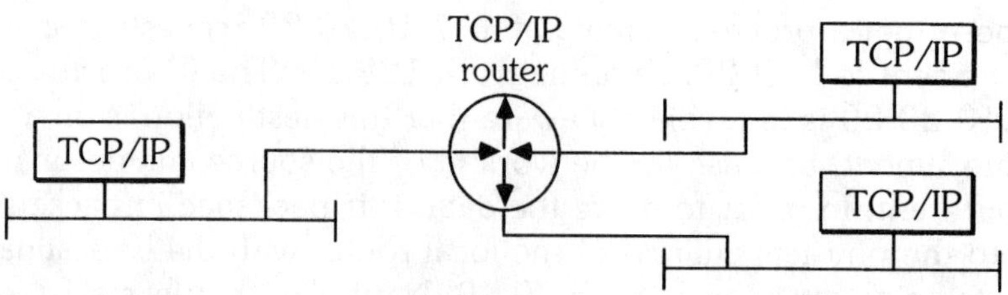

A single protocol router. Learning Tree International

required, one to support each protocol. This is not a good idea: the various routers are unaware of one another, and can make incorrect assumptions about the network topology, potentially leading to broadcast storms and other nonoptimal situations.

This problem can be solved with a device known as a multiprotocol router. In effect, it provides the functionality of multiple single-protocol routers in a single box. This is advantageous, since it eliminates the need for multiple WAN connections and reduces administrative overhead to some degree. It is important to note that a multiprotocol router does not offer any conversion between the various protocols. Each local device perceives the router as being dedicated to its own protocol, which means that multiprotocol routers cannot be used as part of an interoperability strategy. That function is reserved for gateways.

Multiprotocol routers are popular because they provide a "quick fix" for purposes of integration. A single backbone can be shared, with the various protocols coexisting on the wire. They remain completely unaware of one another's existence. As a short-term solution, these devices offer some relief, but the integrator is still faced with the interoperability problem. (See Fig. 3-12.)

Figure 3-12

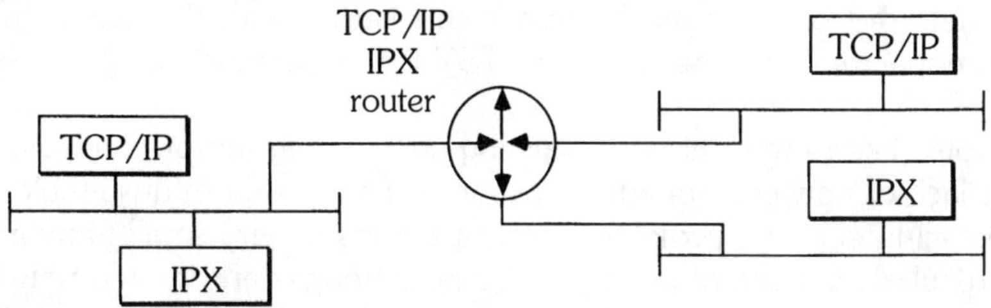

Multiprotocol routers offer some relief in handling concurrent protocols, but do not provide for interoperability. Learning Tree International

 Mixed bridged/routed networks

The well-integrated network anticipates that both bridges and routers will be used. Bridges offer great functionality when building local subnets. They are independent of the network protocol selected, and their transparency makes them ideal for situations where "plug and

play" technology is required. This is most often the case in remote offices, where local technical staff may not be available. Bridges can also be used for larger integration projects, such as linking all the LANs in a high-rise building, provided that:

- All LANs are using the same protocol (Ethernet, Token Ring, etc.)
- The building is considered to be a single addressable subnet for the organization's internet
- There is no requirement for sophisticated network management capabilities, such as security, accounting, and other areas that we have mentioned

If all the LANs are Token Ring and local to one another, then source routing should be considered. It provides excellent functionality when it is used for its intended purpose. It should be avoided on remote links—except in very small networks—because the broadcast overhead can quickly become substantial.

We have also discussed Transparent Spanning Tree as an option for linking Ethernet-based systems. In fact, TST is rarely used in practice. Most vendors offer proprietary extensions to TST that work much better and offer more functions. In most cases, all bridges deployed in a subnet are from a single vendor. If this is the case, the use of the vendor-specific implementation of TST is recommended.

We would like to reiterate our warning about using remote bridges. Their lack of flow control and transparency makes them unsuitable for almost all integration projects, except perhaps in very small networks. It is particularly important not to mix remote bridges and routed networks. Since the router cannot see the transparent bridge, it assumes that all devices can be reached at the high bandwidth offered by the LAN. This leads to timeouts and an overall reduction in throughput. In addition, routers tend to rely on the ability to broadcast in order to perform functions such as advertising routing tables and network address to physical address resolution. Remote bridges have to forward these frequent broadcasts, consuming valuable WAN bandwidth.

Over the last few years, a variety of devices have appeared that combine the functionality of multiple internetworking components into a single unit. There are bridges that offer routing and routers that offer bridges. These devices are often called *brouters*. In addition, hubs can now offer various capabilities, including bridging, routing, and switching. We are not quite sure what to call them, but that doesn't seem to prevent the industry from developing them and bringing them to the marketplace. Regardless of the name of the product, or the functions offered, it is important to remember that repeaters, bridges, and routers describe a view on internetworking. Combining all the functions into a single box does not change this, though it may prove more convenient.

Another trend in internetworking technology is away from shared media LAN systems and back to point-to-point communications. A variety of technologies, such as Ethernet switching, virtual LANs, Asynchronous Transfer Mode (ATM), and switching bridges, have emerged to overcome the limitations of the basic LAN. In general, we applaud these efforts. Shared media systems require that all participants share the available bandwidth. In a switched point-to-point system, bandwidth is dedicated to the individual session. This means that a guaranteed level of service can always be offered.

The integrator may wish to take a wait-and-see approach to these new technologies. Standards are still being developed, and the price of the technology can be expected to drop dramatically over the next few years. Since most of these technologies can take advantage of the existing cable system, it should be possible to deploy these new technologies quite easily. The one exception may be the case in which the current network no longer has the bandwidth required to do the job, or if a major new traffic load is anticipated, such as that offered by multimedia-based applications. In these last two cases, it might be worthwhile to investigate these technologies today.

Regardless of the technology chosen, remember:

Bridge local, route remote!

Gateways

If it's not a repeater, bridge, or router, it's a gateway! Gateways provide a wide range of services. These services may address both

internetworking and interoperability. For example, IBM offers a device called the 8209 Bridge that is really a gateway that allows a TST-based Ethernet environment to be connected to an SR-based Token Ring. Clearly this is an internetworking gateway. On the other hand, a company called Soft-Switch offers a series of products that enable dissimilar e-mail systems to interchange messages. This product addresses interoperability issues. Novell offers a gateway that enables LAN-based microcomputers to access an IBM mainframe as if they were locally attached terminals. This same function is offered by a wide variety of companies, some in hardware, some in software, plus all manner of combinations. All of these products can be classified as gateways.

Given the diversity of gateways, it becomes difficult to approach the subject in a structured manner. In the next chapter we address the issues of interoperability, where you will see that many interoperable solutions rely on gateways to achieve integration.

Internetworking gateways fall into two broad categories:

- Peer-to-peer gateways
- Peer-to-terminal/host gateways

Peer-to-peer gateways are used when we wish to link highly dissimilar networks. Examples might include OSI/TCP/IP or SPX/IPX/APPN. The gateway must act as a local participant in each network and make the remote side appear to be native. If a high degree of integration is required, this can become very complex. This is perhaps why we see so few of this type of gateway.

The conversion of an ETES protocol into an incompatible system raises many questions, such as address translation, naming conventions, data encoding rules, handshaking protocols, and others. Even if this can be accomplished, the gateway remains a potential single point of failure. Integrity issues must also be considered, since the gateway may not perform a "perfect" translation.

Peer-to-terminal/host gateways are much more readily available. They have been in use for years, and cover just about every

conceivable conversion. In general, they provide the ability for peer-type devices (normally microcomputers or workstations), attached to peer-networks (normally LANs), to access minis and mainframes as if they were locally attached terminals. These types of gateways are very popular. In fact, our first major section in Chapter 4 will be on remote terminal access. It examines this issue from an interoperability point of view. But peer-terminal/host gateways may also be implemented in hardware. For example, Attachmate sells a card that can be used to emulate an IBM 3278 terminal. This card enables the microcomputer to be directly attached to an IBM 3174 cluster controller. This is a hardware-based internetworking solution to the same issue that is addressed in Chapter 4.

Peer-terminal/host gateways are often used to access remote information providers, such as CompuServe and America Online. A LAN may be equipped with a modem pool that enables LAN-attached machines to dial out as asynchronous terminals. The reverse is also true. Many mobile users wish to be able to access the file server on their LAN. This is done using access server technology (which is just another way of saying gateway). These devices enable a remote device, using asynchronous communications, to participate in the LAN as if it were locally attached. These access servers are designed for specific file servers, such as NetWare, LAN Server, and Banyan Vines; their functionality and performance vary.

If all these options sound confusing, don't worry too much. In most cases, when attempting to integrate peer and terminal/host networks, there is a great deal of assistance available from the vendors of both the peer network and the terminal/host system. In addition, a wide variety of sources are available that cover these technologies. These include magazines, trade shows, and product sourcing groups such as Datapro. In addition, the integrator is likely to find that whatever the problem, it has likely already been solved by someone that they know. While we strongly recommend that the integrator get solid references before installing an internetworking gateway that connects a peer-to-peer environment to a thermal/host network, we do not feel that it is necessary to spend too much time addressing the issue. In addition, it would not be practical to try to address all the potential combinations of peer and terminal/host connections. It is unlikely that we could even scratch the surface.

In summary, gateways can address both interoperability and internetworking. Interoperability is handled in Chapter 4. Internetworking gateways generally focus on peer-to-terminal/host connectivity. Much information is available from a wide range of sources. Since the integrator's internetworking requirements are much more narrowly defined, we have chosen not to cover this particular aspect of gateway technology in detail. Solutions do exist, and should be used if required.

Summary

Internetworking addresses the binding of network technology. The principal components used by the integrator to accomplish this task are repeaters, bridges, routers, and gateways. Of these, bridges and routers should be examined in detail, since they offer alternative approaches to the same problem.

Bridges are selective transparent repeaters that implement a virtual LAN or subnet. They operate at the subnetwork layer (OSI/RM L2). Their basic function is filter/forward. They enable the integrator to better manage network traffic by keeping local traffic from impacting the entire set of users. Bridges are considered "plug and play" technology, since they do not require sophisticated implementation, and can operate unattended for long periods of time. They are an excellent choice when the LANs to be connected are homogeneous and local to one another. Remote bridges should be avoided, because they lack the flow control functionality needed to handle the bandwidth mismatch problem.

Routers implement an internet. They can be used for both local and remote connections. An important prerequisite to the selection of routers is determining the network (or ETES) protocol to be supported on the backbone. Routers are chosen to match these protocols. The network protocol will specify how network addresses are to be defined, and how packets of information are to be relayed through the internet. (See Fig. 3-13.)

Routers are the key building block of the integrator in building the well-integrated network. They offer the functions and the flexibility

Figure 3-13

	Case 1	Case 2	Case 3	Case 4	Case 5
Problem	Connecting small set of local homogeneous subnets	Mixed subnets	Multiple routable protocol internets	Handling nonroutable protocols	Incompatible protocols
Examples	1. LAN-LAN 2. 2 Remote LANS 3. 10BASE-2 10BASE-T	1. TRN-Ethernet 2. AppleTalk 3. TRN	1. TCP/IP 2. SNA 3. OSI	1. NetBIOS 2. LAT	1. OSI-TCP/IP 2. SNA-DECnet
Proposed solutions	– Bridges – Remote bridges	– Gateways – Routers	– Routers – Multiple-protocol backbones	– Brouters – Bridges	– Gateways
Advantages	– Plug & play – Low cost	– Fast fix	– Control – Scalability	– Solves the problem	– Solves the problem
Disadvantages	– Lacks scalability – Lacks control	– Performance of G/W – Router management	– More management administration	– High overhead	– Performance – Overhead

Sample case studies in internetworking. Learning Tree International

required to implement a standard backbone strategy. They can be scaled to a variety of sizes, and act as the primary interface point between the local autonomous networks and the backbone. If required, they can also handle multiple network protocols concurrently, enabling one to implement a multiprotocol backbone. Multiprotocol backbones do not imply integration; they provide coexistence only.

Emerging technology, in particular switching technology such as ATM, should be considered if current networks do not have sufficient bandwidth, or if major new traffic loads such as multimedia will be deployed in the near future. Otherwise, it is suggested that the integrator wait for the standards to stabilize and prices to decrease. There is always a premium, both in terms of price and complexity, in implementing "bleeding-edge" technology.

Internetworking gateways are difficult to define. Peer-to-terminal/host gateways are common and abundant. The integrator should have little difficulty in finding one that is suitable for any particular situation. These gateways can be implemented in hardware

or software. Peer-to-peer gateways are somewhat uncommon. It is fortunate that there is little demand at the present time for such devices. It may be that in the future, as OSI networks become more common, there will be a greater need, due to a demand for TCP/IP/OSI connectivity. In the near term, it is unlikely that the integrator will require this type of gateway.

In many ways, the next chapter is the study of interoperable gateways. It is time for an examination of the other major component of the well-integrated network, interoperability.

But first, why not take the time for a brief review?

Concepts review

1. What is the major problem facing the integrator in the area of internetworking?

   ____

2. What are the major integration components for internetworking?

   ____________, ____________

   ____________, ____________

3. What addresses do bridges use to forward frames?

   ____

4. What addresses do routers use to forward packets?

   ____

5. Name two types of routing protocols used by bridges:

   ____

6. Bridges that are based on geographic coverage may be called:

   ____

7. What component must a protocol suite have for routers to be a viable integration technology with that suite?

____

8. Why are routers, rather than bridges, generally the preferred integration component?

____

9. Name some examples of gateways:

____

10. When are gateways an applicable integration solution?

____

Interoperability

CHAPTER 4

INTEROPERABILITY deals with applications on separate computers working together across a network. As we saw in earlier chapters, internetworking deals with the lower layers (1 through 4) of the OSI model, and it ensures that data gets across the network from one endpoint to another. Interoperability is concerned with the rest of the OSI model layers, layer 5 (Session), layer 6 (Presentation), and layer 7 (Application). This is somewhat of an arbitrary definition, since the words "internetworking" and "interoperability" are sometimes used by workers in the industry to mean the same thing. We have chosen to give these two terms specific definitions to more easily develop the material.

To complete the job of integration, it is important to have internetworking in place. We might go so far as to say it is necessary. As it turns out, internetworking is not a sufficient condition for integration. The job of integration is not completed until the issues of interoperability are grappled with, and any resulting problems solved. Since interoperability deals with the upper layers of the OSI model, it might be considered the application interface between the processes on hosts at the endpoints of the network, and the network itself. We deploy software components known as application programming interfaces (API) to solve interoperability problems. These are the interfaces to the network services by the end-user applications employing those services.

An enterprise network of medium to large size often has a heterogeneous mix of data and application types. There are myriad forms of e-mail, database access, remote process execution and various types of file transfer and access services. If the requirement in your enterprise network is to keep these disparate resources isolated, then you have no problem and there's little need to consider interoperability. However, in a well-integrated network, we often require these various differing APIs and application services to exchange data and process information. In today's networks, interoperability requirements are the rule, rather than the exception. See Fig. 4-1.

Consider also that interoperability must be provided for, regardless of the underlying transport or subnet protocols, different types of

Figure 4-1

Internetworking + Interoperability = Integration

- E-mail
- Database
- Client/server
- Electronic data interchange
- Distributed filesystems
- Remote terminal access
- File transfers

Today's enterprise networks have a wide variety of interoperability requirements. Learning Tree International

operating systems end hosts, or network operating systems deployed on the network. To provide this level of integration requires a system integrator who is trained in recognition of the issues, and feasible solutions.

The integrator has two primary tasks in the area of interoperability. One task is to develop a long-term strategy to facilitate future interoperability. That is not his main task, but it is an outgrowth of proper planning that should be done in any case. The most important task is the selection of components that address current interoperability problems while keeping in mind future needs and future growth. This chapter addresses the analysis of interoperable situations, and the selection of appropriate network software components at the highest layers of the protocol stack.

When looking for solutions in any technical field, an excellent plan of attack is to divide a given task or problem into manageable components that can be analyzed, dissected, and understood; then, apply the identifiable unique solutions, and solve one problem at a time. This way, the entire issue of interoperability in all its facets can be addressed.

The entire area of interoperability can be subdivided into seven major areas covering all pertinent aspects. We will assume that internetworking is in place. We can then turn our attention to the

interoperability tasks, starting with the simplest and progressing through the more complex ones. This is also the order in which they build on each other. A task studied first becomes a foundation for interoperable solutions to the more complex tasks.

We can name the following areas of concern with respect to interoperability:

- Remote terminal access
- Remote procedure execution
- Interoperable file transfers
- Distributed file systems
- Integrating network operating systems
- Integrating electronic mail systems
- Integrating distributed database access

First, there is the problem of remote terminal access. Users on stand-alone workstations or on host-connected terminals at one end of the network are required to run applications on hosts at the other end of the network. We term this task *remote terminal access*.

The next task might be the execution of a remote procedure on a remote host by a procedure on a local host. For this task, there are two hosts and at least one remote and one local procedure. The first procedure is running on the local host where the user might be working. This local procedure calls upon the execution of a procedure on a remote host. We term this *remote procedure execution*.

The third task might be one of file transfer, where a data file or an executable, binary, text, or data file—or any type of information or program file, for that matter—that is stored on one computer needs to be transferred across the network to another computer.

The fourth problem is the opening, reading, writing, and closing of files on a remote system from the local system, without the need to transfer the file. This is called the *Distributed File System*, or DFS, task.

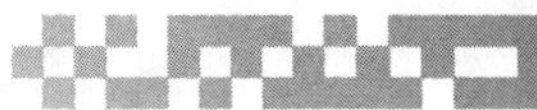

The fifth situation concerns distributed file systems that become the basis for a range of services supporting networked personal computers and workstations. This is classified as *PC network services*. But opening, reading, and writing to remote files is just one of the services that a PC network operating system provides. Record and file-locking features supporting multiuser access to the shared files is another important feature. Other services include remote printing via the use of queues, and some form of user authentication and access control—in other words, security services.

The sixth task requiring an interoperable solution is the integration of the many kinds of e-mail systems often found on enterprise networks. As a matter of fact, the need to provide enterprise e-mail services to all users across the enterprise network is one of the main motivating factors for multivendor integration in today's networks. It is often one of the first integration tasks tackled by an enterprise network.

The last task is the integration of access to remote databases, especially where those databases are of different types. This might be considered the *distributed database access* task.

Throughout our presentation we have kept in mind that we are not comparing solutions to one another in terms of technical merit, efficiency, or ease of implementation, but how they compare in providing interoperable solutions. For example, rather than comparing the various PC network operating systems to one another in terms of their functionality, we will be looking to see what techniques are used to integrate them. In other words, how does each NOS type support integration to other NOSs, and to a variety of workstation types?

Of course, as we look for interoperable solutions and apply them to existing problems on our networks, we are also going to be planning and rating solutions as to their usefulness for future growth and for future network needs.

So, as each of these seven issues is tackled, and we complete the integration job by fulfilling interoperability, we will rate the solutions that we find along the way for each task, and how they fit in with our

blueprint (see Chapter 2). In each case, we will try to address the following questions and considerations:

- Will this component or this solution be a good solution across the entire network?
- Is it a universal solution? Would we deploy it network-wide and consider installing it on all platforms, minis, mainframes and user workstations?
- Is this just a solution unique to a small population of users? Perhaps it has a place in our network, but it is not particularly good for interoperability. Perhaps it is a proprietary solution that we must live with for a while in a legacy portion of the system, for example.
- Does this scheme, solution, protocol, or particular service help us in interoperability, or does it hinder the interoperability environment? Is it part of the solution or part of the problem?

Keep in mind as we go along that our rating system does not say anything about how good the solution is technically, or in any aspect other than interoperability.

Before we begin looking at each of these interoperable situations, we have one more administrative comment. The approach in each interoperable situation is presented in four parts.

1. The first part of each section seeks to identify the interoperable problem that we're trying to solve in that section.
2. The second part of a section lists the major protocols, services, or products currently being used to provide that service or perform that task. In some cases, the list is quite exhaustive. In others, it's a short list, but it basically covers the major solutions or protocols used to solve the problem.
3. Once we're finished enumerating possible solutions, we tackle the important or major solutions in detail. We discuss their features, how they might perform the given task, and in some cases we give references for further information, such as pertinent RFCs. The point in this third section is to understand how the protocol works—

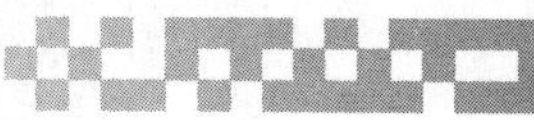

how good or how poor each interoperable solution really is. And it enables us to rate each one with respect to the others in its class.

We answer the following questions: Is it a good universal solution, or is it something we might want to preserve for the future? Is it leading edge (the correct aggressive solution), trailing edge (the conservative approach), bleeding edge (too soon to buy), or forget-it edge technology (a big mistake to purchase and install)? Lastly, we bring it all back together at the end of each section by discussing the impact of the solution on network traffic, and the additional problems the solution might create. In other words, does it have a heavy or a light impact on network loading? How do we classify the traffic: bursty, steady, heavy, light, etc.? What is the impact on network bandwidth? Will we see a lot of traffic or very little? And what impact does this protocol have on server loading and on the workstation?

❹ The fourth and last topic in each section is a discussion of the overall strategy to follow in achieving interoperability for the current area.

At the end of each section there is also a chart summarizing the findings, and rating the major protocols and services against one another from the standpoint of interoperability.

Remote terminal access

Consider a user on a workstation or terminal somewhere at one end of the network. The user desires to execute or control an application on a remote host from his local device. A remote host is another independent computer somewhere else on the network. These two computers are typically connected by a high-performance network. In other words, there is already some form of internetworking in place. This excludes the case of terminals directly attached to their host.

This, then, is the problem of remote terminal access: executing a computer procedure whose input comes from the user's keyboard. It can be block-oriented or character-oriented. In other words, it generates either synchronous or asynchronous traffic. The output of

the procedure is to be displayed on the user's monitor. We must assume that there is a high-performance network between the user and the target. (See Fig. 4-2.)

Figure 4-2

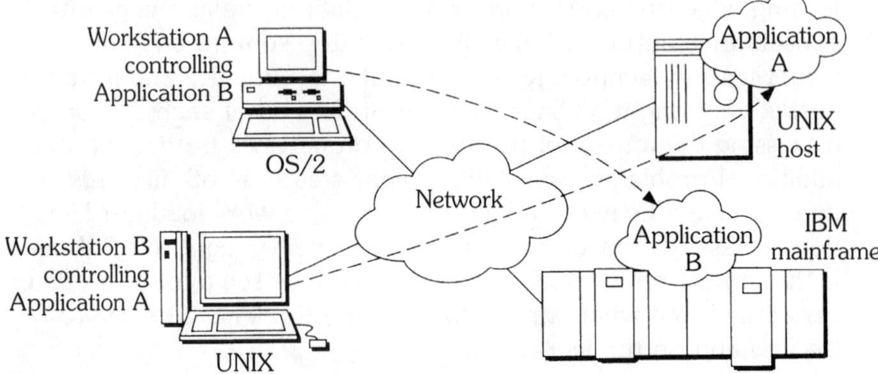

Remote terminal access is typically the first and simplest interoperable requirement to satisfy. Learning Tree International

There are many solutions to this problem. They are divided into those that support asynchronous traffic and those that support synchronous traffic. Although we sometimes use a terminal that communicates asynchronously, and, with the appropriate converter, controls an application on a remote host that requires synchronous input, it is not common. This is the case of a DEC VT100 controlling an application on an IBM mainframe, for example. The VT100 is an asynchronous character-oriented device, and the mainframe host is typically set up to receive block-oriented communications from a block-oriented terminal, such as an IBM 3270.

The typical remote terminal access solutions are the following:

- The most universal solution of all is the use of the TELNET protocol from the TCP/IP protocol suite.
- On certain UNIX networks, particularly the Berkeley UNIX variety, there are the r* Utilities. For example, we have rlogin and rshell as two protocols for this procedure.
- Digital Equipment Corporation uses a protocol in their network of VAX computers for terminals and workstations acting as

terminals to access the remote hosts over an Ethernet network. This protocol and this service is called Local Area Transport (LAT), a very popular protocol in DEC VAX environments.

- A popular protocol in OSI networks (those networks that use the OSI protocol suites) is the VTP or the Virtual Terminal Protocol, VT for short.
- For synchronous hosts, there is the 3270 terminal emulation and a number of local area network protocols that support the transport of 3270 data-stream traffic to an IBM SNA host.
- For those situations where an asynchronous terminal needs to send traffic to a synchronous host we have TELNET 3270. This is a popular protocol for accessing an SNA host from a TCP/IP workstation or host. It is abbreviated as TN3270 protocol.

There are other solutions that support terminal access over high-speed networks, but most of them are proprietary to the various minicomputer vendors. These will not be considered, since the list above is fairly exhaustive, and covers most situations encountered in enterprise networks. (See Fig. 4-3.)

Figure 4-3

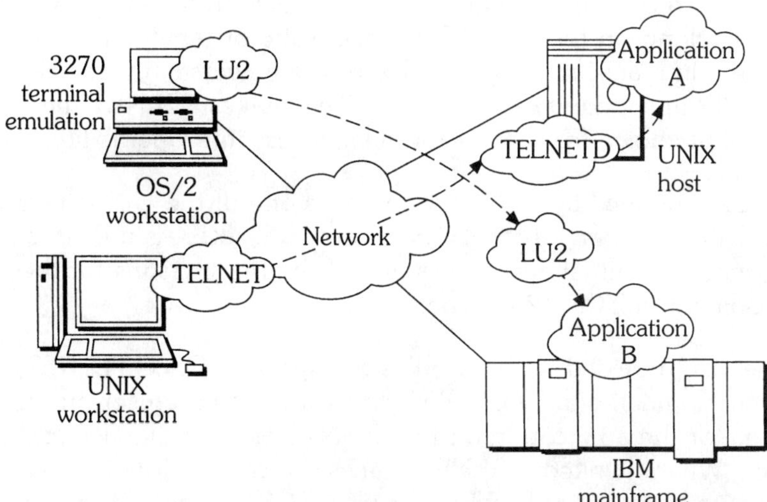

Two possible solutions to the remote terminal-access interoperability problems: TELNET and TN3270. Learning Tree International

The figure above shows an example of how two of these solutions solve the problem of remote terminal access. The TELNET program running on a UNIX workstation connects via a network (typically an Ethernet LAN using the TCP/IP inter-networking protocols) that is serviced at the other end by another UNIX host, a TELNET daemon, which receives the characters from the UNIX workstation and feeds them to the application. It then takes the characters from the application that need to be displayed on the UNIX workstation screen and sends them across the network to the client.

Alternatively, there is a 3270 session being conducted between an OS/2 workstation at one end of a network. It might be the same type of network used by the UNIX workstation in the previous example, or it might more typically be a Token Ring network. The workstation has established a Logical Unit 2 (LU2) session with the application running on an IBM mainframe.

These two examples are the most common remote terminal access situations. Let's look at each solution in detail.

The universal solution: TELNET

TELNET can be considered the universal solution for interoperability. Since it belongs in the TCP/IP protocol suite or family of protocols, it is present in many, many hosts, due to the universality and availability of TCP/IP in so many environments. This makes TELNET the protocol of choice for remote terminal-access interoperability.

TELNET is defined by RFC 854, and was one of the earlier protocols to be defined as part of the TCP/IP protocol suite. As a matter of fact, remote terminal access was one of the first reasons for the development of TCP/IP internets.

The service provided by TELNET is transparent. Once TELNET is invoked, no additional work needs to be done by the user on the terminal or the application on the remote host to make use of the service. When invoked, TELNET makes the connection and manages the session automatically. TELNET uses TCP for reliable delivery of the characters and, of course, IP for connectionless datagram routing. It works through port 21 on TCP.

TELNET provides three basic services: the *Network Virtual Terminal*, a standard ASCII character-based interface for any type of host or workstation the user might be on, and keyboard input and screen output management across the network, irrespective of the local or remote computing platform or operating system type.

These services are:

- A virtual keyboard that supports character input. Limited to the specific characters in the 7-bit IA5 (International Alphabet 5) ASCII code.
- A virtual printer for the display of IA5 characters. Virtual printer output can go to a monitor, or it can go into a file at the other end. This protocol is the basis, for example, of file transfer used by the FTP (File Transfer Protocol) for moving characters in a file from one host to another.
- A symmetric connection. The client side can be either a program or a terminal and the server side can also be either a program or a terminal. (See Fig. 4-4.)

Figure 4-4

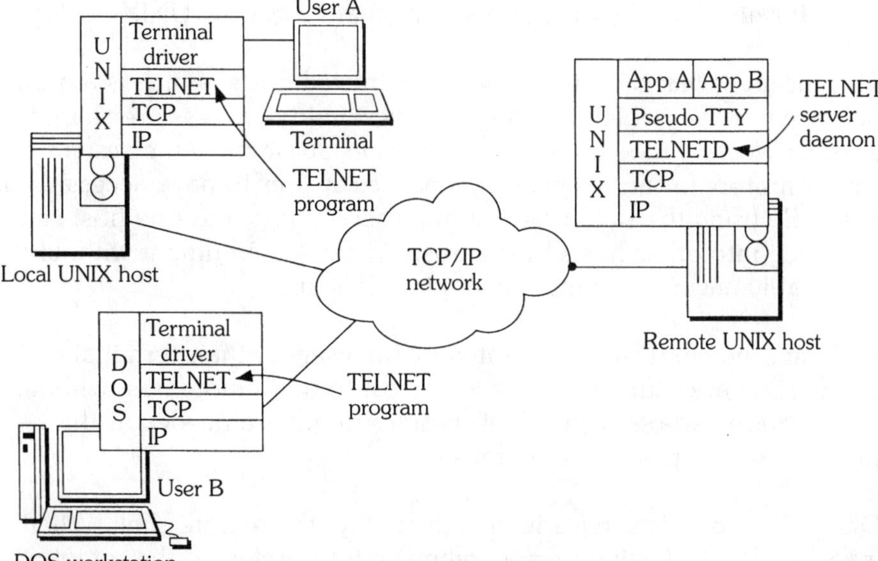

How the UNIX TCP/IP solution of TELNET works for a remote host client and workstation client for remote terminal access. Learning Tree International

TELNET provides two alternatives for a connection, with the decision being made at connection time: either the file is encoded in 7-bit ASCII, or it is encoded in an 8-bit character set. Choosing the latter alternative enables the transport of EBCDIC-encoded characters, in order to support 3270 terminal traffic, or the transfer of binary files under FTP. The options can be negotiated by either the client or the server.

Interoperability Rating: Best

No matter what the client or server operating environment is, if both support the network Virtual Terminal environment of TELNET, interoperability is assured. It is the preferred protocol when interoperability is being considered in an enterprise network that has a variety of hosts and a variety of terminals and client types.

The Berkeley UNIX solution: rlogin.

The rlogin facility is the 4.3 BSD UNIX remote login capability. It employs TCP port 514 and requires the use of the TCP/IP protocol suite. It is used mostly in environments that have BSD UNIX hosts.

This facility is based on the concept of trusted hosts, which is a group of machines over which user login names and file access are shared. In other words, an equivalency has been established by the system's administrators for authenticating a particular user to have access to all hosts. By using the rlogin facility, the user logging into one host can easily log into other hosts in the trusted-hosts environment without necessarily having to authenticate himself again.

In rlogin, the characters generated by the users at their terminal or workstation are sent from their local host to the remote host without modification. No other protocols besides rlogin are needed at the process level to produce this transfer.

One variant of this service is the rsh facility: the remote shell. When invoked, the rsh facility runs a command interpreter on the remote machine and passes the command arguments issued by the user on the local machine to the remote machine, also without modification.

The resulting effect is essentially the running of a remote shell on the remote machine. To the local user, it appears that he is running the remote shell at his local machine. Of course, just as in rlogin, the user skips the login to the remote machine because he's automatically authenticated as being a part of the trusted-hosts environment.

From the standpoint of security, rlogin and rsh often provide "back door," or unsecured, entry to UNIX hosts, creating a security liability or hole. This is something system administrators should avoid.

Interoperability Rating: Poor

Although rlogin, rsh, and other facilities of this nature have been ported to other environments besides BSD UNIX, their use is not widespread. They will not be readily available on hosts as often as the TELNET protocol, for example. They will also not be seen on as many workstation types as TELNET. For that reason, they are rated much lower than TELNET in interoperability.

Of course, if your enterprise network contains machines that all have this facility, then by all means, rlogin or rsh are appropriate choices. There is no need to install TELNET everywhere for remote terminal access. But they are definitely not as universal as TELNET, so we would have to rate them in the poor category. They are certainly not something to be sought out in providing future growth capacity for a network, or as part of an overall integration strategy in enterprise networks.

The OSI solution: The Virtual Terminal protocol

Networks that specify and employ the OSI protocol suite have the Virtual Terminal protocol (VT) available for remote terminal access. OSI protocols might be seen in networks specified by governments under the GOSIP specification, or in manufacturing environments under the MAP (Manufacturing Automation Protocol) specification.

VT supports a variety of terminal types: scroll-mode terminals, such as the VT100 type; screen- or page-mode terminals, the 3270 type;

and form-mode terminals. VT is not a direct terminal interface like TELNET. It is an *Application Service Element* (ASE), in the nomenclature of the OSI protocol suite. It is actually an application process support protocol, which means that a program or a local application process is required to interface the terminal's keyboard and screen to VT itself. In other words, the VT program does not interface with the terminal or screen directly. The interface program is host-dependent, and implements the support needed for terminals unique to that host. VT itself is host-independent.

In a VT environment, client and server negotiate user-element mapping functions at connect time. One such negotiated element might be the use of a mouse and the support of mouse control signals. Another might be the type of monitor (color or monochrome), the type of keyboard, or the type of printer (a display device or character printer). The fact that VT supports varied display and user input devices makes VT more flexible than TELNET. (See Fig. 4-5.)

Figure 4-5

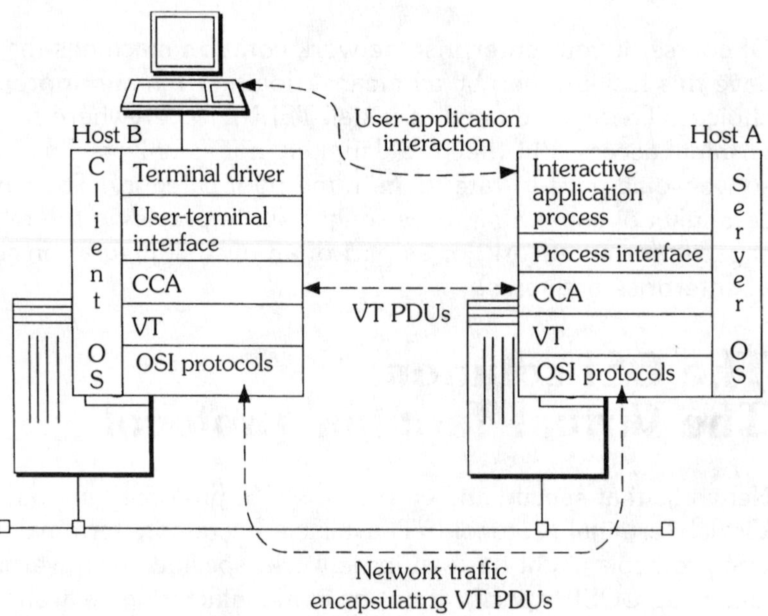

CCA = Conceptual Communication Area
PDU = Protocol Data Unit

The OSI solution for remote terminal access: VT. Learning Tree International

The figure above shows the protocol stack employed in the VT environment. Note the typical OSI network environment of host-to-host operation. Although we might occasionally see a workstation as the user platform, the more typical situation in OSI networks is a user terminal connected to a local multiuser host running an application on a remote multiuser host.

Internetworking is provided by the OSI lower-layer protocols, such as TP-4 and ISO IP, over some data-link protocols, such as Ethernet or Token Ring. At the application level, we would have the VT protocol running on both machines under the specific operating systems in those hosts. There is a conceptual communication area in each computer, which is the buffer where input and output are transferred between VT in each of the devices and the user terminal interface, which is host-specific. On the user host, there is a terminal interface. There is also a process-to-VT interface program for the application running on the remote host.

User authentication and session initiation is not handled by VT but by ACSE, the Association Control Service Element, a support protocol.

Interoperability Rating: Mixed.

As the OSI protocol suite gains popularity, and more and more workstations and hosts employ it, this solution will become more desirable and more widespread. Since the abandonment of strict adherence to the GOSIP mandate by the U.S. Government, this increase in popularity has come into question. At the moment, due to the scarcity of deployment of OSI protocols, we would have to rate this solution as being not as good as TELNET. On the other hand, if the OSI protocol suite were to become more popular, this would become a more desirable solution, perhaps rivaling TELNET at some future time.

The DEC solution: LAT

One of the DEC solutions to the problem of asynchronous DEC terminals communicating with hosts over Ethernet is the *Local Area Transport* (LAT) protocol. DEC came up with an ingenious method of transmitting characters from a VT-100 terminal to a VAX host in an environment with multiple VAXes and a terminal on every desk.

LAT is an extension of Ethernet, and it is used as the shared transport mechanism in an Ethernet network. To access the Ethernet network, asynchronous terminals (mainly VT100's) connect to a LAT terminal server. The terminal server receives the asynchronous terminal traffic over serial ports, creates a LAT packet with it, and encapsulates it in an Ethernet frame in its Ethernet adapter. The frame is then sent to the appropriate host. The terminal server manages the flow of incoming and outgoing data characters. (See Fig. 4-6.)

Figure 4-6

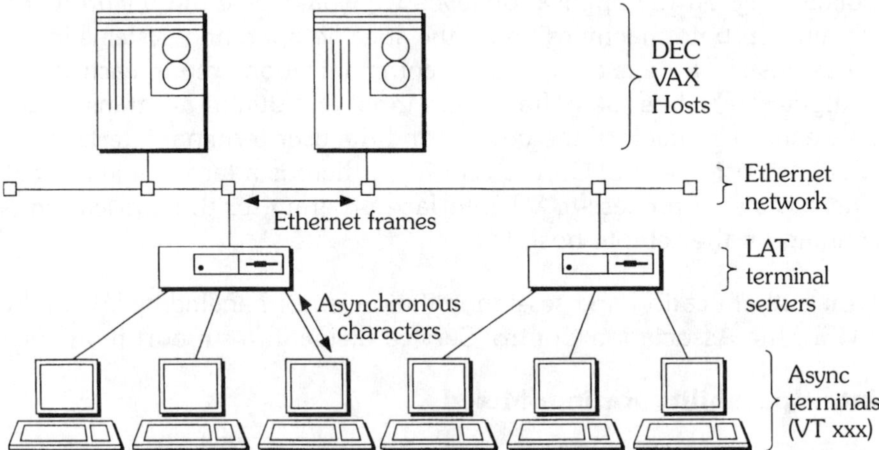

The DEC LAT network solution: attaching character-based terminals to a high-performance Ethernet network in order to run programs on VAXes.
Learning Tree International

LAT is a three-layer protocol with Ethernet at the lowest layer, a slot layer above it for identifying which terminal any one character belongs to, and a virtual circuit layer that manages the connection between each terminal and the host.

LAT is a very simple and nonroutable protocol. It is meant to be employed in a local network, with at most bridges as the internetworking devices, not routers. It is from this simplicity that it derives its efficiency. It also increases efficiency by stuffing characters bound for the same host from many terminals into one common frame. But because of the nature of the Ethernet frame, plus the slot-layer and virtual-circuit protocols, it has a considerable amount of overhead.

The key element in a LAT network is the LAT terminal server, which assembles and disassembles frames and manages the handshake with locally connected terminals. Workstations can replace asynchronous terminals by emulating them. The workstation can manage its own LAN/LAT connection without need of a terminal server, but this usage is uncommon.

Interoperability Rating: Poor

We have to rate this protocol as being poor in interoperability because it is too vendor-specific. Essentially, the only environment that makes widespread use of this protocol is the DEC VAX environment. LAT is more efficient than TELNET, and is preferable for a VAX-only environment. Also, it's very difficult to build large enterprise networks that have to support LAT because it is nonroutable. It may be encapsulated in TCP/IP to be routed across an enterprise network, but this increases its inefficiency.

Connecting to an IBM mainframe

The classic method of running a program on an IBM mainframe is for a user on a 3270-type terminal connected to an SNA network to establish a Logical Unit 2 (LU2) connection with the host. The SNA network handles establishment and maintenance of the circuit, and the data transfer between the terminal and the application.

Traffic from the 3270 data stream represents a large fraction of overall enterprise network traffic, due both to the traditional interaction of large mainframes with a large population of terminal users, and to the emergence of the new environment of client/server software. We tackle the client/server scenario and its network requirements in Chapter 6. Here we look specifically at 3270 terminal traffic support. (See Fig. 4-7.)

There are many possible ways for a workstation to emulate the operation of a 3270 terminal:

- A personal computer, Intel-based DOS machine, or Macintosh, with an appropriate SDLC adapter plugged into the

motherboard and appropriate software for 3270 emulation, can be directly connected to the IBM SNA network. There are a number of products on the market that do this very well.

➢ Moving up in sophistication, a personal computer running OS/2 as its operating system can be outfitted with a Token Ring adapter. Then using 802.2 (logical link control) software, it can move 3270 traffic from the workstation to a classic 3174 on the token ring. Alternatively, the 3174 communication controller (also known as a *cluster controller*) can be a PC emulating a 3174.

➢ A third alternative might be an OS/2 PC attached to an Ethernet network employing logical link control (802.2) to get to an Ethernet-attached 3174. We can also use a high-level protocol such as NetBIOS or IPX to encapsulate and transport the 3270 traffic. In the latter case, the workstation establishes a session with the gateway via the upper-level protocol, and the 3270 traffic is then handled over that session and converted into SNA at the gateway. The proper SDLC handshake is handled by the gateway with the host over the SNA network on the other side. The connection by the PCs to the gateway can be made via any Data Link protocol—for example, Ethernet or Token Ring, which are the two most popular.

➢ A popular interoperable solution, which is also universal in scope, is to run 3270 traffic encapsulated in TCP/IP packets. The protocol is TN3270, which can transport 8-bit encoded EBCDIC traffic to a gateway that then converts and transfers the traffic to the mainframe.

➢ Another technique is to run TCP/IP and TELNET 3270 on the mainframe itself. This requires the direct attachment of the mainframe to the high-performance network. Either Token-Ring or Ethernet network controllers can be employed for this purpose. The latter approach seems to be growing in popularity. It offers a universal solution, since the trend is to run TCP/IP on gateways, hosts, and workstations, and to install an enterprise-wide TCP/IP network infrastructure. The use of TCP/IP in the mainframe also provides solutions for remote program execution (RPC), file transfer (FTP), and distributed file system (NFS), all of which are popular for a wide range of workstation clients. Providing solutions to all these needs brings

with it TN3270, which is also a part of the TCP/IP protocol suite. Having this protocol at the mainframe essentially for free (riding the coattails of the other protocols), we can naturally turn to it for our remote terminal needs. (See Fig. 4-8.)

The diagram in Fig. 4-9 explores using TN3270 over a TCP/IP network. There is a DOS workstation running TCP/IP for internetworking, the TELNET protocol and 3270 terminal emulation software, and perhaps even a script handler such as the *High-Level*

Figure 4-7

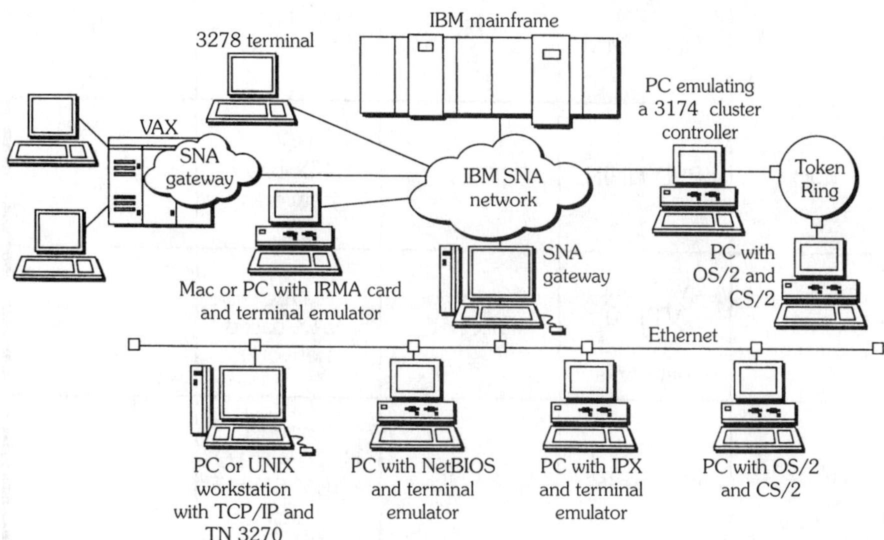

The IBM SNA network, and the many possible ways to connect a terminal to it. Learning Tree International

Figure 4-8

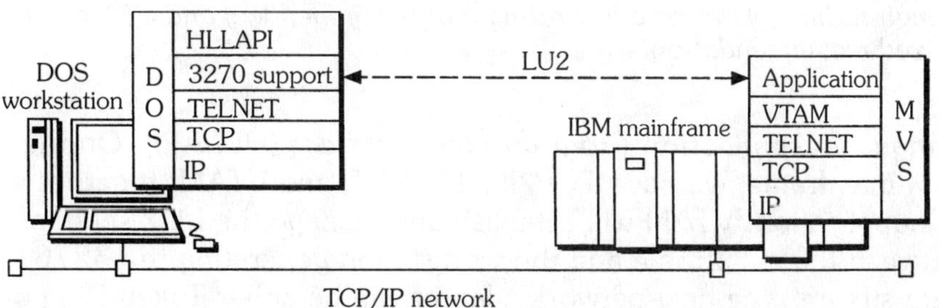

The TN3270 solution of attaching an asynchronous terminal to an IBM SNA network via a TCP/IP enterprise network. Learning Tree International

Figure 4-9

Remote terminal protocol	Common platforms	Typical protocol suite	Typical use	Interoper-ability rating
TELNET	UNIX DOS OS/2 Windows NT	TCP/IP	Universal connectivity mostly used in UNIX world	+
VT	UNIX DEC VMS	OSI	OSI-based networks	0
rlogin	BSD UNIX	TCP/IP	BSD UNIX-based networks	–
LAT	VAX and VT100 terminal networks	DECnet	DEC VAX-based networks	–
TN 3270	UNIX IBM hosts	TCP/IP	Connect UNIX WS TO IBM SNA host	+

The table reviews the major remote terminal access protocols, the most common platforms they are found on, the typical protocol suite they are a part of, a typical use, and the interoperability rating. Note that a (+) means a high rating, (–) means a low rating (not recommended) and a (0) means mixed recommendation. Learning Tree International

Language Application Programming Interface (HLLAPI). On the IBM mainframe, we have TCP/IP, TELNET, and VTAM, together with the application. VTAM will establish and manage the LU2 session between the mainframe and the workstation generating the 3270 data stream. The host network adapter can be either Token Ring or Ethernet, with Token Ring being the odds-on favorite.

The TN3270 data stream uses both TCP and IP because it is a TELNET connection, and of course the connection is made over TCP port 21. The network Virtual Terminal traffic is 8-bit EBCDIC in this case. We're very happy with the use of TELNET to move 3270 traffic, because there's such a great variety of platforms that support it on the client side. TCP/IP, TELNET, and the use of TELNET 3270 are very popular on a variety of workstation types: UNIX, DOS, Windows NT and OS/2 workstations. A variety of hosts such as VAXes, HPs, and numerous others also support TCP/IP, as pointed out earlier in this section. TN3270 then becomes a natural gateway for users on the larger hosts to get to the mainframe from their respective terminal types.

Network resource utilization

Moving remote terminal access traffic across a network can place a very heavy load on that network. For connectionless high-performance LAN situations, such as TELNET running over Ethernet, there may be a very high protocol overhead. For example, one byte of TELNET data carried over Ethernet, using a 64-byte minimum frame, requires 18 bytes of Ethernet overhead, 20 bytes of IP overhead, 20 bytes of TCP overhead, and 5 bytes of padding. This represents a 1% utilization of the available bandwidth!

Protocol overhead is just one problem. The other is that even though the traffic from any one terminal is fairly light, there are typically many hundreds of terminals requiring remote access that are connected to the network. This is especially true where the use of terminal servers is the norm, as in VAX networks. In this case, the aggregate traffic from all these terminals can fill up the available bandwidth very easily, leaving no room for other types of traffic, such as file transfer, RPCs, and distributed file access.

Typically, remote terminal access traffic is characterized as steady and heavy. It is important to note the characteristics of this traffic because, at a later point in this chapter, we will consider the burden other protocols place on network bandwidth. The characteristics of these two situations are markedly different, and will cause problems

when they both use the network simultaneously. File transfers, for instance, have a bursty characteristic. File transfers can heavily load a network segment, which can cause the sessions supported by remote terminals on that segment to time out.

In any event, remote terminal access has a high protocol overhead, is typically an inefficient consumer of network resources, and places a heavy loading requirement on the network. For that reason, we tend to prefer higher-performing networks in which there are many terminals. Connectionless low-performance networks should be avoided, or their use minimized.

For example, AppleTalk's LocalTalk protocol can easily get swamped when many Macs send TELNET traffic over a particular LocalTalk segment. The data rate of LocalTalk is 230 kbps. We don't recommend the use of LocalTalk in large, heterogeneous networks, and we urge a switch to high-performance Ethernet and Token-Ring options when connecting Macs for remote terminal access.

When using connection-oriented wide-area network protocols to support connectionless traffic, one has to be aware of the high overhead engendered by call set-up and closing, which adds to the inefficiency and further degrades performance. For example, an X.25 link might be used to connect IP routers. It can be configured to close a virtual circuit after a certain amount of idle time, and reestablish that circuit when additional traffic warrants build-up at the router. This adds to the degradation of performance, and should be seriously studied.

Whenever possible, the use of high-performance WAN protocols, such as frame relay, should be considered. Certainly, the use of dial-up access or switched virtual circuits, with their additional frequent call-setup overhead, should be minimized.

Integration strategy

TELNET is the protocol to be recommended for success in interoperability, as it is overwhelmingly superior to all of the others considered here. Because of the nature of TCP/IP and the marketplace, TELNET is available on many different hosts, many different platforms, and many different types of data-link protocols. A strategy that moves

the enterprise network toward universal acceptance and use of TCP/IP and TELNET should meet with a high degree of success.

Please bear in mind that TELNET does not support the newer computing environments, such as graphical user interfaces and the use of pointing devices such as mice. That kind of environment has to be supported with other protocols, such as the X11 protocols supporting the use of X-Windows. As the use of more sophisticated terminal types increases, changes to the protocol have to be made to accommodate them. For now, we seem to be happy with character-based and block-oriented terminals.

To summarize, here are a few rules of thumb to use when designing interoperable remote terminal access:

- As with any network use, match the traffic load to the available bandwidth. In the case of remote terminal access, it is a relatively easy task to model and predict the traffic load on the network. We can therefore estimate both the required bandwidth and the bandwidth that will actually be used on our enterprise networks, backbones, and subnets fairly well.
- We urge the use of higher-performance (above 1 Mbps) LAN protocols whenever possible. That is, switch Macintoshes from LocalTalk to TokenTalk and EtherTalk, for example.
- Segregate the steady, heavy terminal traffic from bursty, distributed file-server traffic by subnet. In other words, put hosts and the terminals that need remote access to them on the same subnet, and keep that terminal environment separate from file transfers or a distributed file system environment, which is bursty by nature. Keep the two segments connected with bridges or routers, but have the subnets separated. Terminal users can experience a significant slow-down or time-out in their connections during the burst periods of a file transfer protocol if they are on the same subnet.
- Employ lower-overhead wide area network protocols whenever possible to improve efficiency. Choose frame relay over X.25. Choose statistical multiplexers over static multiplexers—they improve performance, and tend to lower monthly phone costs, even when there is a higher initial equipment cost disparity.

Remote program execution

Consider the situation of a UNIX computer in which a particular process is running. Suppose that the process requires the execution of another procedure to complete processing. Typically, in that situation, the client process will make a call from within the program to another procedure (the server procedure) outside the program and pass it some parameters. When the calling program receives a reply, it continues operating. The procedure being called in this case executes in the same host, but as the calling program.

An identical situation can arise in which the client process is on one machine, and the responding server process is on a separate computer. The two machines are typically connected via a high-performance network. There is an application service that makes it appear to each process as if the other were on the same computer, yet they might be across the network from one another. In the case of UNIX computers, for example, this service software is often called a remote procedure call.

Since we might have a variety of situations in which such a service is performed, but that are somewhat different from the UNIX RPC environment, we generally classify these similar services as remote procedure execution. One of the characteristics of these services is the transparency of the network to the requesting process. In other words, the remote process serving the request typically appears to be local to the requesting client process.

Another characteristic of remote procedure execution is the presence of a high-performance network and the use of a high-performance protocol suite, such as TCP/IP, to connect the two hosts. The two hosts might be two minicomputers, a mainframe and a minicomputer, or more typically in today's environment, a workstation, DOS, UNIX, Macintosh, or OS/2 machine, and a large host.

In multivendor networks, it is often the case that the client and server host operating systems are different. This requires that whatever remote procedure solution we implement be available under the two operating systems. It also requires that both machines run the same protocol suite for internetworking. It may require the use of gateways

for protocol conversion, if the protocols at the two endpoints of the transaction are different.

Typical solutions to the remote program execution problem are

- the Sun UNIX RPC (Remote Procedure Calls);
- the IBM CPI-C solution using the APPC and APPN protocols;
- the DCE (Distributed Computing Environment) RPC (Remote Procedure Calls);
- the OSI ROSE (Remote Operating Service Element) protocol;
- the sockets protocol.

In reality, the choice of remote program execution protocol is only of particular interest to programmers. System and network integrators have a passing interest in what remote procedure call environments are being employed. It can be useful to the integrator to know the type of remote procedure calls being employed, though, as it is then possible to effectively estimate the loading on the network given a particular implementation.

Figure 4-10 demonstrates a client process making local and remote call to external processes. As we shall see, the call to remote processes needs to be handled by a network protocol we call the remote procedure call.

Figure 4-10

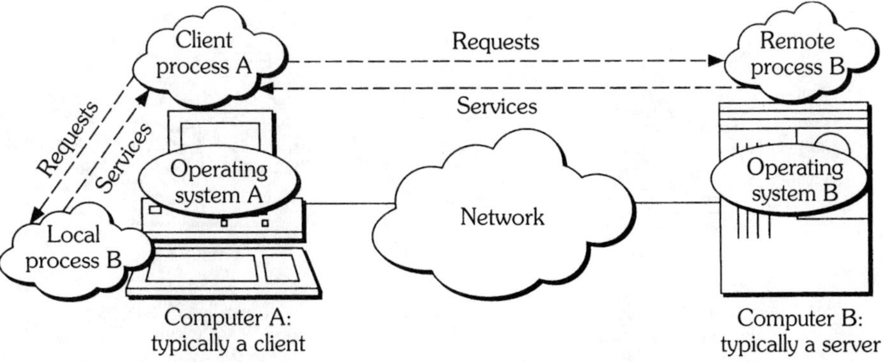

Remote procedure execution scenario. Typically, workstations are clients making the request, and minicomputers are the servers executing the remote procedure. Learning Tree International

The Sun UNIX solution: RPC

Sun Microsystems developed a protocol which it published to the Internet community in 1988, and which was subsequently adopted as RFC-1050. This Sun protocol is part of a series of protocols used to support a distributed file system called NFS (the *Network File System*). The unique thing about the NFS protocol family is that Sun made the source code for the RPC available for a minimal license fee. This caused the protocols to be ported to many other computer environments by many vendors besides Sun, proliferating their use and turning them into a fairly widespread solution for remote program execution and file access. It has been accepted by the industry and codified in RFC-1050.

Sun RPC was designed to execute remote procedures by emulating the pipelining of requests and replies between two procedures within a UNIX host along the principles discussed earlier. It was originally intended to be used with NFS and UNIX. Today RPC is a major tool in its own right, and has been successfully used to support client/server program development on a variety of client and server platforms.

To illustrate how an RPC works, consider the case shown in Fig. 4-11. Here the RPC protocol operating in a client workstation supports NFS

Figure 4-11

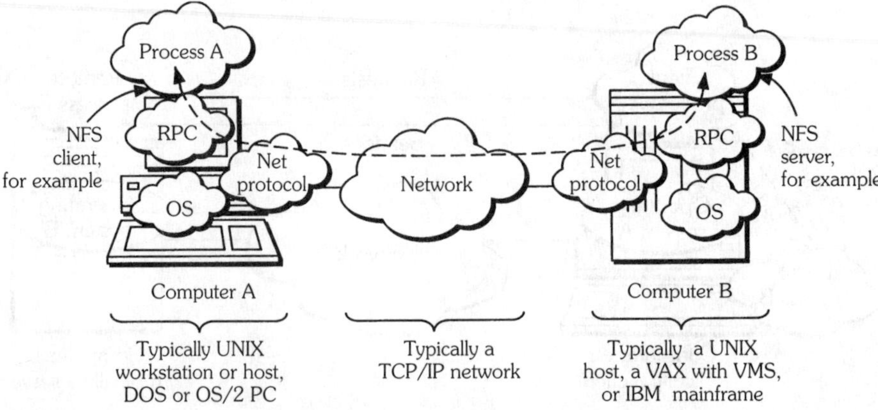

RPC is the network function between an application service such as NFS and the internetworking services. Note that there are RPC server and client components. Learning Tree International

calls made by the workstation application. The NFS server host also uses RPC as the interface between NFS and the network.

Due to the popularity and proliferation of Sun NFS, the RPC protocol is available on many different platforms and operating systems, making it the protocol of choice when considering interoperable remote procedure execution services.

Typically, RPC is considered a session protocol, and it is placed within layer 5 of the OSI model. It should be noted that a number of authors and workers in the field place it in other layers. RPC provides some of the functionality of the session layer of the OSI model, and thus it naturally can be thought of as belonging there. Please bear in mind that the TCP/IP protocol suite was not developed with the OSI model in mind, and it is often hard to fit protocols that were not developed under the guidance of the OSI model layer definition to those layers themselves.

Interoperability Rating: Best

Due to the popularity of RPC, it has been widely used and implemented. There are many interoperable variants, making it a very popular protocol as well as a universal interoperable solution. It is thus favored as the solution of choice.

Consider how a remote procedure call might be executed, as shown in Fig. 4-12. There are two computers in this scenario. A client process is running on computer A and begins executing. At some point, the process in computer A needs to run another procedure to complete execution. The client process suspends execution and sends a request to execute a procedure. The RPC program in the client receives the procedure call and forwards the call over the network. The request from the client process is forwarded to the server computer via the rest of the protocol stack. It is received by the server RPC processor. The server has been idle up to this point, waiting to receive requests. The server remote procedure execution services (RPC server) start up, and feed the received remote procedure call to the indicated local procedure on the remote server. The remote server executes the procedure and returns the output to the RPC server program on the remote computer, which in turn

Figure 4-12

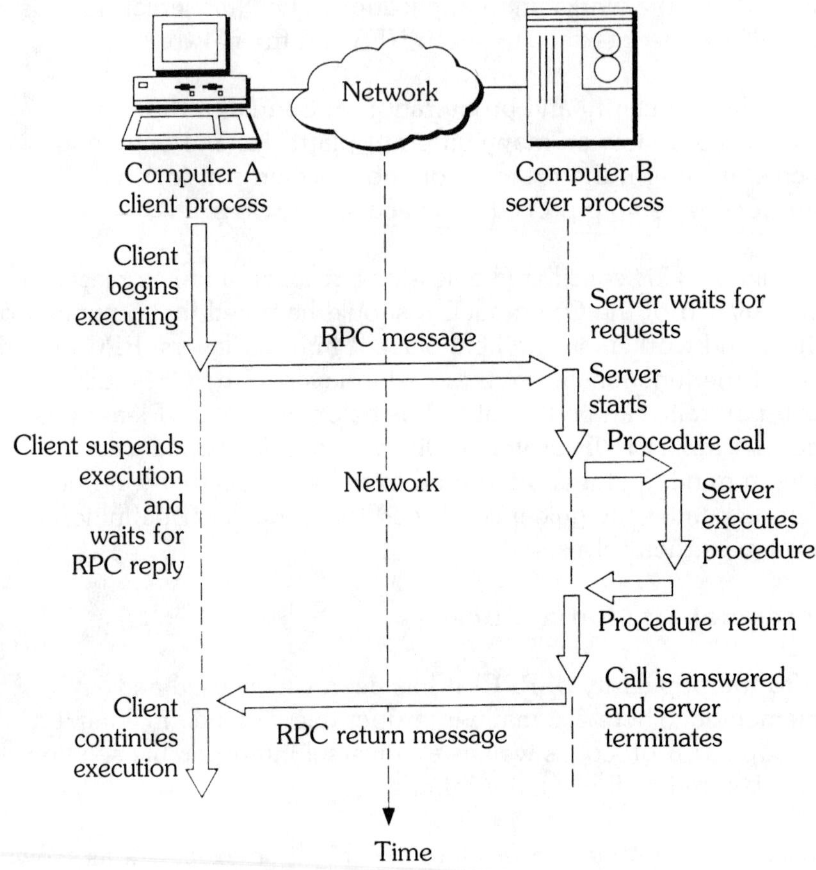

A typical RPC call request by client, and procedure execution and reply by server. Learning Tree International

forwards it across the network as an RPC message reply to the client. The RPC processor in the client passes it on to the client process. The client process then continues execution after having received the reply. It is a one-time transaction. There's no continuing conversation between client and server. If another RPC message is needed in order to continue operating at the client process, that message will be sent as a separate message distinct from the first.

The Remote Procedure Call protocol requires that the information transmitted across the network be encoded in a universal format. Bear in mind that the operating systems of the client and server, and their local

storage and manipulation of data, can be quite different. To expedite communication between the two machines, the RPC protocol specifies the use of a universal representation of the data as it goes across the network. This representation of the data is specified in a protocol called the *External Data Representation* (XDR). XDR was developed by Sun in conjunction with RPC, and it is defined in RFC-1014.

XDR defines an implicit format for the data. This means that the two machines have agreed ahead of time on the format of the data, before it goes across the network between one machine and the other. No format specification is sent with the data itself.

This implicit specification in XDR differs from other protocols that perform the same function. For example, an explicit data representation is the *Basic Encoding Rules* (BER) specification of the OSI protocol suite. BER causes every communication from one host to another to carry with it a specification of the formatting of the data, as well as the data itself.

Data types represented by XDR are 8-bit bytes with the least significant bit at the highest memory address (big-endian format). XDR defines 4-byte integers and 64-byte hyper-integers. A Boolean structure is defined, as well as the IEEE format for floating-point numbers with 23 bits of mantissa, 8 bits of exponent, and 1 bit for the sign.

In the process of transmitting RPC requests and replies, each host is responsible for converting the data between the local format and the XDR implicitly-defined format. XDR has a limited range of format options, but it has great appeal due to its universality.

The OSI solution: ROSE

The OSI protocol suite defines the *Remote Operations Service Element* (ROSE) for the execution of remote procedures. This is a very basic protocol used by other application-level protocols in the OSI protocol suite. It is not intended for programming client-server remote procedure executions, but its presence in a host may be used to make remote procedure calls. ROSE is known as an *Application Service Element* (ASE), and it is connection-oriented. In other words,

the calling application has to establish a virtual circuit at the application level before its request can be forwarded.

The connection-oriented nature of ROSE should be contrasted with Sun RPC, which is connectionless. In RPC, a request goes across asynchronously, i.e., without having a connection established beforehand. It must be pointed out that although the two protocols, Sun RPC and OSI ROSE, perform similar tasks, they're not compatible. The data representation in the OSI ROSE environment is defined by the ASN.1 protocol. Specifically, the rules for formatting the data fall under the ASN.1 protocol called *Basic Encoding Rules* (BER), as previously described.

In this case, the definition of each data element being sent across goes with the data element. In other words, we have explicit representation. The BER protocol specifies many more data types than those represented by XDR.

Note that ROSE does not interface directly with an application or process; it does so through an application-specific service element, as shown in Fig. 4-13.

Figure 4-13

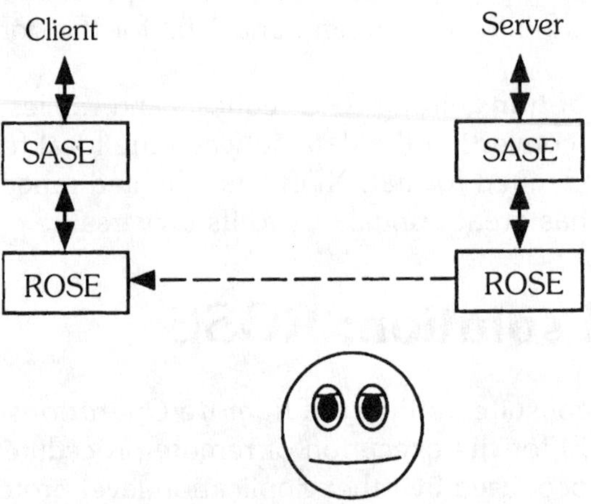

The operation of OSI ROSE in a remote-procedure execution mode. Learning Tree International

Interoperability Rating: Mixed

Although there is great promise for the OSI protocol suite and many organizations are adopting it, specifically government organizations that require an interoperable protocol suite, it is still not very popular. Thus, there are very few implementations of this protocol. In other words, there are few products on the market that support it. It is not as widespread as the Sun RPC, and for that reason cannot be counted as a major solution in our interoperable protocol bag of tricks. We look forward to greater acceptance of the OSI protocol suite and greater use of ROSE in the future.

The DCE solution: DCE RPC

The DCE Remote Procedure Call protocol is very similar in operation to the Sun RPC, although it is not compatible with it. The Open System Foundation (OSF), in its Distributed Computing Environment (DCE), has defined a method by which client processes can make calls to server procedures, and have those procedures execute and return responses exactly in the manner of the Sun RPC.

The DCE RPC has been defined and made part of the DCE environment. There is source code available for RPC runtime libraries. A number of procedure libraries are also available to respond to remote requests from programs. Interestingly, there are many different network infrastructures that support the DCE Remote Procedure Call environment. This is not explicitly specified in DCE. Typically, it is a TCP/IP stack, but it could also be an OSI stack, or in fact any other. See Fig. 4-14.

Interoperability Rating: Mixed

This protocol should be rated *mixed* primarily because it is still too new, there are very few implementations, and it is not as widespread as the Sun RPC. On the other hand, it has a great future, because it promises to be integrated with many different protocol stacks, and is part of the DCE suite of solutions. We intend to use it more in the future, as it becomes more widely supported, and as it is adopted across different vendor platforms, perhaps eventually replacing Sun RPC.

Figure 4-14

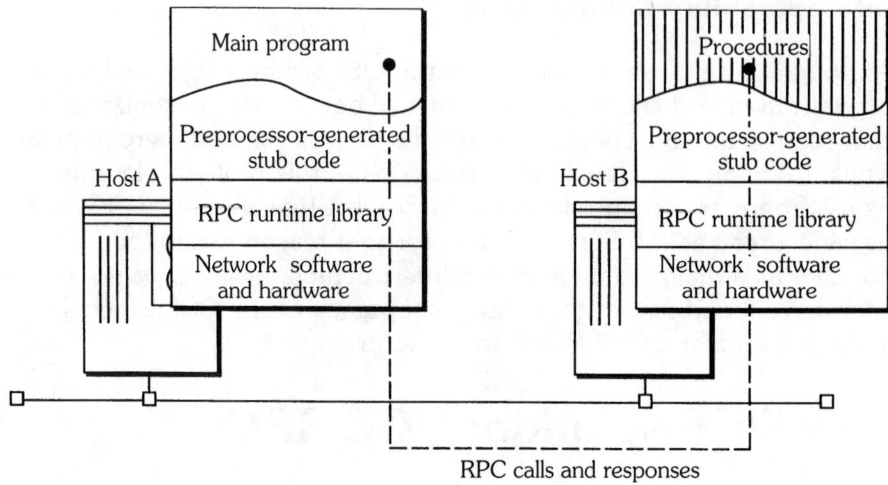

The DCE remote procedure functions. Learning Tree International

Part of its great promise, of course, is that the rest of the DCE suite will deliver all of the necessary services to develop client/server applications. Thus, it will make sense at some future time to migrate toward greater use of DCE RPC.

A Universal Alternative solution: The Sockets Network I/O

UNIX and a number of other operating systems use the Sockets I/O system for remote procedure execution. The UNIX I/O paradigm supports the ability of an application to make *open*, *read*, *write*, and *close* operations to files in the local file system. When an *open* call to the file system is made by a process, UNIX returns a file handle. If the UNIX system is configured properly, a file handle is similarly returned when making an *open*, *read*, *write*, or *close* request to a remote file.

In a remote call, the Sockets I/O system under the local operating system returns an integer that identifies the assigned socket. This integer is termed the *socket handle*. The socket handle is

```
socket# = socket (af, type, protocol)
```

In this socket handle, the first parameter (*af*) is the protocol family, for example, TCP/IP, Xerox, PUP, or AppleTalk. The second parameter (*type*) is the type of communication process: stream- or datagram-oriented. The third parameter (*protocol*) is the transport protocol in the suite used to support the transfer of data. If we're using TCP/IP, for example, we specify whether we're using TCP or UDP.

For operations of the Sockets I/O on a network, we require the association of two sockets, the one on the local machine and the one on the remote machine. The associated sockets are termed a *socket pair*. Therefore, interprocess communication in the Sockets environment is accomplished using this socket pair. This provides functionality similar to the UNIX *pipes* service.

The *BIND* and *CONNECT* commands are used to connect, or bind, the two sockets across a network. The BIND command binds each process to sockets locally. The CONNECT command then takes the local socket in the machine which is making the CONNECT call and associates it with the socket on the remote machine as follows:

```
Connect (socket, destaddr, addrlen)
```

The first parameter is the local socket number on the computer making the call associated with the process. The second parameter is the socket number, TCP port, and IP address of the machine being called. The last parameter is the length of that address. Once the connection is made between the two sockets, i.e., once the socket pair is established, the connection can be used bidirectionally by either machine, irrespective of which process made the original call.

Figure 4-15 demonstrates the operation of Sockets I/O and the use of the socket pair.

Once the socket is opened, *open*, *read*, *write*, and *close* operations can be performed directly to the socket pair. This can only be done after the connection is made.

Figure 4-15

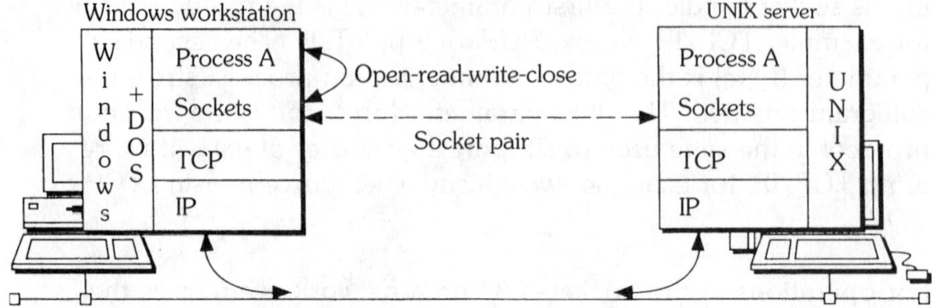

The use of a socket pair by two communicating hosts in a TCP/IP network.
Learning Tree International

Socket I/O is a very popular protocol in the Windows environment as well as the UNIX environment. The protocol implementation in the Windows environment is called *WINSOCK*. A UNIX server and Windows client is a popular client/server pair exercising Sockets I/O between them for remote procedure execution.

Interoperability Rating: Best

This protocol environment has been widely implemented. The sockets protocol has been ported and is available on a number of platforms, principally UNIX, Windows, Windows NT, OS/2, and even in the Macintosh environment. It is highly interoperable among these platforms. We therefore have to rate this protocol as high in interoperability, and very useful for the purposes of integration.

Applications that make calls to Sockets I/O have a high degree of interoperability with services across a network that can receive those calls.

Be aware that the number of actions supported through a socket pair is limited to the *open*, *read*, *write*, and *close* operations. Even so, it allows us a wide range of solutions.

Also, be aware that the socket pair only provides a pipe, and does not in any way define the formatting of the data. Interoperable formatting of the communications is left to the client and server processes using the socket pipe.

The IBM solution: CPI-C

IBM has developed a solution to support remote program execution. IBM calls this environment the *Common Programming Interface for Communication* (CPI-C). It provides for asynchronous conversational peer-to-peer program support. It requires the use of the APPC (*Advanced Peer-to-Peer Communications*) protocol, part of the new peer-to-peer IBM network services. To expedite the execution of APPC between workstations and a mainframe or workstations and minicomputers, IBM encourages the use of the new APPN networking environment (*Advanced Peer-to-Peer Networking*), also available from IBM.

The main difference between IBM's solution and those discussed previously is that CPI-C supports multiple-path conversations. The CPI-C services are invoked in a similar manner to Sockets I/O, binding client and server to a pipe. Conversations are carried on with APPC primitives. But, as can be seen in Fig. 4-16, client processes typically continue to execute while invoking and waiting for replies from server processes. They can also carry on multiple transaction conversations with server processes during the execution of a client process.

CPI-C procedures require the client process to request and allocate a connection to the server and a certain number of resources, which are then available for the client to invoke and use. At the end of the operation, the client process must also request the deallocation of those resources, and close the connection.

This is an extension of IBM connection-oriented application services, and closely follows their connection-oriented networking services. The CPI-C protocol has been widely implemented and fully supported by third parties, and adopted by protocol-making bodies such as IEEE in X/Open, POSIX, and ISO.

Interoperability Rating: Mixed

Although standards-making organizations and many third-party vendors have adopted this protocol and fully support it, the number

Figure 4-16

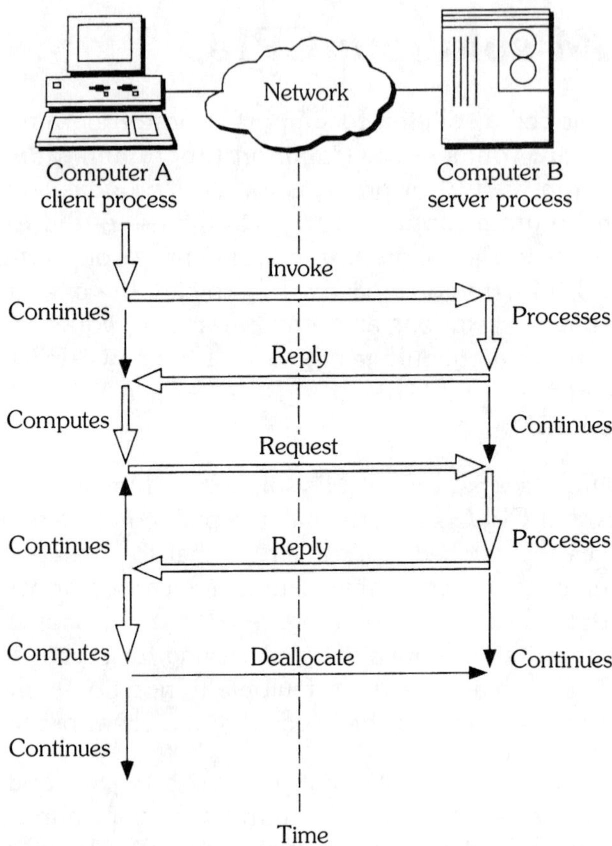

The CPI-C protocol in action, supporting a client-server conversation. Learning Tree International

of server products presently available from vendors other than IBM is very small. This causes us concern about using this protocol as a network-wide solution. You should realize that the IBM environment is highly connection-oriented, and mostly transaction-based, pointing largely to the mainframe as the server platform.

Definitely, if you have an IBM mainframe environment with many IBM solutions already in place, including workstations of the OS/2 variety, DOS, or Windows, interoperating with an IBM mainframe, this may be the protocol of choice for you.

On the other hand, when faced with a multivendor environment where there are a great number of UNIX hosts, Macintoshes, and DOS machines—as well as an IBM mainframe and many other IBM compatible devices—you should be greatly concerned about making this solution the cornerstone of interoperability for remote procedure execution. You would probably be better off deploying something like Sun RPC. In such a varied environment, you might wind up having two or more solutions widely implemented, which certainly makes life very interesting.

Network resource utilization

Remote Procedure Calls, by their nature, are efficient consumers of network resources. Sun RPC, for example, is not connection-oriented, it has minimal protocol handshaking at the time of request, and there is no closing of a connection. The others, such as OSI ROSE and IBM CPI-C, are connection-oriented and may be less efficient.

The aggregate total of all RPCs being made on a network, on the other hand, can severely degrade its performance. It is difficult to predict overall loading by RPCs, but one should attempt to estimate this traffic, as in any other network design situation.

In some cases, RPC-originated traffic may be bursty, as when RPCs are used within a distributed file system (NFS is a case in point). On the other hand, traffic may be steady when it occurs in high-volume short transactions. File access and transactions are situations that make loading of the network by remote procedure execution difficult to predict, and they must be studied on a case-by-case basis.

Integration strategy

In most cases, one does not have a choice of which remote procedure execution protocol to use. Application protocols in use normally bring with them a certain RPC type specified for that application service. For example, Sun NFS uses Sun RPC. We are therefore more often concerned with the application-level protocol interoperability than with the RPC interoperability.

Typically, the choice of RPC type is a concern in the development of applications that use the RPC directly. Thus, it is mostly of great interest to programmers. For OSI networks, we typically use ROSE. More commonly, in TCP/IP networks, we use Sun RPC. Therefore, the application service in use often dictates the RPC technology of choice, as well as the type of network used to support these services.

A rule of thumb might be that whenever possible, we should use the RPC protocol that has the widest support—in other words, the one that has been ported to the most operating systems. Right now, that appears to be the Sun RPC service. In the future, we look for widespread use of the DCE RPC, and perhaps more use of OSI ROSE. We should not discount the growing popularity of IBM solutions, including CPI-C. Figure 4-17 summarizes and rates the RPC protocols just discussed.

One thing that should be noted is that there are no gateways from one RPC type to another.

Interoperable file transfers

The movement of an entire file from one computer to another is generally known as file transfer. This task can be accomplished in a variety of ways, with varying degrees of performance. Certainly, we can move a file by copying it to some intermediate medium such as a floppy disk or a magnetic tape, transferring to the receiving system, and then uploading it through a similar device on the receiver. This process is commonly known as "sneaker net."

In today's organizational computing environment, we perform this file transfer via high-performance network connection; in some cases, it might still be done over a serial low-speed link.

When the sending and receiving hosts have the same or a similar type of operating system and hardware, the task is fairly simple, and there are many solutions, some generic and some proprietary. The problem we would like to tackle here is the one in which the two endpoints of a file transfer, sender and receiver, are radically different. We might have an MS-DOS personal computer at one end, and a multiuser UNIX host at the other. These two might be connected over a high-

Figure 4-17

Remote procedure protocol	Typical platforms	Protocol suite	Major developer/ promoter	Major use	Interoperability rating
Sun RPC	UNIX Windows OS/2 DOS Apple	TCP/IP	Sun Universal	NFS client/ server development	+
DCE RPC	Varies	TCP/IP	OSF	New client/ server development	0
OSI ROSE	Varies	OSI	ISO	None	–
UNIX sockets	UNIX Windows Windows NT NetWare	TCP/IP	Universal	Client/ server connection-less	+
IBM CPI-C	OS/2 IBM Host MVS	SNA	IBM	Client/ server connection-oriented	0

The table above summarizes the information on the most common remote procedure execution protocols. It gives the most common platforms the protocol can be found on, the typical protocol family it is deployed with, the major developer or promoter, and some major uses. It also rates each protocol for interoperability, with (+) being the most interoperable, (–) the least, and (0) a mixed bag. Learning Tree International

performance network such as Ethernet with TCP/IP, or over an asynchronous low-speed link.

No matter what the type of connection, we are expecting available protocols to assist in interoperating between the two machines. To begin to understand the scope of the task at hand, let us break it up into a series of known problems. (See Fig. 4-18.)

Figure 4-18

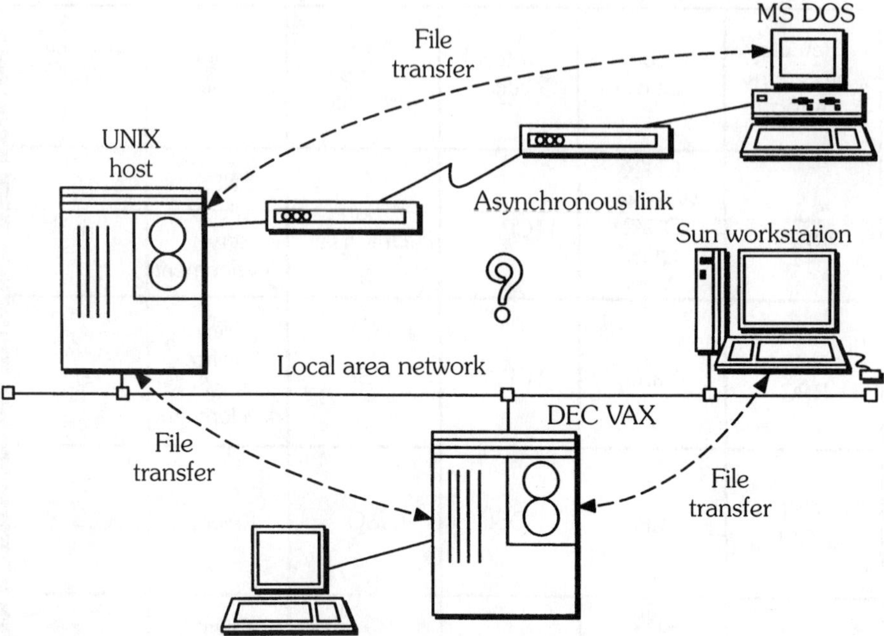

The various facets of file transfer: high-performance network transfers and low-performance transfers over an asynchronous link. Learning Tree International

If we have a connection between workstations and hosts over a high-speed or high-performance network, there are a number of file transfer protocols that can assist us.

In contrast, if we have two machines connected point-to-point via an asynchronous link, we require a different set of protocols. These two machines might be collocated and connected by a short cable from serial port to serial port, or they might be many miles apart, in which case you might need to use modems and a telephone link to connect them.

Interoperable solutions, then, fall into two categories: those that assist us with the point-to-point low-speed file transfer connection, and those that are employed in the high-performance LAN and WAN environment.

A number of protocols, such as XMODEM, ZMODEM, KERMIT, are popular in the point-to-point, low speed, asynchronous connection

area. In an IBM/SNA synchronous network, we might employ a protocol such as IND$FILE to accomplish a similar file-transfer task. The protocols employed in the high-performance LAN and WAN environments are the File Transfer Protocol (FTP) and the Trivial File Transfer Protocol (TFTP) from the TCP/IP protocol suite. The OSI family of protocols uses the FTAM (File Transfer, Access, and Management) protocol.

There is a second solution that can be employed in the high-performance LAN and WAN networks; it is often overlooked as a very practical and effective method of file transfer. This is the use of the file copy command which comes as part of the distributed file system. For example, UNIX NFS (Network File Transfer System) and any PC network operating system allow the transfer of a file from one host to another by copying. In the case of NFS, the transfer occurs from host to host or from workstation to host. In the case of the PC NOS, it's from the client workstation to the file server.

We can classify all of these methods of file transfer as *store-and-forward* systems.

Store-and-forward systems

Store-and-forward systems are used to move files between highly dissimilar machines—for example, from an IBM mainframe to an Apple Macintosh and vice versa. These can operate over both synchronous and asynchronous links, or over high-performance networks. One of the key characteristics of store-and-forward systems is that the file to be forwarded from one host to another can be stored in an intermediate location and then forwarded to the destination.

For example, we might transfer a file from one workstation on a PC network to another via the file server. That file will be uploaded to the file server from the sending workstation, and stored there until the receiving workstation logs in and downloads the file. The characteristic of a store-and-forward file system is that the entire file is moved between end systems. This process is called *file transfer* for short.

Store-and-forward systems require user involvement. The user on the sending machine or the receiving machine needs to invoke the protocol and manipulate parameters to obtain the desired result of transferring the file. Often, the file name and file location have to be indicated during the course of the file transfer. We may need to change directories on the receiving and sending hosts, and commands are available to do so. Parameters of the file transfer program—indicating whether the file to be transferred is binary or text, for instance—may need to be adjusted. All of these parameters and actions require direct user intervention, another key characteristic of store-and-forward systems.

Store-and-forward file systems often require data translation, especially when the endpoints are highly dissimilar. Once the file is transferred, it may not be usable by the receiver's applications as is, and may require further action to make it usable, such as conversion of the file format.

Another important characteristic of store-and-forward file systems is that the transfer occurs in block mode. This is true even when we use an asynchronous link and an asynchronous, character-based connection to support the file transfer. The entire file is broken up or fragmented into manageable pieces or blocks. The blocks must fit the frame size of the data link protocol being used. An appropriate protocol is used for retransmission of any blocks that might not have arrived error-free.

Typically, store-and-forward file systems are programmed to work as active, foreground tasks. This is the case in most single-tasking microcomputers. File transfer systems have typically been implemented in software, not hardware. There is a program that embodies the protocol, and it is required on the receiver and the sender. It is typically a symmetric process, and a cooperative process, in that neither the receiver nor the sender controls the other, but instead they actively cooperate in the process of sending and receiving the file.

There are a couple of very good examples of systems that use store-and-forward file transfer protocols. The first is e-mail transfer protocols, which move messages from one host to another. We can go further and classify this type of store-and-forward file system as a message-switching system as well, because we're moving files that are actually messages.

Our second example of a store-and-forward system is the way in which we employ PC file servers, where we typically store application programs that users might wish to execute on their workstations. When a user wants to run a program, after making the appropriate connection to the file server over the high-performance network and being validated as a user, the binary code of the application program is downloaded from in the file-server disk to the user's workstation memory. It is then executed from the workstation memory. We have thus transferred a file from one computer to another. The process, though, required the transfer of that binary program representing, which is a copy of the application program that was in the file server's disk. This type of file transfer occurs transparently to the user, who is not even aware that such a file transfer has taken place. Nevertheless, we are employing a store-and-forward technique, and it may be one of the few cases where the technique did not require explicit user interaction, but was instead performed transparently for the user by the network operating system services.

Asynchronous file transfer protocols

A simple way to transfer a file from one computer to another is to provide a link between the respective I/O ports (typically serial ports) of the two machines. This is a point-to-point connection. Many times this might be the only way to move a file from one computer to another. The protocols used over this type of link come from the public domain. They have been created by individuals working in the computer field, and receive very wide public support.

The common characteristic of asynchronous protocols is the encoding of the transmitted file into ten-bit characters. Each character has a stop and a start bit, and seven or eight data bits. This results in very high overhead, typically 30%. The most common of these protocols is XMODEM, which might even be termed the lowest common denominator of interoperability, because it is so popular and available on so many different systems. It has very high overhead, and a major drawback is that it does not detect all errors with its error-detection scheme. Also, if for some reason the protocol fails in the middle because of a lost block, it will stop and not recover. You will have to begin again at the start of the file transfer.

More sophisticated protocols are available in place of XMODEM. KERMIT and ZMODEM are good examples—they retransmit blocks that have been received in error, improving the performance.

Figure 4-19 demonstrates the typical way in which a character-encoded file grouped into blocks can be transferred by an asynchronous protocol. The protocol shown is KERMIT. We might or

Figure 4-19

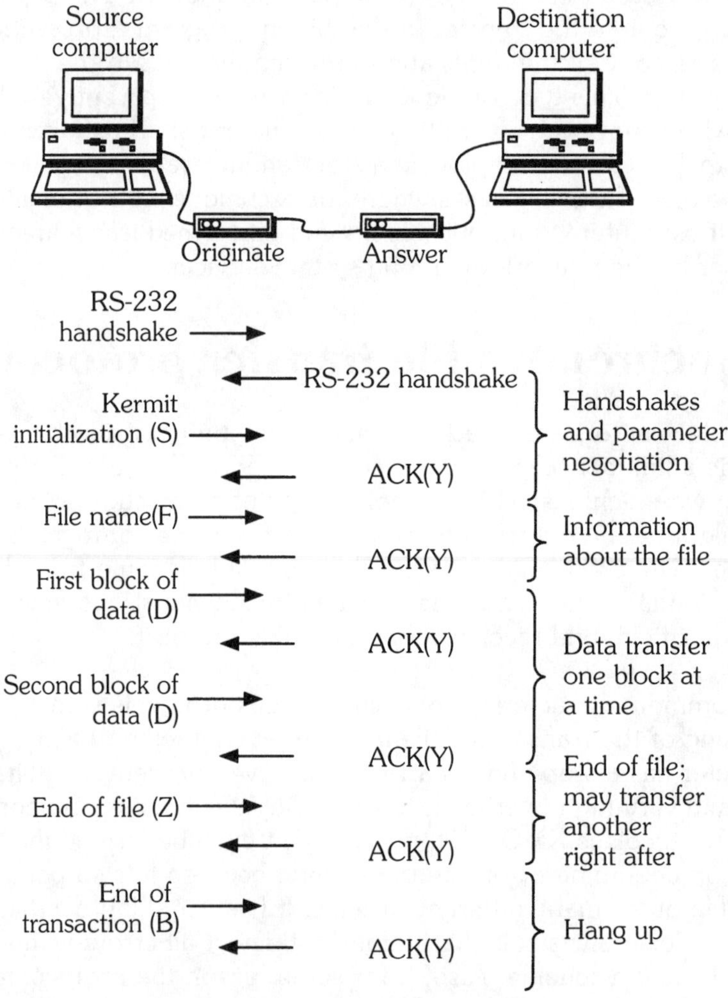

A typical KERMIT store-and-forward file transfer group of transactions. Learning Tree International

might not have modems between computers as shown in the picture, but it is a point-to-point protocol in any case.

There's an exchange of signals that represents the handshake at the physical layer, the RS-232 handshake. Once the two machines recognize that they have a serial connection between them, the KERMIT protocol at the sender sends an initialization block containing identifying parameters and any parameters to be negotiated. The receiving computer sends back an acknowledgment block. The type of each block is identified. The initialization blocks are known as *S* blocks, and the acknowledgment blocks are known as *Y* blocks, for example.

The next block is from the sender, and contains the name of the file and other parameters concerning the file to be transferred. It is acknowledged by the receiving computer. If at any point in this exchange the sender does not receive an acknowledgment within a certain time-out period, it will retransmit the last block sent. This is a stop-and-wait protocol, in that only one block at a time is sent. Each time a block is sent, the sender waits for an acknowledgment of that block before sending the next one.

After acknowledgment of the file information block, the sender begins sending the file one block at a time, each block being acknowledged in the manner just described.

At the end of the file transfer, there is an end-of-file (*Z*) block. The KERMIT session might contain a number of files to be transferred, with a corresponding Z block at the end of each. In this case, there is only one file to be transferred, so the end-of-file block is followed by an end-of-transaction block, which is basically an invitation for both computers to close the KERMIT session.

Asynchronous file transfer protocols have very high overhead. Besides the 30% bit overhead in each character, there are handshake blocks in KERMIT that must be exchanged before any data is transferred. There are blocks that need to be exchanged at the end in order to gracefully close the KERMIT session. Also, within data blocks, there are a number of characters at the beginning of the block—the header characters—and a number of characters at the end

of the block, such as error-check and CRC characters, further adding to protocol overhead.

Additionally, each character of the file requires the transmission of ten bits. The data portion usable for carrying file information out of the ten bits may be seven or eight, depending on whether we're transferring an ASCII file or a binary file.

Putting it all together, we see that we can typically have 35% overhead in this type of protocol, which translates into having to send an additional 3.5K in order to transmit a 10K file.

Of course, it's a penalty we gladly pay when there is no other recourse in transferring a file between dissimilar computers. But it is much more desirable and appropriate, when available, to use a high-performance (lower overhead, higher speed) protocol over a local- or wide-area network connection. These are discussed next.

High-performance file transfer protocols

Higher-performance file-transfer protocols typically require a packet- or frame-oriented internetworking infrastructure. Transmission can occur over a local area network or wide area network links. The protocols are classified into two types: those that require user intervention and the execution of a program containing the protocol, and those that use the copy command of a network operating system.

The TCP/IP solution: FTP

The FTP (File Transfer Protocol) is perhaps the most popular and the most interoperable protocol for this purpose. Its popularity is due to its inclusion in the TCP/IP protocol suite. It was added to the TCP/IP protocol suite very early on, and is defined by RFC-959. It requires the use of TCP over IP.

A user on a local host, which may be a workstation (single user) or a multiuser host, needs to establish an FTP connection with the remote

host. The remote host does the authentication of the user. In other words, the user has to have a valid account on the remote host. Once authenticated, the user can perform directory listings on the remote host and upload and download files from it to the local host. This is not a transparent protocol—it requires user interaction.

The FTP protocol uses the TELNET Network Virtual Terminal protocol to accomplish this purpose. How does FTP work? Figure 4-20 shows the components of FTP and several typical applications.

Figure 4-20

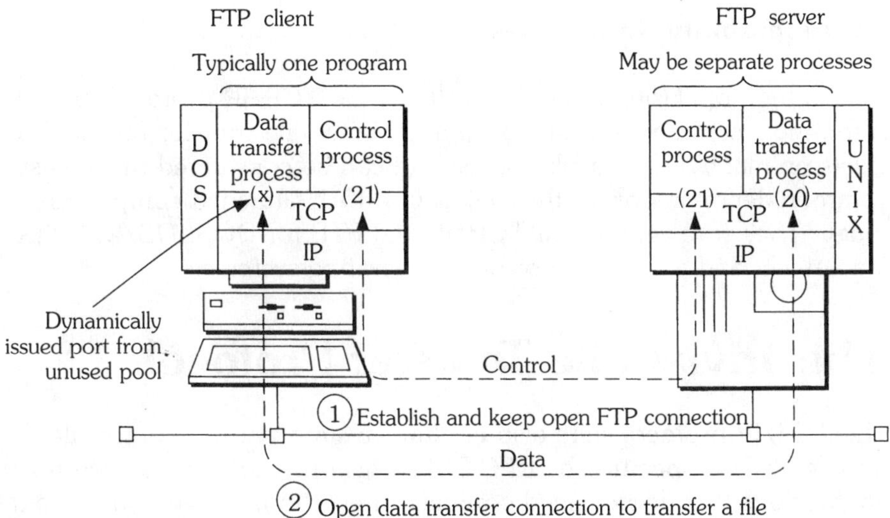

The FTP protocol in action. Learning Tree International

The local host has a client process, a program, that implements the FTP functionality. In reality, the program has two processes: the control process and the data-transfer process. They are often parts of the same program. The remote host also has two FTP server components: the control process and a data-transfer process.

When the user requests an FTP connection, the two hosts establish a TCP connection via the control processes. FTP employs port 21 on both machines for the control processes to establish the connection. This connection remains open for the entire duration of the FTP session between the client and the server. At this point, the user can

issue requests for directory listings on the remote host. These are handled under the control process. Once the desired file is located, the user issues a request for a file transfer, using the FTP *get* command.

This initiates the opening of an additional port on the FTP client for the data-transfer process. The port is dynamically issued from the pool of unused TCP ports on the client. On the FTP server, the data-transfer process uses TCP port 20. The two data-transfer ports are bound to each other, providing a pipe. The file is then transferred in blocks or TCP segments.

Interoperability Rating: Best

Due to the popularity and universality of the TCP/IP protocol suite, it is the most commonly used high-performance file transfer protocol. It exists on many different platforms, and can be considered the lowest common denominator on the majority of LAN-attached computers today. There are versions of TCP/IP and FTP for DOS, OS/2, UNIX, Macintosh, and a multitude of other operating systems.

The Trivial File Transfer Protocol

The TCP/IP protocol suite also contains a simpler protocol for file transfer, TFTP, specified by RFC-783. This protocol can be used for applications that do not need complex interactions between the client and server. It is typically used over LANs, not WANs, and has a number of uses in that environment. For example, it is used in diskless workstations' BOOTP bootup process.

TFTP is functionally simpler than FTP. TFTP runs over UDP, not TCP, using UDP port 69. The block size is restricted to 512 bytes. A major drawback is the lack of a user authentication requirement at the remote host. This is a good protocol for the dissemination of public information to a wide variety of users without necessarily creating accounts for all of them. TFTP can also become a security breach or security hole on certain systems; therefore, it must be carefully monitored and managed.

Interoperability Rating: Mixed

This protocol is appropriate for a few situations. The BOOTP process is one. It is not a universal solution, in that it might not be available on all TCP/IP implementations, and the host administrator might not have allowed it to exist or be used on certain hosts, due to security concerns. It is often missing on a host after it has been removed by a network administrator trying to improve host security. In any case, because of its limitations, it is probably not as appropriate as FTP.

 Using a DFS for file transfers

An often overlooked method of performing file transfers is the use of a distributed file system. For example, we might use the *cp* command of NFS, or the NetWare NCOPY command. The details of distributed file systems are discussed in the next two sections of this chapter, and will not be repeated here. We merely point them out here because they are often available in high-performance network environments that have network operating systems or distributed file systems already in place, and are overlooked as a file-transfer service.

As an example, FTAM is an OSI solution that embodies both a file-transfer facility and a distributed file system. For that reason, we don't discuss FTAM details here, but leave them for the next section.

Interoperability Rating: Best

If a DFS, or an NOS, has been installed on a network and the file-sending and file-receiving machines can make use of the services of this DFS, then there's a very high potential for successful integration. By all means, in that case, do not seek other methods of file transfer. Make use of the facilities that are already installed.

 Putting it all together

Consider the example of transferring a file from one host to another via an intermediate host, and requiring the use of both a high-performance file transfer protocol and an asynchronous file transfer protocol. See Fig. 4-21.

Figure 4-21

• Example of downloading a copy of an RFC from an Internet host to a local PC

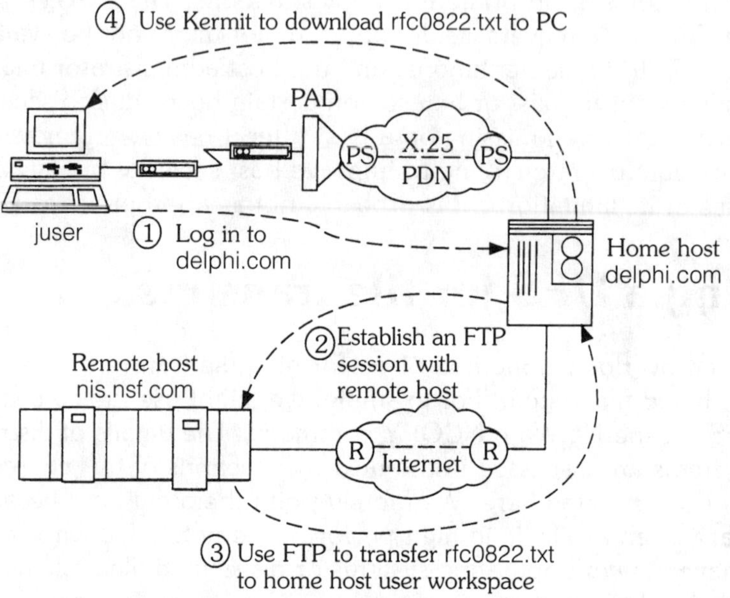

• juser@delphi.com FTPs to nis.nsf.com and downloads a file to delphi.com, and then uses Kermit to download the file to a local PC

The use of FTP and KERMIT to complete the file transfer between two hosts. This situation often arises when we need to do file transfers from an Internet host to our PC, and only have terminal access to the Internet. Learning Tree International

The task at hand is a very common situation nowadays in downloading an RFC text file from an Internet host to a local personal computer. You might have a connection between your personal computer and your "home" host, which is where you have an Internet account, via an X.25 PAD connection. The host, of course, has a high-performance TCP/IP connection to the worldwide Internet. The repository of RFCs on the Internet might be a remote host such as nis.nsf.com. Your home host might be delphi.com.

Let's see how we might go about downloading this text file (the RFC) from the remote host via the home host to your personal computer. The first step would be to log in to the delphi.com host via a local connection into a PAD, and then through the X.25 packet data network into the home host. Once you are authenticated as a user, you request an FTP session with the remote host. Once granted that FTP session, you can do a file transfer of the RFC in question.

Now in the remote host, you may not have an individual account for yourself. You may be coming in through a public account with no password or a very simple password. Often that account has a user name of "anonymous," with no password or all passwords allowed.

The RFC documents are in directories accessible by the anonymous user. Once the FTP session is under way, one can request the transfer of an RFC, for example, as shown in the diagram. Let's say it is the RFC0822.TXT file. That file will be transferred from the nis.nsf.com to the delphi.com host and remain in the user's work space on the latter. At this point, the file is still not in the user's personal computer, but resides totally on the home host—a typical example of the store-and-forward technique, where a file is temporarily stored on the home host after being transferred from the remote host.

At this point, the FTP connection with the remote host can be concluded, and we can move on to the next step, which is to download the RFC0822.TXT file to our workstation. This can be accomplished by an asynchronous protocol such as XMODEM or KERMIT.

We see, then, that to accomplish the transfer between these three machines, we've actually made use of two different protocols and a number of networking technologies, and have done it in an interoperable manner, since the three platforms in question might be completely different. Our personal computer might be a Macintosh, the home host might be a DEC VAX, and the remote host might be an IBM mainframe or some other type of large platform. Yet by using interoperable file transfer protocols commonly available on all three platforms, we have been able to complete our file-transfer task.

Network resources utilization

File transfers are "bursty" by nature. They require and absorb all of the bandwidth that a particular link has to offer, but they do so for a very short time unless the file is very large or the link is very slow.

If we mix file-transfer traffic with terminal traffic on the same subnet or link, the terminal users will experience delays. As a matter of fact, many of the delays, in certain cases, will be so severe as to cause some terminal sessions to be lost due to timeouts. The mixing of these two traffic types is to be avoided if at all possible.

We should also avoid file transfers across slow WAN links, or increase the bandwidth of the link appropriately to improve performance. Certainly, we do not want to be transferring binary executable files from a remote file server across a WAN link to workstations on a local network. If application programs are to reside on a file server to support users on workstations, then by all means provide a local file server for the storage of those application files. Use the WAN link for the transfer of data only.

Whenever possible, one should avoid using asynchronous links for file transfers. They have a high protocol overhead and are very inefficient. With certain protocols, the entire transmission has to be restarted if a block is lost in the middle of transmission.

Integration strategy

Support for file-transfer interoperability follows a few basic principles. First, one should support a variety of asynchronous file transfer protocols on all hosts. In particular, and as the protocol of last resort, one should have at least a version of XMODEM available on all hosts. Certainly it is desirable to have ZMODEM and KERMIT versions on these hosts as well, to improve performance, but XMODEM is the bottom line.

As for a high-performance strategy, at least FTP should be supported on all servers. Allow workstations that often need to transfer files to and from those servers to use FTP. FTP is certainly the universal solution. Network administrators should encourage users to employ any distributed file system currently in use when they need to transfer files.

Network administrators should go farther and expand the use of DFS, and even provide gateways between the various types of DFS in use on a network. The gateways will also avoid the use of dual protocols on clients. This subject is discussed in more detail in the next section, as well as in Chapter 5.

Lastly, when considering file transfers, one must also consider security. Remember that users must have an account on the remote host (even if that account is an anonymous account) to access files on the remote host. Use TFTP with care, since it often represents a security risk at the remote host. See Fig. 4-22.

Distributed File Systems

A workstation can open, read, and write to local files. Using the workstation operating system, how can it also open, read, and write to files on other computers on the network at the same time? The solution is to use a distributed file system. See Fig. 4-23 and 4-24.

A distributed file system has two components: one that works in the workstation, which we might term the *client portion* of the distributed file system, and one that works on the remote host, where the file to be opened resides.

The remote host component of the distributed file system is called the DFS server. Working together with network protocols that move the request from client to server and back, the client and server components make remote files appear as if they were local.

Typically, this is done in two ways. One is to make the directory structure, or a portion of the directory structure, on the remote host appear to be part of the local directory structure on the application. The NFS protocol employs this technique. In the DOS environment,

Figure 4-22

File transfer protocol	Character-istic	Protocol suite	Promoter/ developer	Major use	Interoper-ability rating
FTP	High-perf. universal solution simple	TCP/IP	Internet	Host-to-host file transfers	+
FTAM	High-perf. complex solution	OSI	ISO	U.S. gov't. EDI + X.400 E-mail systems	0
NFS cp	Use of a DFS. needs NFS	TCP/IP	Sun	Host-to-host file transfers	0
XMODEM	Universal async protocol	ASCII async connections	Public domain	Character based file transfer host-to-PC	+
Kermit	Higher performance async protocol	ASCII async connections	Columbia University	Character based file transfer host-to-PC	0

The table summarizes the characteristics of interoperable file-transfer protocols. The horizontal double line separates high-performance protocols (above) from asynchronous point-to-point protocols (below). Note that a (+) rating typically means the protocol is ubiquitous or universal, rather than high-performance. Learning Tree International

directory structure on the remote server is made to appear to be part of a local drive—in other words, a local logical drive that has no physical existence on a local machine. A local directory points to a directory on the remote server. This service is usually part of the PC network operating system function.

Let's enumerate some of the typical distributed file-service solutions. First, we have what might be called the universal solution, which is

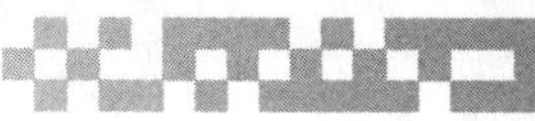

Figure 4-23

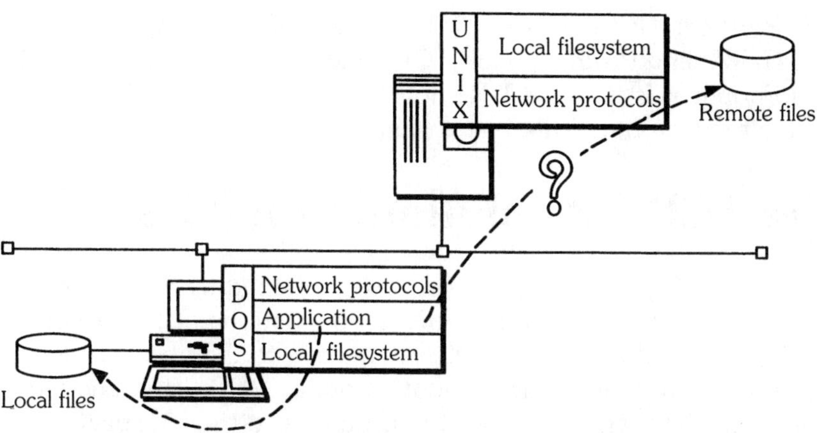

The problem addressed by a DFS: an application on a workstation needs to open, read, write, and close files on a remote file system as well as on a local file system. Learning Tree International

Figure 4-24

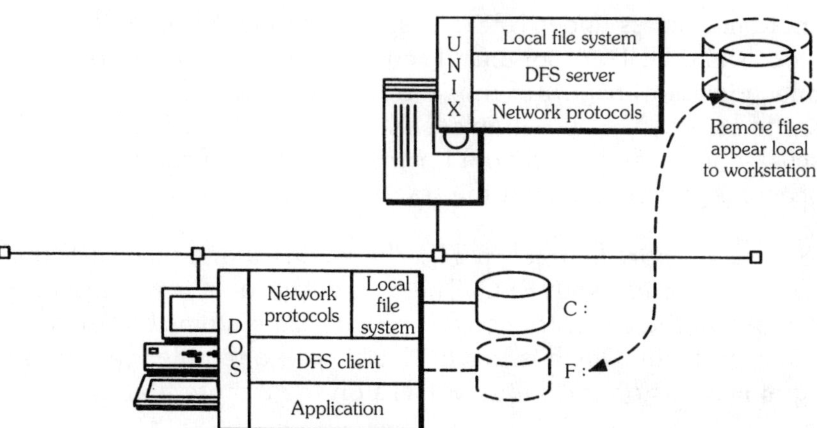

The distributed file system solution: for a PC, the DFS makes the remote disk appear as a local disk; for a UNIX workstation, portions of the remote directory structure appear local. Learning Tree International

part of the TCP/IP protocol suite: the Network File System (NFS). Then there is the OSI protocol suite's File Transfer, Access, and Management (FTAM) facility. A new facility, part of the Distributed Computing Environment (DCE), is the Andrew File System (AFS).

In the PC networks environment there is the IBM/Microsoft Server Message Block (SMB) protocol. Novell uses the proprietary NetWork

Core Protocol (NCP). The AppleTalk protocol suite employs the AppleTalk Filing Protocol (AFP). One alternative to these, which is not so popular, is the AT&T Remote File System (RFS). All of these provide the distributed file system function.

The TCP/IP solution: Sun NFS

The Network File System protocol was developed by Sun Microsystems for use with their UNIX workstations. It has become a de facto standard in other UNIX systems due to the liberal licensing practice of Sun. Many other vendors, notably DEC, IBM, and HP, have ported NFS and its associated protocols to their respective computing environments. There has been movement of the NFS protocol to the PC network environment. Today, NetWare file servers and Windows NT file servers, for example, also support NFS clients.

Remote file access under NFS relies on a number of other TCP/IP protocols and UNIX programs. It requires UDP (not TCP) for connectionless transport of the NFS requests and responses from client to server. NFS also employs the RPC and the XDR protocols (discussed in an earlier section) for the execution of remote procedures and encoding of the transmitted data.

NFS relies on *port-mapper*, a TCP/IP routine that issues UDP ports dynamically for use with RPC. The *mount* protocol is employed, which allows the client to mount a remote file system. *Lockd* is another protocol, which issues file locks on the remote file. The *statd* program checks on the status of locks on the remote file.

How does NFS work? NFS permits the creation of what is termed "a virtual file system." It provides for a way of attaching a portion of the remote directory structure on the server to some point, called the *mount point*, on the local directory structure. Under UNIX, it is simply done as a portion of the remote directory tree attached to the local directory tree. Under DOS, a local logical drive, such as F or G, is assigned to a directory on the remote file structure. In either case, we have an NFS client program that runs on the UNIX or DOS workstation, and an NFS server program that needs to run on the remote host. NFS clients do not share their local file structures with other clients. That job is left to NFS servers only.

Interoperability Rating: Best

NFS is the most widespread of the distributed file system solutions, not only because of the popularity of TCP/IP, but also because it has been ported to many different platforms. The server portion exists on large hosts and on PC network file servers. NFS clients are also very popular, spanning the range of workstations and hosts from PCs to large machines. See Fig. 4-25.

Figure 4-25

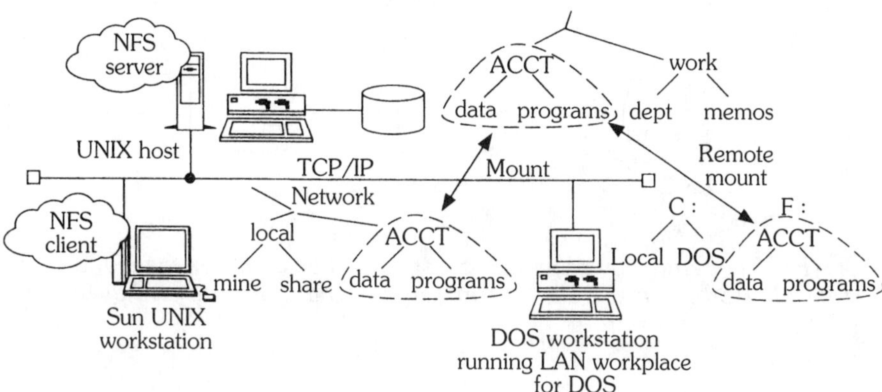

Sun NFS creates a "virtual file system." Learning Tree International

The OSI solution: FTAM

The OSI protocol suite contains the *File Transfer, Access and Management* facility (FTAM), which provides for file transfer (copying a file from one system to another), file access (reading or writing an individual record), and file management (changing the attributes of a remote file).

FTAM operates by providing a network-wide common view of the file. This is termed a *virtual file store*. Under FTAM, there is an FTAM client, which is called the *initiator*, and the FTAM server, which is called the *responder*. It is the responder's, or server's, responsibility to translate the local file format and file view by mapping the server's file system into the virtual file-store format understood by all of the FTAM clients. The FTAM initiator, or client, then views all the responder's file systems on the network in one

common format that it can work with. It is the responsibility of the initiator to translate the common virtual file-store view of the file system into its local file format. For example, under DOS, we would be pointing logical drives to FTAM file sets. Under UNIX, on the other hand, FTAM file sets are viewed as part of the local directory tree. See Fig. 4-26.

Figure 4-26

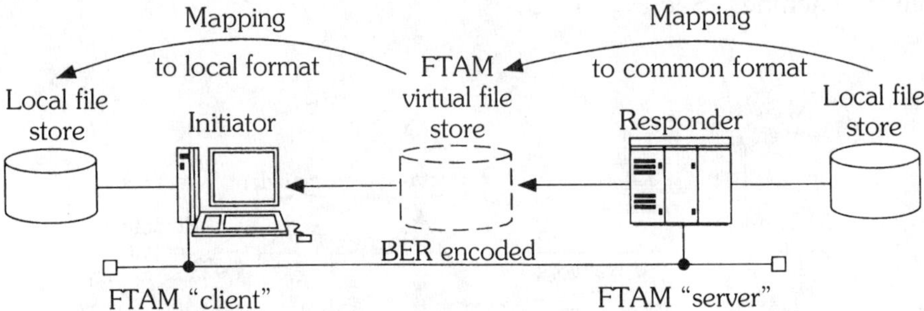

OSI's FTAM operates by providing a network-wide common view of a virtual file store. The end systems map the virtual file-store format to their local file system format. Learning Tree International

Interoperability Rating: Mixed

FTAM is a substantial distributed file system with many features. If the OSI protocol suite were more popular, this would be a strong competitor to NFS, although some feel it is a poor performer. Unfortunately, it has very few efficient implementations that can be relied upon as good interoperable solutions.

Certainly, major vendors such as DEC and IBM, as well as some of the PC network file-server vendors such as Novell, have FTAM implementations. These are all interoperable, by and large. But it is hard to recommend FTAM as an interoperable solution because of the scarcity of products.

The DCE solution: DCE DFS

The Distributed Computing Environment (DCE) protocol specification has been produced by Open System Foundation consortium of vendors. More detail on the DCE suite of protocols can be found in

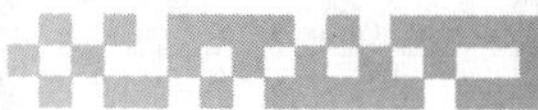

Chapter 6. Our purpose here is to put the DCE offering of a DFS in perspective with the rest of the DFS solutions.

The distributed file system specified by the DCE protocol suite is an adaptation of the Carnegie-Mellon University specification: the *Andrew File System* (AFS). It is called the *DCE DFS* in the rest of this section.

One of the unique features of this DFS is its replication service, which makes data readily available across the network. DCE DFS keeps replicated sets synchronized through disruptions of connectivity.

The DFS is implemented as a distributed client/server application. The client portion opens, reads, writes, and closes remote files on servers via requests. The server portion, which runs on the remote host, manages data on the remote machines, and services requests from clients. The client portion also acts as the redirector, and screens calls to the local and remote file systems. All of this, of course, is the classic way in which a DFS operates.

Interoperability Rating: Mixed

DCE is becoming popular because of the current thrust to implement client/server systems. It is not yet popular enough to recommend as a universal solution, or even partially as a good interoperable solution, because of the scarcity of products. Many vendors have adopted DCE, and are working hard to produce products to offer to the marketplace. Thus, it is a strong candidate for future networks. We recommend you keep an eye on this solution as a contender for the spot currently held by NFS. The DCE DFS has more features than NFS. As it is implemented more widely in client/server systems, it may well become the future universal solution.

The IBM/Microsoft solution: SMB

The *Server Message Block* (SMB) protocol is very popular for peer-to-peer PC networks and PC network operating systems. It was produced by collaboration between IBM and Microsoft, and it was to be used for some of the original LAN products. It has become a *de*

facto standard in some PC networks. Major PC network products that use SMB are

- Microsoft LAN Manager
- OS/2 LAN Manager
- DEC Pathworks
- Microsoft Windows for Workgroups
- IBM LAN Server
- Microsoft LAN Manager for UNIX
- LANTASTIC
- Microsoft Windows NT Advanced Server
- many other peer-to-peer networks

SMB defines four message types that implement the following commands:

- Session control, which starts and ends a redirection connection to a shared resource on the server
- File access, which enables the open, read, write and close of a remote file
- Printer access, which enables the printing of a file on a remote computer's printer
- Message command, which sends and receives messages between network computers. The latter is typically employed for one-line, simple messaging from one workstation to another.

SMB messages are structured in what is called *Network Control Blocks* (NCBs), and sent by the redirector to the underlying protocol for transmission.

The typical transmission protocol supporting SMB is NetBIOS. NetBIOS is also a very popular protocol, having been implemented on many different workstations and been made a part of many different types of PC network operating systems. Unfortunately, NetBIOS has a major weakness. It is not routable, so it cannot be

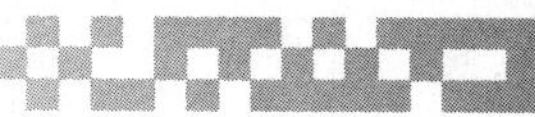

deployed on very large networks—principally, enterprise networks. It has been relegated to small departmental networks. This is therefore a weakness of SMB as well. See Fig. 4-27.

Figure 4-27

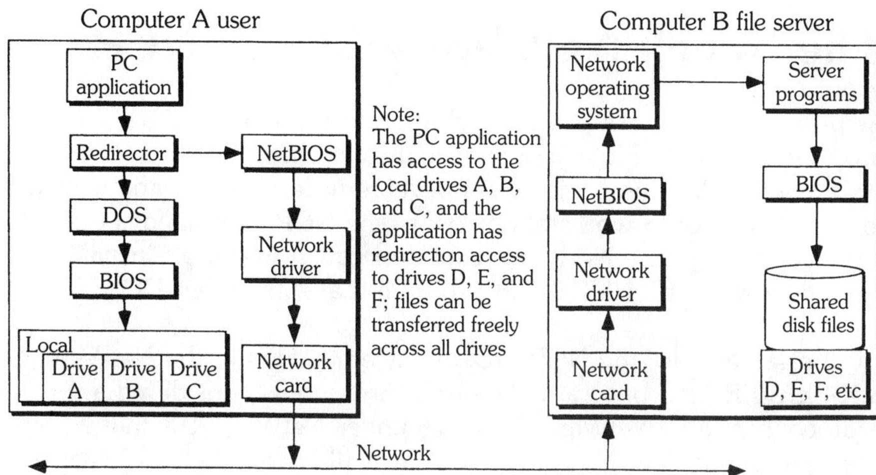

The IBM/Microsoft SMB redirector on a client, and the SMB server program at work. Learning Tree International

SMB requires a server program and a client program. The client program runs on the user workstation, and the server program runs on the remote machine whose file system is to be accessed. SMB enables requests from the application on the user's workstation to be redirected to the network, sent via a transmission protocol (typically NetBIOS) over a high-performance network to the server. The server receives those requests, acts on them, and responds. The server might be asked to read a portion of the server disk and return the data to the client using the same transmission protocol, for example.

The SMB program on the client is typically termed the *redirector* because it is the traffic cop that directs application printing and file-read requests to either the network or the local operating system. For the DOS environment, the redirector makes the remote file system directory structure appear as if it were a local drive.

Interoperability Rating: Mixed

SMB is extremely popular in PC networks. Therefore, it is an excellent solution for that environment. Unfortunately, because of its strong dependence on NetBIOS, it is limited as an enterprise-wide interoperable solution.

The Novell NetWare solution: NCP

As part of the network protocol, Novell has implemented a proprietary remote file access protocol called the *NetWare Core Protocol* (NCP). This protocol supports remote printing, and is meant for support of DOS workstations only. The NCP command set is available through the NetWare redirector called the *Shell*, usually implemented in the NETX.EXE program that runs under DOS.

On the server side, the NCP program is embedded in the NetWare 3.X SERVER.EXE program. The NCP protocol is considered a protocol native to NetWare. It is only under NetWare 4.X that NCP has been unbundled. Now a NetWare 4.X file server can be installed without necessarily installing NCP. NetWare NCP provides functionality similar to that of SMB; its major drawback is that it is a proprietary protocol, and is only deployed on NetWare networks.

Interoperability Rating: Poor

NCP is only valuable for NetWare DOS clients. The only vendor that has implemented the NCP protocol on servers has been Novell. It is very popular in today's networks because of the high penetration of the Novell NetWare product in the PC network marketplace (somewhere between 60% and 70% of all networked DOS workstations use NCP). It is a Novell proprietary protocol, and there has been no major effort to port it to any other platform. We therefore have to give this protocol a poor rating for interoperability.

It is true that one has to support this protocol in most of today's enterprise networks, but that is only because of the popularity of the Novell solution, not because it is an interoperable solution. See Fig. 4-28.

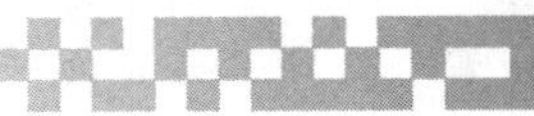

Figure 4-28

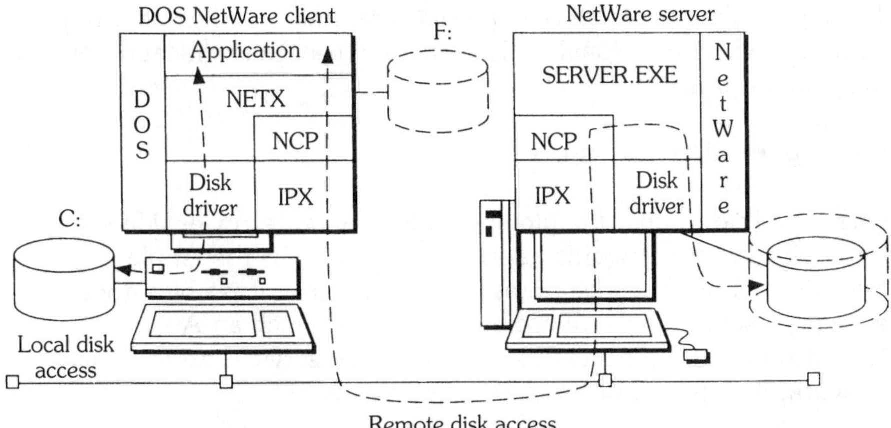

Novell NetWare NCP protocol in action between a NetWare server and a DOS workstation. Note that NCP is part of the NETX.EXE program on the workstation, and part of SERVER.EXE on the server. Learning Tree International

The Apple solution: AFP

Apple has implemented the *AppleTalk Filing Protocol* (AFP) as part of its AppleTalk protocol suite. It is the service that provides transparent access to remote files from one Macintosh to another, or from a Macintosh to a file server. AFP enforces secure access to the remote files via the file server's security functions. It has very similar functionality to NFS, XER, and RPC, but implemented the Apple way. See Fig. 4-29.

Figure 4-29

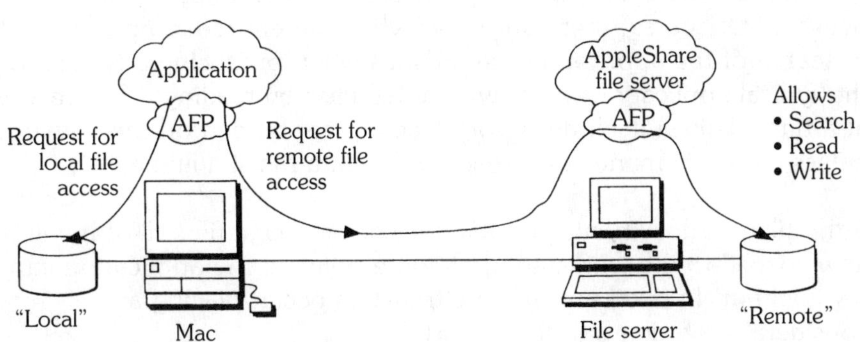

The AppleTalk Apple Filing Protocol in action. Learning Tree International

The Apple filing protocol only provides remote search, open, read, write, and close functionality. Printing and user authentication are left to separate protocols.

Interoperability Rating: Poor

One would not think of deploying AFP as an enterprise-wide solution for a wide variety of workstation types. It only fits the Apple Macintosh class of workstations, so in that sense it is not a good enterprise solution. There are a few ways in which an Apple Macintosh can have access to distributed file services around the network; AFP is one of them. The other is NFS.

On the other hand, many hosts and PC network operating systems have implemented AFP on the server as an alternative DFS to allow Macintoshes to have a DFS service. So while AFP is a poor choice as a universal solution, it is definitely the choice to support Macintosh workstations on a network. As you will see in the next section, almost all PC network operating systems support AFP, just to be able to support these Macintosh clients.

Network resources utilization

Distributed file systems typically place a heavy burden on network bandwidth. They are employed in two ways. The first is to read and write to remote data files. If the data file is a database of organized records, the network traffic will probably be minimal, in that only a few records at a time will be read by the application across the network. Of course, there are times when the entire file may be read in search of the appropriate records, as when producing a report. But the typical application works with a database by reading a few records at a time. This in itself does not place a great burden on the network, unless there are many users reading the data file simultaneously.

If the file is a document, for example, or an image file, all of it will be downloaded to the workstation at some point so the application can work with it. In that case, full file transfers occur, placing a considerable burden on the network.

Furthermore, distributed file systems often supply part of the functionality of a network operating system. Network operating systems support workstations by storing necessary executable programs on the file server. When a user wishes to execute an application at the workstation, a copy of the binary file is downloaded across the network from the file server. It is in this latter case where DFSs, as part of NOSs, produce the heaviest traffic load. The traffic is bursty because this is not continuous, but while the application is downloading we will see 100% network utilization.

Consider some of the following guidelines to minimize the impact of using a DFS on a network:

- Strive to give all LAN users a “home” server on their respective LAN segments. In other words, do not put their applications on servers on the other side of the enterprise network.
- Minimize DFS traffic across WAN links.
- Keep the DFS traffic on a corporate backbone to a minimum.
- Do not mix DFS traffic and remote terminal access traffic on the same LAN segment, if possible. That is, segregate PC NOS users from TELNET users.
- Employ remote routers rather than remote bridges for DFS traffic whenever possible. Routers are more efficient than bridges in their use of bandwidth, especially for DFS traffic.

Integration strategy

Integrating different distributed file systems requires two perspectives: the client's and the server's. Integration is typically carried out at the client. That is, a client might need to access two different types of DFSs simultaneously from within an application. This would require the use of multiple protocol stacks at the client. Support for multiple protocol stacks at the client is provided by a network operating system vendor. The DFS characteristics are also tailored to fit the client on a client-by-client basis. For example, NFS directories appear as logical DOS drives for a DOS client.

We leave the discussion of the client-side integration issues and techniques to Chapter 5. Of interest here is the integration of the server side of DFS, which is a more difficult task. The server DFS is very dependent on the host operating system and file structure. We examine specific examples in the next section, under *network operating systems*.

There are a few guidelines for integration that you might consider. As with any integration technique, consider reducing the variety of DFS types. Also consider standardizing to just a few popular DFS types if restriction to only one is not feasible. For example, NFS, a popular universal solution, might be your solution for enterprise-wide interoperability. Consider the use of DCE in future networks, and migrating from a great variety to DCE. Minimize the variety of DFS types in the departmental PC networks. Unfortunately, you will probably not be able to reduce dependence on Novell NCP, Microsoft SMB, or Apple AFP. Over time, these might be migrated to the TCP/IP environment using NFS if enterprise-wide interoperability with only one protocol is a strong requirement. See Fig. 4-30.

PC network integration

Going into any bookstore today, you will find shelves and shelves of computer books describing the operation, administration, and use of PC networks and PC network operating systems. Our purpose here is not to add to that store of knowledge on the use or installation of PC network operating systems, but rather to show how one goes about integrating network operating systems of different types within the enterprise network.

We are concerned with interoperability when we have a variety of workstation types, all simultaneously accessing data and files on file servers of different types. It is important to understand some of the characteristics of PC file server-based local area networks first, and then to put the *Network Operating Systems* (NOSs) in perspective within the enterprise network. See Fig. 4-31.

PC networks operate in one of two modes. In the first, classified as peer-to-peer, there isn't an identifiable machine that acts as a

Figure 4-30

Distributed filesystems	Characteristics	Typical server platforms	Typical client platforms	Typical protocol suite	Developer/ promoter	Interop-erability rating
NFS	De facto universal interoperable solution	UNIX NetWare VMS LAN Man	DOS UNIX Windows OS/2	TCP/IP	Sun	+
DCE DFS	Becoming popular with client/server systems	UNIX VMS OS/2	UNIX DOS	Varies	OSF	0
FTAM	Large OSI, GOSIP nets, Gov't. nets	UNIX VMS NetWare	UNIX DOS	OSI	ISO	0
SMB	Peer-to-peer PC nets	LAN Server LAN Man Peer-to-peer	Windows Windows NT DOS OS/2	NetBIOS	IBM Microsoft Many others	–
NCP	NetWare-based PC networks	NetWare	DOS Windows NT DOS OS/2	IPX/SPX	Novell	–
AFP	Mac nets only	NetWare AppleShare MVS Banyan LAN Man	Mac	AppleTalk	Apple	–

A review of the characteristics, typical platforms, protocol suites, and promoters of each DFS discussed in this chapter. Learning Tree International

repository of the common files or the file server. We call this type of network a *peer-to-peer* PC LAN. The workstations share files with one another. They can also send one another small on-line messages. From one workstation to another, they can send print jobs to be printed on the locally attached printer.

Typically, this type of LAN is economical. There is no dedicated file server. It has frequently been deployed by small businesses in an effort to connect a few microcomputers together.

Figure 4-31

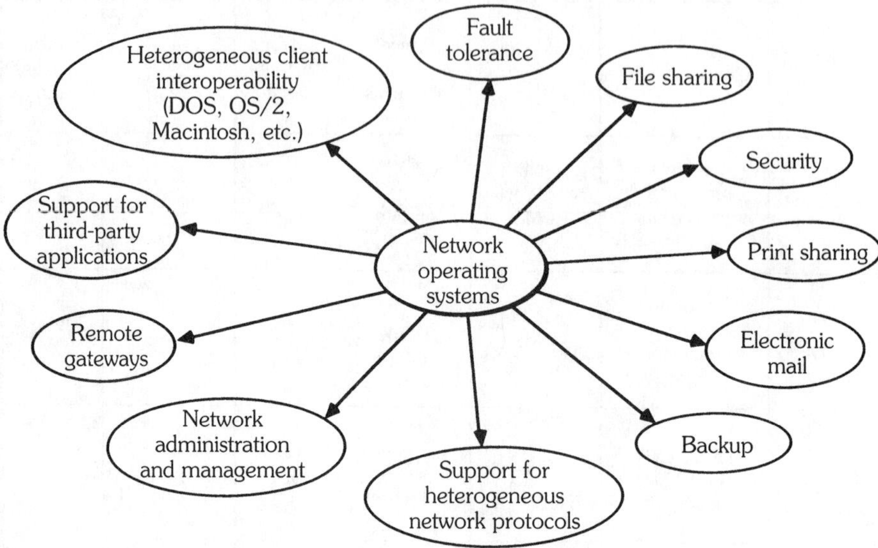

Features of a network operating system. Learning Tree International

In an enterprise network, we have the second type of operation, which is called *file-server based*. In this situation, one machine is identified as the file server, and dedicated entirely to that task. It is the common repository of data and program files. A program called the *network operating system*, or NOS, is installed on that machine. It is connected by a high-performance local area (and sometimes a wide area) network to a number of client workstations. This type of PC network is deployed at the departmental level.

In the file-server based PC network, applications run on the workstation. The workstations, called *clients*, also have some form of protocol or data communication software operating in them that enables them to reach the file server. This program is called the *redirector*. On file servers and clients, there is also a data communication protocol to send and receive requests and responses. The file server typically runs some type of operating system—a local operating system. Additionally, we run a program on this file server that provides printing and file access services, which we call the *network operating system*.

For DOS machines, the server drive and the shared printing devices are mapped to DOS drives and DOS printer ports by the workstation redirector software. The characteristic of the network traffic in this environment is burstiness.

Network operating systems provide a variety of functions. First and foremost, they enable remote file sharing. They implement some form of DFS, which was described in the previous section. Typically, file sharing is accomplished in a multiuser manner, with record and file locking.

Network operating systems also provide for printer sharing, in which locally attached devices or devices attached to other workstations can be shared via the file server. Different versions of network operating systems all provide some form of queuing facility for print jobs. These are the two core functions of a network operating system.

Additionally, servers provide security facilities and user authentication mechanisms. They might provide some simple electronic mail, such as messaging. Servers can also be the focal point for remote gateways to hosts, such as an SNA network. Servers support extensive network administration management facilities. They support heterogeneous client populations, including DOS, OS/2, Macintosh, and UNIX.

Backup facilities might be included as part of the network operating system, although backup is usually not done at the file server, but remotely on a workstation and over the network. Recent versions of network operating systems tend to support backup being done at the file server as an option.

The enterprise role for PC file servers

Traditionally, the file server platform has been a powerful microcomputer. The task of file sharing and printer sharing kept this file server very busy, with few resources left to execute any other type of application at the server. As these file servers have grown more powerful, with extra capacity to execute additional applications in support of other user needs, we have seen PCs become platforms for the server side of client/server software.

For example, a PC file server can also perform the task of a database server by having an SQL server program running on the file server at the same time as the NOS. This makes PC file servers viable players in the client/server software area.

Considering this new environment of client/server architecture, where do PC LAN servers fit in? Typically the client/server architecture has three tiers. The first consists of network workstations, what might be termed the *clients*, or consumers of server resources. This is where client programs and the client side of a split client/server application run.

The clients are connected via high-performance LANs or WANs to a second tier, where most of the PC LAN file servers are found. PC LAN file servers usually provide only file and printer sharing, with an occasional database server function. The server side of client/server software runs on minicomputers found at the second tier—multiuser hosts. Minicomputers are the preferred platform to support the server side of a client/server application.

The second tier can be connected via LANs or gateways to the enterprise mainframe or a larger host, a server of servers. Typically this third-tier platform has the legacy applications running on them, as well as some of the new server-side components of client/server applications.

The PC marketplace

Looking at the PC LAN file server marketplace, we see that we have a number of vendor offerings. It is useful to enumerate these, and to see what percentage of the market each of them holds, just to get a bird's-eye view of the magnitude of the interoperability problem that we face in dealing with network integration. See Fig. 4-32.

Novell offers NetWare 3.x and 4.x to support DOS, UNIX, and Macintosh workstations, and it has the largest market share. Microsoft has a number of current offerings. The first, the LAN Manager under OS/2, has been in the marketplace for some time. This program is being retired. The second and more recent offering is the Windows NT

Figure 4-32

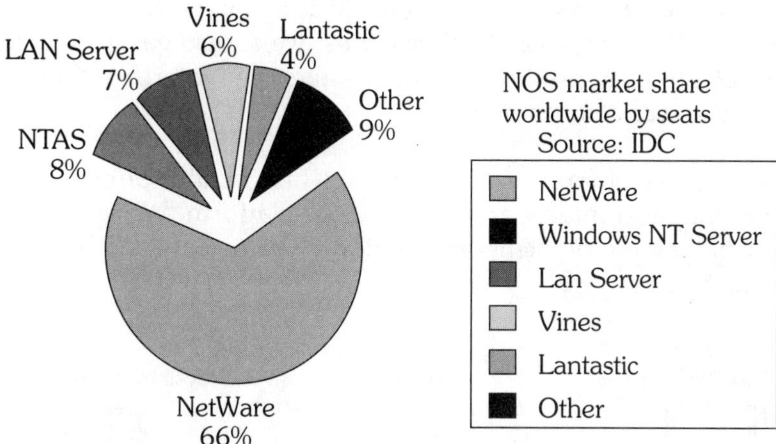

The PC network operating system marketplace.

Server program. It is useful to note that Microsoft also offers a peer-to-peer LAN for Windows workstations called Windows for Workgroups.

IBM offers LAN Server under OS/2. Banyan offers the VINES network operating system. Sun supports MS-DOS PCs with their NFS program. Hewlett-Packard has ported the LAN Manager program from Microsoft to the UNIX environment, and offers the LM/X network operating system to support DOS clients from their UNIX hosts. HP also has ported NetWare to run under UNIX on their line of computers.

Digital offers their port of LAN Manager called PATHWORKS, which has many additional features that enhance the original Microsoft product. DEC has also ported NetWare to run under VMS on a VAX. There are numerous NFS server implementations which support clients that use the NFS client software. NFS should not be neglected as a viable PC file-server solution. Lastly, Apple offers AppleShare as a form of network operating system.

Our task in this section is not so much to compare these network operating systems feature by feature to see which is better, or which has important features that we might use, or which has better security and so on. Our task is to be sure that from any client

workstation, we have a way to open, read, and write files on file servers of different types. In other words, we should easily be able to access remote files, print to remote printers, and interoperate in a mixed file-server environment.

Our criterion for interoperability will be which vendor offering provides the most interoperable file-server platform. We will also look to see what variety of client workstation types each vendor product supports. See Fig. 4-33.

Figure 4-33

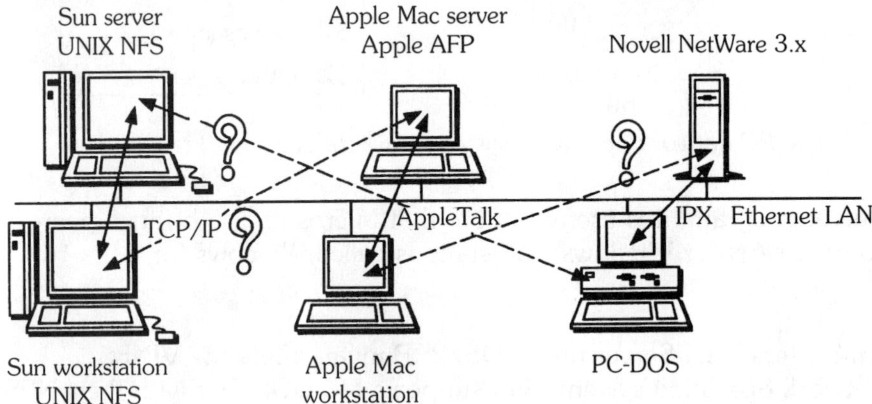

The classic problem of interoperating among file server types: sharing files and printer services from one PC network operating system to another.

Learning Tree International

Interoperability in a mixed PC NOS environment requires solving two problems. The first is that each file-server type typically uses a different DFS, so there has to be some translation of a DFS request at the client for the proper server NOS. This includes both file access and file printing requests.

The second problem is that each DFS typically requires that a particular protocol stack be used underneath for transport services. For example, NFS usually requires TCP/IP, NetWare NCP usually requires IPX, and SMB for Microsoft usually requires NetBIOS. When looked at from the client point of view, not only will there be multiple shells or redirectors operating under our application to work with two or more different file-server types, but there will probably also have to be multiple protocol types underneath each redirector.

Classification of solutions

Interoperable solutions to the problem of accessing multiple file servers simultaneously can be classified into two areas. The first we might term the *client-side solution*: running a multiple-protocol stack on the client. See Fig. 4-34.

Figure 4-34

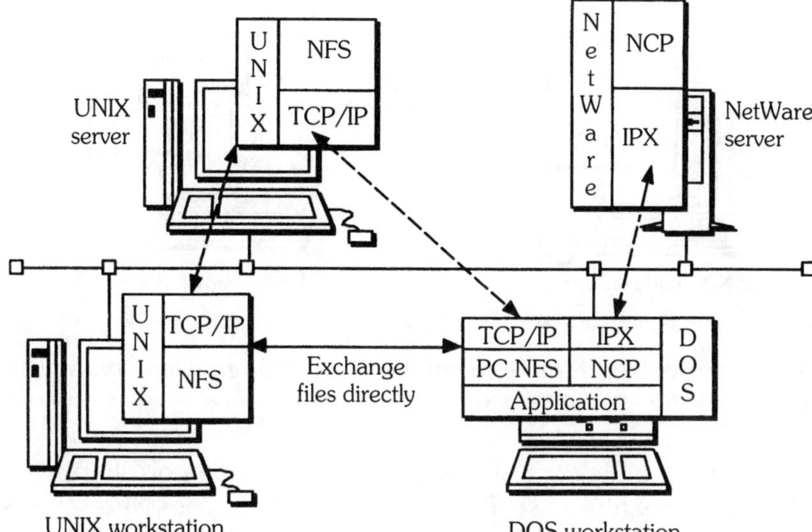

The client-side solution to the multiple PC NOS problem: multiple stacks on the client. Learning Tree International

In this case, a DOS machine might run IPX and the NCP redirector under the application to access a Novell file server, while simultaneously running TCP/IP and PC NFS under the application to access a UNIX server. This solution is explored in detail, client type by client type, in Chapter 5.

Of greater interest in this section is the second type of solution. It is called the *server-side solution*, with multiple application services and protocol stacks on the server. See Fig. 4-35.

Let's assume that we have a variety of client workstations in our client population requiring access to file services, and let's have the

Figure 4-35

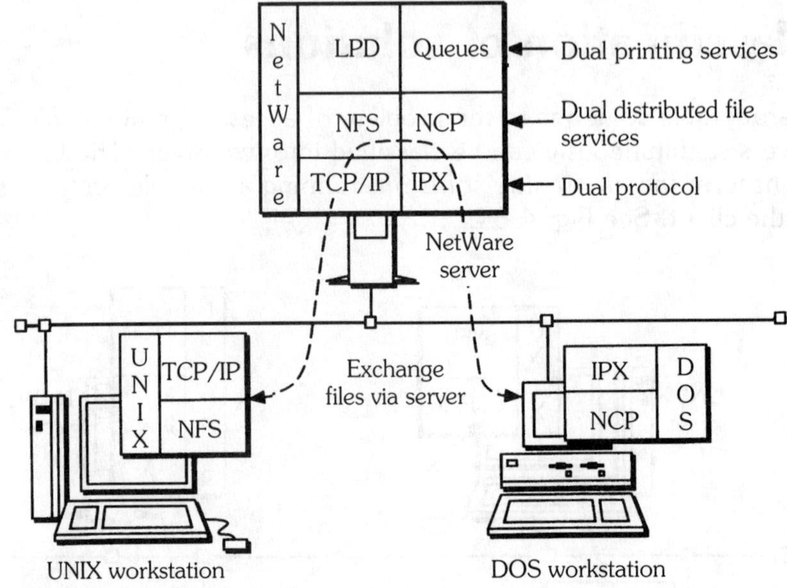

The server-side solution to the multiple PC NOS problem: multiple protocol stacks and application services on the server. Learning Tree International

clients access the file services with their native protocols. In other words, UNIX machines use TCP/IP and NFS to access NFS servers, DOS machines use the NCP redirector with the IPX protocol to access Novell servers, and Apple Macintosh uses AppleTalk to access AppleTalk servers with AFS.

The solution is not to have multiple servers, but to have one server that supports all three protocol stacks and file services. It is this server's task to present a view of the server's file system in the native environment to each client-station type.

For example, a Novell 3.x file server will not only have the native NCP/IPX protocols, but also the NetWare for NFS and the NetWare for Macintosh programs. This enables the Novell file server to serve UNIX workstations in their native NFS/TCP/IP environment, DOS workstations via IPX and NCP, and Apple Macintoshes via AppleTalk and AFP.

The problem with this type of solution is that not all network operating systems support all client types in their native environment, so our search for interoperability and our classification of interoperability leads us to look at each network operating system from the point of view of what client types it supports and how it supports them. Once we found the universal server, would we consider moving all of our file servers to this one type throughout our enterprise, so that all the needs of our client workstations could be properly satisfied without having to run multiple-protocol stacks on our client's workstations?

One drawback of providing the server-side solution is that file transfer between clients can no longer be easily done peer-to-peer, but now must be done as an upload to the server with a subsequent download by the other client type.

The server-side approach to PC file-server integration can be organized as a series of decision steps.

First, there is the analysis phase. Enumerate the mix of client types in your enterprise network. Then look at the type of network operating systems being used by the departments. Are any clients currently running dual protocol stacks? Look at what mixed-protocol services are supported by each network operating system currently in use.

One approach would be to consider switching all network operating systems to a single type, and clients to a single type. It might not be a realistic solution, but it's certainly one option. If this approach is not viable, then look at which clients need to talk to which servers outside their local native environment. In other words, do DOS clients currently talking to a Novell server with IPX and NCP need to access files on a UNIX NFS server on another LAN?

If the dual protocol stack on the client is not a viable solution—in other words, if the client-side solution is not an option—then consider loading the appropriate protocol stack on the foreign, or remote, server to support the new client. Alternatively, install new servers that support multiple-protocol stacks as a focal point of integration among the different client populations.

Whichever approach is taken for integration, it is important to know all the possible ways in which a client can reach all possible servers outside their LAN. This is covered in detail in Chapter 5. It is also important to know what integration components, multiple-protocol stacks, and services are offered by the major network operating system types. This is the subject of the present section, and will be covered in detail NOS by NOS.

✻ The Novell solution: NetWare 3.x

Novell NetWare 3.x provides native support for DOS clients. When server.exe is loaded on the file server, IPX and NCP are automatically loaded and available. To support UNIX, OS/2, and Macintosh clients, Novell provides additional products that must be added to the file server to support these client types. Support is provided via programs called *NetWare Loadable Modules* (NLMs). See Fig. 4-36.

Figure 4-36

NetWare 3.x File Server

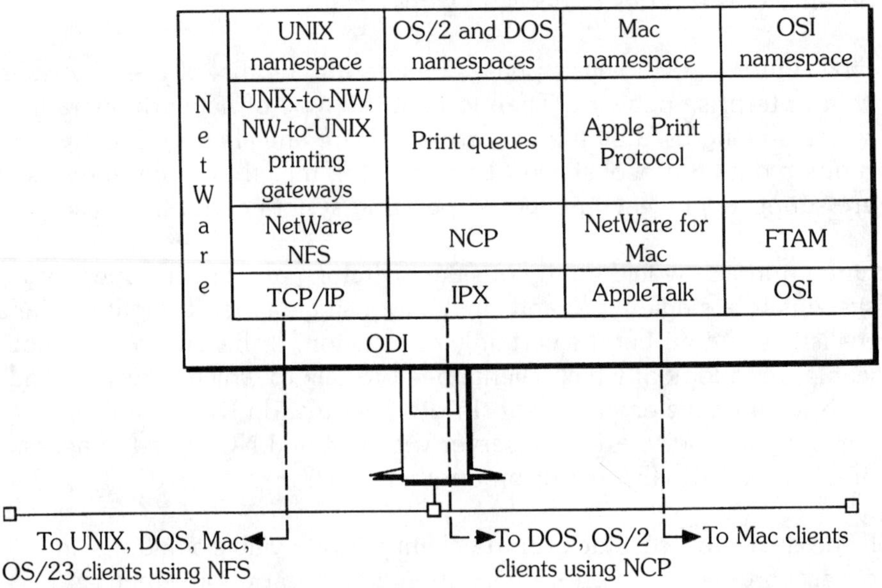

NetWare	UNIX namespace	OS/2 and DOS namespaces	Mac namespace	OSI namespace
	UNIX-to-NW, NW-to-UNIX printing gateways	Print queues	Apple Print Protocol	
	NetWare NFS	NCP	NetWare for Mac	FTAM
	TCP/IP	IPX	AppleTalk	OSI
	ODI			

To UNIX, DOS, Mac, OS/23 clients using NFS

To DOS, OS/2 clients using NCP

To Mac clients

NW 5 NetWare
ODI 5 Open Data-link Interface

The Novell solution, showing the many protocol stacks that can be loaded on a NetWare server. Learning Tree International

NFS clients are supported via TCP/IP NLMs. The set of programs is called NetWare NFS. It enables the simultaneous maintenance of a UNIX namespace next to the DOS namespace, NFS calls via RPC and XDR to be supported on the Novell server, and, through a line-printer daemon, the bilateral execution of print jobs from and to a UNIX workstation or UNIX server. In other words, DOS print jobs can be sent to UNIX printers, and UNIX print jobs can be sent to Novell queues to be printed by the Novell file server.

Macintosh clients are supported by adding the AppleTalk protocol to the Novell file server. The NetWare for Macintosh NLM needs to be loaded at the file server, which also enables the creation of a Macintosh namespace. As far as a Macintosh workstation is concerned, a NetWare server looks just like another Mac that has been set up as a common repository of files and a common print server.

This environment also provides a bilateral printing capability where DOS print jobs submitted to Novell queues can be printed on remote Apple printers on the AppleTalk network, and AppleTalk print jobs submitted to AppleTalk queues can be rerouted to Novell queues.

OS/2 clients are supported via IPX or TCP/IP, and can access files on the NetWare server as if they were DOS files. If the OS/2 namespace has been loaded on the server, the files can alternatively be accessed as OS/2 HPFS (High Performance File System) files.

Interoperability Rating: Best

Novell has worked very hard with NetWare 3.x to provide a platform supporting a heterogeneous mix of clients. To Macintoshes, it looks like a Mac file server. To DOS machines, it looks like extensions of DOS drives. To UNIX machines, it looks like an NFS server. Most client types are supported in their native environment, without the need for multiple protocol stacks at the client.

The one glaring omission is the lack of support on the NetWare server for NetBIOS and SMB protocols, and NetBIOS clients. This is a consequence of the Novell-Microsoft feud in the marketplace. It is understandable, but leaves us with a hole in our interoperability bag of tricks that needs to be handled. It forces NetBIOS stations to use

dual-protocol stacks or, following the course of action we advocate throughout this book, the reduction of enterprise dependence on the NetBIOS protocol.

In general, NetWare appears to provide a very good all-around solution for interoperability in a mixed file server and client environment.

✻ The Microsoft solution: LAN Manager and NTAS

To understand the programs offered by Microsoft as network operating system software, it's useful to look at the progression of NOS offerings by the company. See Fig. 4-37.

Figure 4-37

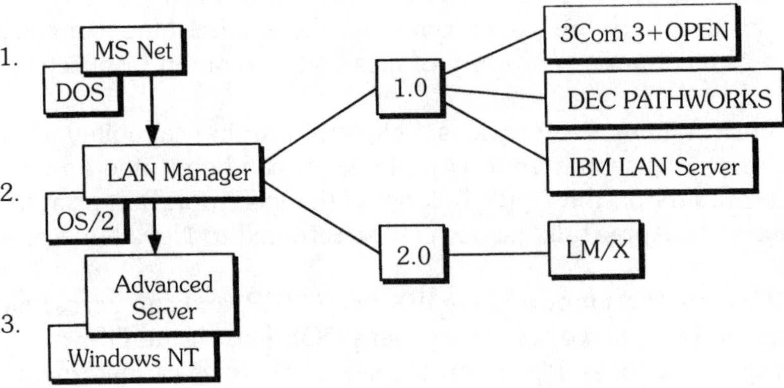

Progression of network operating systems offered by Microsoft, culminating in their Windows NT Server product. Learning Tree International

The latest offering by Microsoft, NTAS, is a third-generation product. The family of NOSs started with the first-generation product, MSNet. It ran under DOS and was a fairly unsophisticated program. With the advent of OS/2 and the IBM— Microsoft joint effort to develop it came the second-generation product, LAN Manager.

Early versions of LAN Manager were picked up by OEMs and turned into their own PC network operating systems. Out of that effort came 3+Open from 3Com, now no longer sold. DEC also picked up LAN Manager and ported it to the DEC VAX, offering it under the name PATHWORKS. IBM had its own LAN Manager, which was termed LAN Server. LAN Manager evolved into version 2.0, which was

picked up by Hewlett-Packard as an OEM and ported to the UNIX environment, where it became LM/X. All of these products competed with Novell NetWare, but didn't make much headway in taking market share away from Novell. See Fig. 4-38.

Figure 4-38

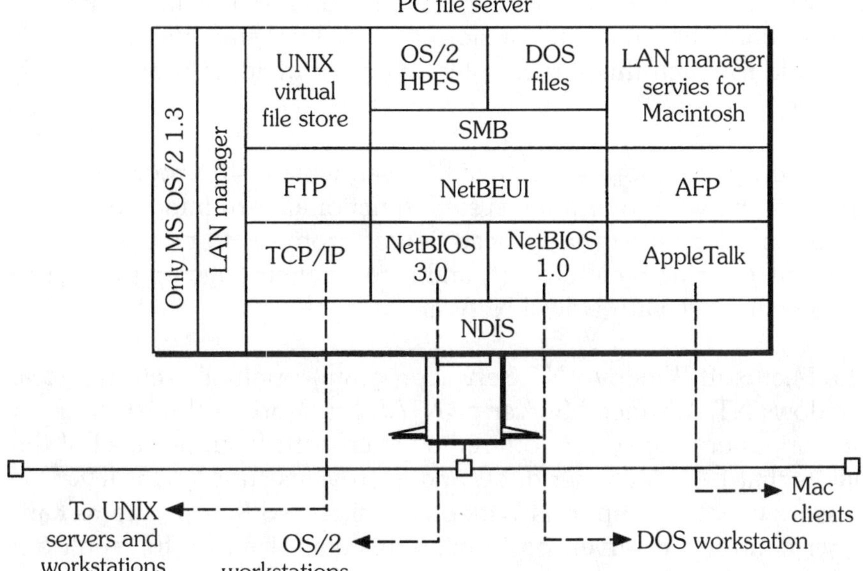

Microsoft LAN Manager, now no longer sold, but with a respectable installed base. Learning Tree International

The Microsoft LAN Manager solution supports a variety of client platforms: Apple Macintosh workstations via AppleTalk AFP, and the LAN Manager services for Macintosh products from Microsoft. It supports DOS and OS/2 workstations with NetBIOS and NetBEUI, and the SMB protocol for file sharing. The Microsoft LAN Manager program also supports UNIX clients by converting the file server into an NFS server. The NFS product set from Microsoft also supports FTP, and uses the TCP/IP protocol suite for transport. It does not support Novell's IPX and NCP.

It is important to note that the Microsoft LAN Manager runs as a program under Microsoft (not IBM) OS/2. It uses the NDIS program to resolve contention by multiple protocols at the server for the use of the network card.

Interoperability Rating: Poor

LAN Manager from Microsoft can be considered an older legacy system, and it's on its way to being abandoned by Microsoft. It also relies on NetBIOS, which makes it a poor choice for an enterprise network. It has no native IPX support, making it difficult for IPX workstations, which are the majority of the workstations on the network, to reach this server unless they run multiple-protocol stacks to client.

The latest generation is Advanced Server, which is a program providing network operating system functionality running on a Windows NT platform. Advanced Server promises to be a very powerful offering by Microsoft, and is to be their flagship product set to compete well with Novell NetWare.

The Microsoft Windows NT Server program, originally referred to as Windows NT Advanced Server or NTAS for short, is the latest network operating system offered by Microsoft. It supports all of the clients that LAN Manager does, and it provides support for IPX network clients. It supports Windows sockets, so that programs can be written for the server, positioning this platform as a full application server for client/server environment. See Fig. 4-39.

Interoperability Rating: Mixed

This should be a very powerful platform for integration of many client types in the future. It is too new to rate it higher, so it must be kept as a product to watch, rather than one to fully embrace for interoperability at present. As NTAS gains market share, it could be a very important product in our arsenal of integration tools.

✳ The IBM solution: LAN Server

LAN Server from IBM has gone through two evolutions. The first product offering by IBM was derived from the Microsoft LAN Manager product, and IBM became an OEM for LAN Manager under the guise of LAN Server from Microsoft. Since the split-up of IBM and Microsoft, IBM completely rewrote LAN Server and the newest, LAN Server 4.0, is completely owned by IBM. It runs under IBM's version of OS/2, and supports a variety of client workstations. See Fig. 4-40.

Figure 4-39

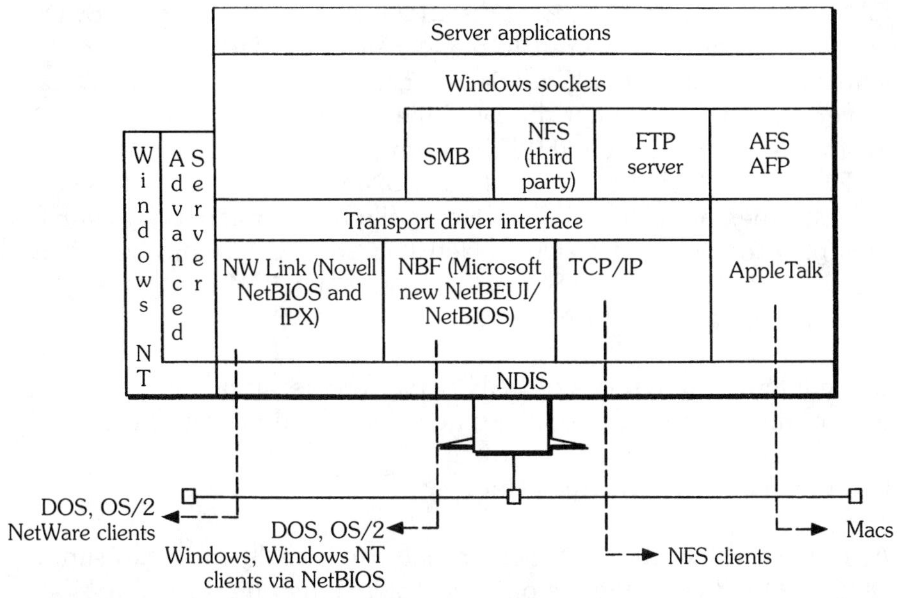

Microsoft Windows NT Advanced Server. Learning Tree International

Figure 4-40

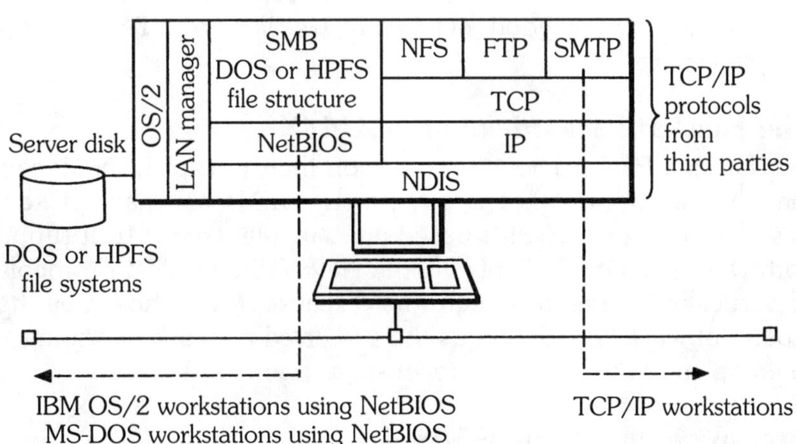

HPFS = High-Performance File System
NDIS = Network Driver Interface Specification

IBM's LAN Server network operating system running on an OS/2 machine. Learning Tree International

LAN Server supports OS/2 workstations via NetBIOS and SMB, DOS workstations also under NetBIOS and SMB, and NetWare OS/2 workstations using LAN Workplace for OS/2. On the LAN Manager server, we can have either the DOS file system or the OS/2 High-Performance File System (HPFS). Third-party products support TCP/IP under LAN Server to support UNIX workstations. These are full implementations of NFS, including file transfer capability via FTP. Multiple-protocol stacks such as NetBIOS and TCP/IP on the server are supported via the NDIS specification, which IBM has implemented on their LAN Server.

Some interesting features of LAN Server are its support of remote boot of diskless workstations, a time service, and file replication.

Interoperability Rating: Mixed

As of this writing, the LAN Server platform is really meant to support DOS or OS/2 workstations only, complementing the rest of IBM's product offering in the area of networking. No attempt has been made by IBM to provide native UNIX client support, which is left to a third party to implement, or to support Macintoshes, for that matter. It is therefore not as good a choice as NetWare or NTAS as an integration platform.

✻ The Hewlett-Packard solution: LM/X

The Hewlett-Packard solution relies on their port of LAN Manager to the UNIX environment. There are many OEMs who have picked up this converted product. It's based on LAN Manager 2.0. It supports both DOS and OS/2 clients via the NetBIOS and SMB protocols. It is of particular interest to integrators, because it was chosen by the Open Software Foundation as the preferred network operating system to complement the DCE protocol suite. See Fig. 4-41.

Interoperability Rating: Mixed

LM/X provides a limited client-base support. There is definitely no native Mac client support, or at least it hasn't been widely implemented. Of course, because of the UNIX platform underneath, it supports UNIX clients very well. But there is no IPX native support either. For that reason, we have to give it a mixed rating.

Figure 4-41

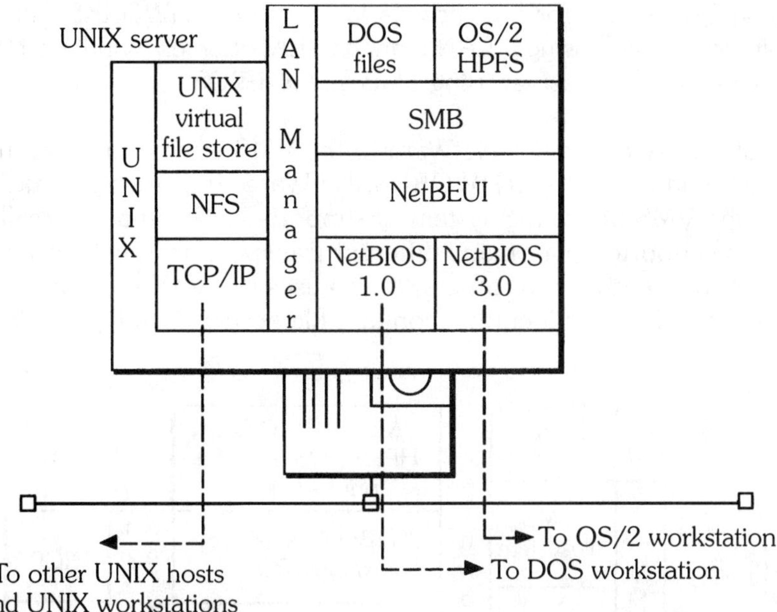

The Hewlett-Packard LAN Manager for UNIX (LM/X) server solution. Learning Tree International

The only bright spot is that it has been picked up by OSF as part of the DCE protocol suite, which may popularize it in client/server environments.

❊ The DEC solution: PATHWORKS and NetWare for VMS

DEC has ported a number of network operating systems from the PC world to run as programs under VMS on a VAX. The first is the port of LAN Manager from Microsoft, called PATHWORKS. The second is a port of NetWare from Novell. Both solutions provide interoperability between VAX users and PC LAN users.

PATHWORKS, a LAN Manager derivative, is supported under many different transport protocols. One can reach the VAX via DECnet, IPX/SPX, NetBIOS, and even LAT. It uses VAX security mechanisms for user authentication and file access. It also employs the SMB protocol for file sharing requests. It provides bidirectional printing services from the VAX environment to the PC network environment. Finally, it has compatibility without a LAN Manager server. That is, a NetBIOS client

running SMB redirector can access a VAX with PATHWORKS while simultaneously accessing a LAN Manager, LAN Server, Windows NTAS, or any other PC file server using SMB and NetBIOS.

DEC also ported Novell's NetWare to the VMS environment even before it had created PATHWORKS. NetWare VMS runs as a task under the VMS operating system. It supports NCP file-access calls, and it is supported over many different transport protocols: IPX/SPX, DECnet, and a variety of others. It also uses the VAX security mechanisms for user identification and file access. See Fig. 4-42.

Figure 4-42

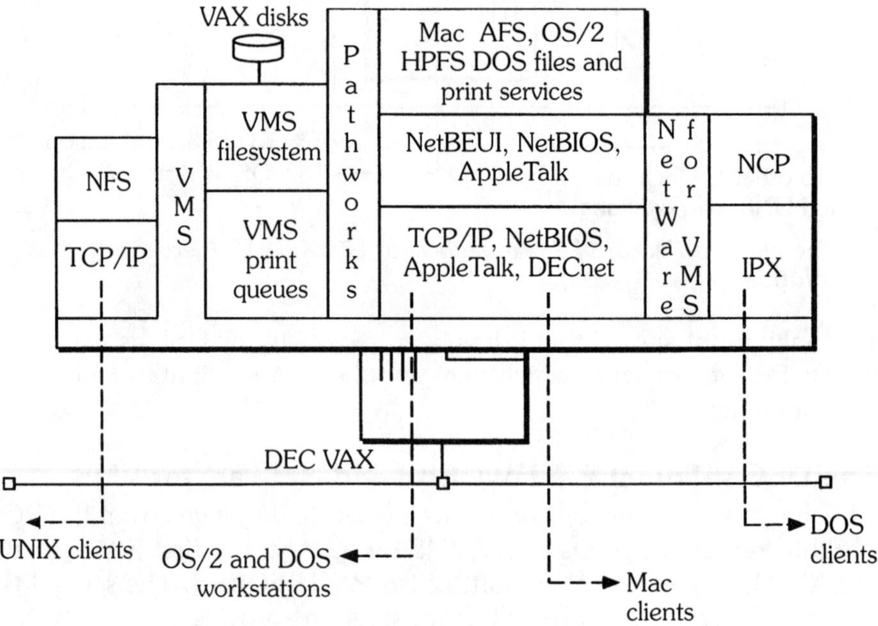

The Digital VAX server solutions. Learning Tree International

Interoperability Rating: Mixed

If you have a VAX, then by all means use it. It turns out to be a very good integration platform. If you don't have a VAX, we don't suggest that you run out and buy one just for its integration properties. Therefore, forget this as a solution.

✻ The Banyan solution: VINES

VINES is Banyan's PC network operating system. It runs under AT&T UNIX on the server, which was modified by Banyan. It is meant to provide file and print services for PCs, mostly DOS workstations, which are the majority of clients on PC networks. It also supports other types of clients: Macs, UNIX workstations, and OS/2 workstations.

The client side for PCs is proprietary to Banyan. It uses a proprietary version of IP. One also could run normal TCP/IP to connect to a Banyan server from DOS or UNIX workstations.

VINES supports NetBIOS clients on the server. One can connect from DOS workstations to the VINES server via FTP Software PC/TCP product. This is a full implementation of TCP/IP for the DOS client. For Macintosh clients, Banyan has implemented a full AppleTalk protocol stack as well.

There are two major features of VINES that differentiate it from other network operating systems. The first is that VINES has traditionally provided very good WAN connectivity right at the server. One can connect VINES servers together over a wide-area network link without resorting to external routers or bridges. This is due to VINES' implementation of TCP/IP, or the use of TCP/IP as a protocol stack. TCP/IP, as we saw in Chapter 3, has very good wide-area network capabilities.

Banyan has also implemented in VINES a very sophisticated directory service called StreetTalk, and more recently expanded to Enterprise Network Systems (ENS). See Fig. 4-43.

Interoperability Rating: Mixed

VINES is a very good network operating system, especially due to its excellent support for wide-area network connections. On the other hand, it has very small market penetration. It has too much dependence on proprietary protocols, and it's lacking some very important native protocols, such as IPX. Thus, we don't suggest that this be used as an integration tool in your enterprise network.

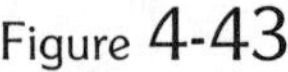
Figure 4-43

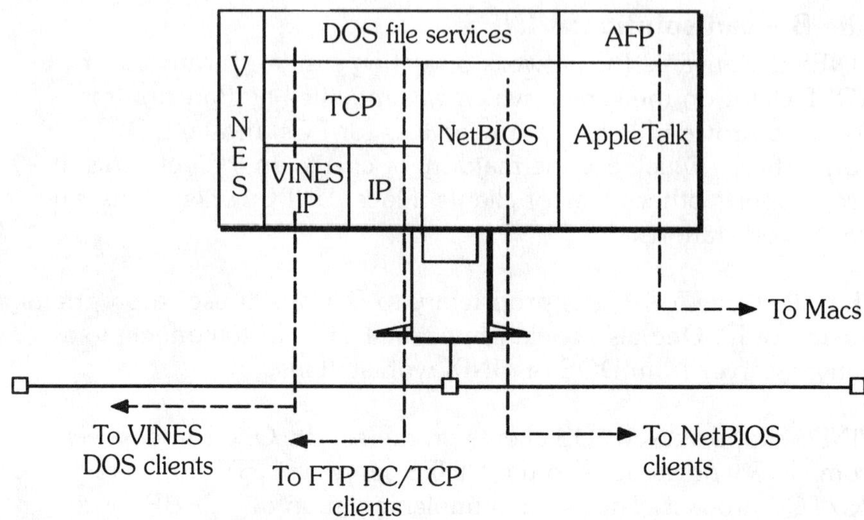

The Banyan VINES solution for PC network integration. Learning Tree International

✻ The UNIX solution: NFS

One should not neglect the possibility of supporting non-UNIX clients via NFS. There are many fine products that enable DOS, Macintoshes, and OS/2 workstations to reach and work with an NFS server. In reality, this is a client-side solution, not a server-side solution. For this solution to work, we need only support NFS under TCP/IP. Chapter 5 treats the NFS solution, client by client, in detail.

There are many products that implement NFS on the client side. Sun has the PC NFS product for DOS machines. Chameleon/NFS is a product by NetManager that supports Windows workstations. LAN Workplace for DOS by Novell is a product that supports DOS access to NFS file servers. FTP Software PC/NFS also provides supports for DOS. Lastly, Windows NT clients have NFS support as well.

Interoperability Rating: Mixed

If your clients are all running NFS client workstation software, then using an NFS server is a good integration point. Considering that there are so many NFS implementations for all the various client types, this is a fairly good solution, although it should be treated as a client-side solution rather than a server-side solution.

 Network resources utilization

In providing server-side solutions to interoperable problems, we must consider two aspects of the utilization of resources. The first is the additional traffic on the LAN the server is on, which will now be coming from clients outside the LAN segment. This is traffic on the LAN segment that was not there before integration. Very large "foreign" print jobs can be especially resource-intensive on the network.

Additionally, the file server used for integration will be further burdened. First, it has to run multiple-protocol stacks. Then it has to support a larger number of clients, which means more client sessions, more disk access, and maybe the loading of additional programs to provide all the added functionality, further increasing the load on the server.

Bridges and routers on our enterprise network might see additional traffic after the integration has taken place. This additional traffic might be over WAN links, and the capacity of these links versus the new offered load must be studied carefully. WAN links can become a bottleneck, and their bandwidth might have to be increased. This, of course, can be an expensive proposition. This problem merits careful study.

We suggest that the impact of connecting multiple PC file servers to our enterprise network and enabling workstations from around the network to access servers not on their local environment be carefully assessed before integration takes place, and then monitored soon after integration, to be sure that network resources are adequate and are enhanced as needed.

 Integration strategy

To improve interoperability, the first technique should always be to minimize the number of network operating system types that have been deployed. Barring that, one should look for and standardize a few network operating system types, focusing on those that support file printing and file sharing for each client community. One should look to use an NOS that supports all the major client types—OS/2,

UNIX, DOS, Windows, Windows NT, and Macintosh—in their native environment, if at all possible.

Also, to improve interoperability, seek to install network operating systems that support the major protocol transport types: TCP/IP, IPX/SPX, AppleTalk, and NetBIOS. See Fig. 4-44.

Figure 4-44

Network operation system	Operating system on server	Native clients supported	Protocol suites supported	File structures supported	Developer
NetWare 3.x	NetWare, VMS, or Windows NT	DOS Windows Windows NT OS/2 (HPFS) UNIX, Mac	TCP/IP IPX/SPX AppleTalk OSI	OS/2 HPFS DOS, NCP UNIX, NFS OSI, FTAM Mac AFP	Novell
LAN Manager	OS/2 UNIX	OS/2 (HPFS) DOS Windows Windows NT Mac	NetBIOS	OS/2 HPFS DOS, SMB	IBM/ Microsoft
LAN Server	OS/2	OS/2 (HPFS) DOS Windows NT Windows	NetBIOS	OS/2 HPFS DOS, SMB	IBM
PATHWORKS	VMS	DOS Windows	NetBIOS DECnet	DOS, SMB Mac	DEC
PC NFS	Any NFS server UNIX (typ.)	DOS Windows Windows NT	TCP/IP	DOS, NFS UNIX, NFS	Sun
VINES	Banyan UNIX	DOS Windows OS/2	Banyan TCP/IP	DOS, NFS UNIX, NFS	Banyan
Advanced Server	Windows NT	DOS Windows NT Windows	NetBIOS IPX/SPX	DOS, SMB Mac, HPFS	Microsoft

Summary of the interoperable characteristics of the more popular PC network operating systems. Learning Tree International

E-mail

Enterprise-wide electronic mail (which is sometimes written as e-mail or email) has become a business-critical tool. For maximum benefit, it is important to tie together all the e-mail systems within an organization. Electronic mail is the passing of electronic messages from computer to computer via computer networks or via modems over common-carrier phone lines.

Traditionally, e-mail meant ASCII text messages. Today, one can e-mail just about any kind of information, from images to voice to videoclips, as attachments to the traditional text message. To understand how to integrate the various types of e-mail systems present within organizations, it is important to first understand the general types and the various brands of e-mail systems. These are discussed in this section.

It is also useful to know the magnitude of the integration problem. Two general integration approaches are discussed next, together with recommendations.

Lastly, we put it all together by looking at a case study.

Some definitions

We will refer to some generally accepted terms in the industry in this section, so we will define them for you. The definitions come from a comprehensive compilation of computer technical terms called the *jargon file*. The entire jargn10.txt file (be careful, it's over a megabyte in size) can be obtained as follows:

```
ftp mrcnext.cso.uiuc.edu
login: anonymous
password: your@login
```

Then change directory to

```
etext/etext91
```

or

```
etext/etext92
```

E-Mail: /ee'mayl/ 1. n. Electronic mail automatically passed through computer networks and/or via modems over common-carrier lines. Contrast snail-mail, paper-net, voice-net. 2. vt. To send electronic mail. Oddly enough, the word "emailed" is actually listed in the Oxford English Dictionary; it means "embossed (with a raised pattern) or arranged in a network." A use from 1480 is given. The word is derived from French "emmailleure," network.

snail-mail: n. Paper mail, as opposed to electronic. Sometimes written as the single word "SnailMail." One's postal address is, correspondingly, a "snail address." Derives from earlier coinage "USnail" (from "U.S. Mail"), for which there have been parody posters and stamps made.

paper-net: n. Hackish way of referring to the postal service, analogizing it to a very slow, low-reliability network. USENET signature blocks not uncommonly include a "Paper-Net:" header just before the sender's postal address; common variants of this are "Papernet" and "P-Net."

voice-net: n. Hackish way of referring to the telephone system, analogizing it to a digital network. USENET signature blocks not uncommonly include the sender's phone next to a "Voice:" or "Voice-Net:" header.

Electronic mail systems

In general, there are two approaches to creating an e-mail system. The first approach is what might be called host-centered or host-based e-mail. Central to this case, there is a multiuser, multitasking computer. On it, some form of e-mail program is running that enables people to open and read their mail, and compose replies and original messages. Users can also peruse mailing lists or compose private mail lists to use to mail out messages to groups of people. A program that enables users to do all these things is called the e-mail interface program, also referred to as the User Agent (UA).

On the same machine there is an e-mail program that sends messages to users' mailboxes on other machines or post offices.

Both the UA and the post-office-to-post-office programs, together with the directories to store user messages (called mailboxes), is called the post office.

In the host-centered approach, there is one post office, and everybody logs into the post office host to get mail. The upper diagram in Fig. 4-45 shows an e-mail post office. It shows users accessing mail via terminals. Users could be coming over the network from a workstation emulating a terminal, or using a modem over dial-up phone lines to do the same thing. The latter is how subscribers get their e-mail messages from an on-line service such as CompuServe or America Online.

On a PC network, the situation is somewhat different. Consider that the e-mail program to open, read, compose, delete, and post mail (the user agent) is running on the user workstation. If there is a file server on the network, that's where the user mailboxes are found. Although

Figure 4-45

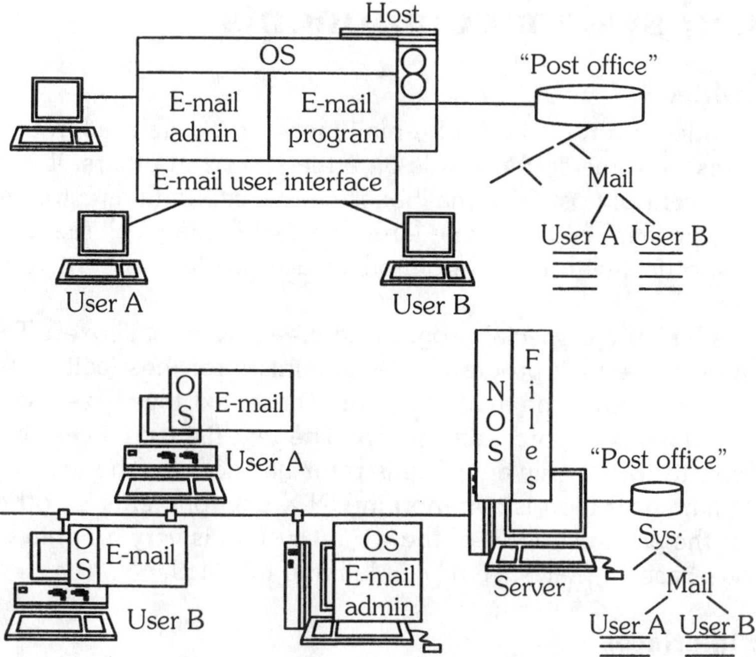

The e-mail paradigms: host-centered and PC network-centered.
Learning Tree International

there is no e-mail program *per se* running in the file server, it is still called the post office, because it is the repository for e-mail messages. All the programs are on the user workstation, including the administration programs that enable the administrator to create mailboxes, edit mailing lists, clean up files, create users, etc. The lower diagram in Fig. 4-45 shows the PC network-based approach to e-mail.

There is a third technique, based on a peer-to-peer PC network such as Windows for Workgroups. In this approach, there is no centralized file server. To send e-mail in this case, the other user has to be available to receive it. There is no drop box or post office to deposit mail. As you can well imagine, this is difficult to operate when people can turn their machines on and off. It can still be done, but both users must have the e-mail program running on their machines, and their computers must be on for them to receive mail. Users are reduced essentially to messaging, a term often used to refer to this type of e-mail, as opposed to true e-mail.

E-mail system components

Post Office

A post office is a repository of e-mail messages, organized in user mailboxes. In some systems, it is on a file server; in others, it is at the multiuser central host. The mailboxes can be separate directories, one per user, or there could be one large file that contains all the mail in that post office, with record pointers as user mailboxes.

Most systems use a special program to create user mailboxes. To administer some mail programs, the administrator must edit certain text files that set up the mail program. These text files have user lists with accounts, passwords, and so on. The text file is in a certain directory, to which only an administrator has access. The text-file type of administration is common in UNIX environments. In other systems the administration of the e-mail system is very graphic, as in Windows-based e-mail such as cc:Mail and MS Mail.

User interfaces

The program used to open, read, delete, reply to, and create new mail is called the *user interface*, or *user agent* (UA). Some e-mail user interfaces allow for the creation of messages within an application; one can then send the message via e-mail right from within that application. This is very common with sophisticated word processors. Applications that have this function are called *mail-enabled*. One of the best examples, of course, is the Microsoft Office suite of applications; WordPerfect Office is another.

✻ E-mail message types

The simplest kind of e-mail is text-based, where the document to be sent is encoded in seven-bit ASCII (just the Roman alphabet, along with numbers and symbols.) Messages get more sophisticated when we send documents that have a variety of embedded fonts and formatting. These must be sent as documents attached to the e-mail message.

A lot of e-mail nowadays comes in the form of attachments. Sometimes the e-mail message is just the header to the attachment. Some mail systems allow for the attachment of very fancy, complex documents that have images, voice, and even video clips in them.

E-mail post office types

Here are a few of the commercially available solutions for post offices. Before we take on the integration problem, we need to understand the various brands of systems that might have to be integrated. The post office types will be classified and details of their operation given later on in the section.

✻ UNIX solutions

The UNIX *sendmail* program is very popular (because it usually comes with the UNIX operating system). Sendmail is actually a program that enables one post-office host to send mail to another post-office host. It is not used to create, read, or administer mail, but it is actually a post-office-to-post-office connection program. A popular user UNIX e-mail user interface is the mail daemon maild; the POP protocol is also used in UNIX workstations to pick up mail from a UNIX server mail drop.

The HP solution Open-Mail is a complete environment in the UNIX world from Hewlett-Packard. It is very sophisticated, and a good example of a client/server program.

An early e-mail solution A very early UNIX e-mail program is MMDF (Multichannel Memo Distribution Facility). It is mentioned here because some of the old U.S. Army and U.S. Air Force facilities actually still use some SCO UNIX MMDF systems.

The IBM solution

In the IBM mainframe environment, the preferred post-office environment is PROFS, which also runs on some of the mid-range machines. IBM also has UNIX solutions, of course, since they sell UNIX AIX platforms. They also have an integrated environment called OfficeVision.

The Digital solution

There are several DEC solutions. One is VAX Mail, a much older program. The other more recent product is All-in-One, a popular word processing and database environment with e-mail capability.

PC network solutions

Here is where there is the greatest diversity in solutions. On the PC networks, the largest share of the e-mail market has been captured by Lotus cc:Mail, with about 20% of the market. Microsoft Mail has captured another 20%. WordPerfect Mail has about 11%. The rest belongs to a legion of contenders.

Public e-mail programs

There are some popular public e-mail programs, such as MCImail and Sprint TeleMail. Some of the online services such as CompuServe offer business e-mail solutions for enterprise-wide e-mail.

Our intention in the rest of this section is to classify the approaches and their details, and then discuss how we go about integrating these e-mail systems into one enterprise-wide e-mail system.

The diversity problem

One of the major problems with e-mail is the proliferation of e-mail systems that users have to deal with on a daily basis. Some people find themselves dealing with about five different e-mail programs.

They might have cc:Mail or Microsoft Mail at the office. They might use TeleMail corporate-wide for international e-mail. They are on the Internet and use an Internet e-mail account to communicate with colleagues at other companies. They might also have a CompuServe account for technical research, and to communicate with other colleagues who are not on the Internet. They might have an America Online account for home use. This user has five accounts, and five passwords to remember. In some cases, like cc:Mail, one has to log into the server as well, which requires a separate account and user name.

It's a big problem keeping track of so many accounts. Wouldn't it be easier if this user could pick up all his e-mail in one place? This can be done by integrating all the systems, and that's a subject taken up in Chapter 5.

✻ The e-mail format diversity problem

Another problem is not only the multiplicity of accounts, but the fact that most message formats are very different from one another. The way the programs attach files, the types of attachments they allow, and how the e-mail is encoded are just some of the differences. The addressing schemes, for example, are completely different in most cases. As a message goes from one system to another, it can lose information as the message is converted. Some systems can't transmit embedded images, others won't attach any files at all, and others don't do video, just to name a few discrepancies.

Mixing e-mail systems

The problem with enterprise e-mail systems is that the post office types are not the same across the enterprise. To integrate, we must cross mixed e-mail systems. How did we get into this problem of having multiple, largely incompatible e-mail systems?

The answer is that enterprise e-mail systems grow from the bottom up: first departmental systems, then corporate-wide integration. Some departments have a host with PROFS or sendmail. The users on a departmental PC network probably have either Lotus cc:Mail or Microsoft Mail. Those disparate solutions fitted the computing

environment and mail needs at the time they were purchased and installed, so everybody has a good e-mail solution for their local departmental system, if they've got any e-mail at all.

The first thing in the corporate computer culture is to convince everybody that e-mail is a good thing. In some cases, we are still fighting that battle, especially if the organization is small and has a tight budget. E-mail is a hard thing to sell, so companies install e-mail wherever they can, and hope to be able to integrate later. Then, when later comes, they find themselves scratching their head and reading this book. That's how we got into this mess.

Once you get over the barrier of convincing everyone to use e-mail, they buy into using their local departmental e-mail. Then they say, "Wait a minute. Wouldn't it be great to send mail across the entire organization?" And e-mail leads the way in causing the organization's networks to be integrated. E-mail is the tail that wags the integration dog. In an ideal world, integration flows from the top down, and e-mail interaction just happens naturally.

✳ The problem

The question then is: "If the user is logged into a NetWare server and is using cc:Mail, and the other person is using POP3 mail via a UNIX workstation against a UNIX host, how do we get mail from one user to the other?" There are a number of solutions for that and other similar cross-system situations.

Figure 4-46 shows the problem graphically. User A can create and send e-mail within his own post office to User B. User B's post office is on a separate system. How do you get mail from A to B? Consider all the differences. Each of these users might have different network protocols to communicate with their respective servers—in other words, different kinds of connectivity. Their e-mail mailing lists are different. The maintenance of those lists by network administrators is different. Their message formats may be different: one may use an X.400 message format, the other one may be 822 encoded (that's the UNIX SMTP mail format).

There may be different attachment types. One might create an image and want to send it to the other, but the other can't receive it because

Figure 4-46

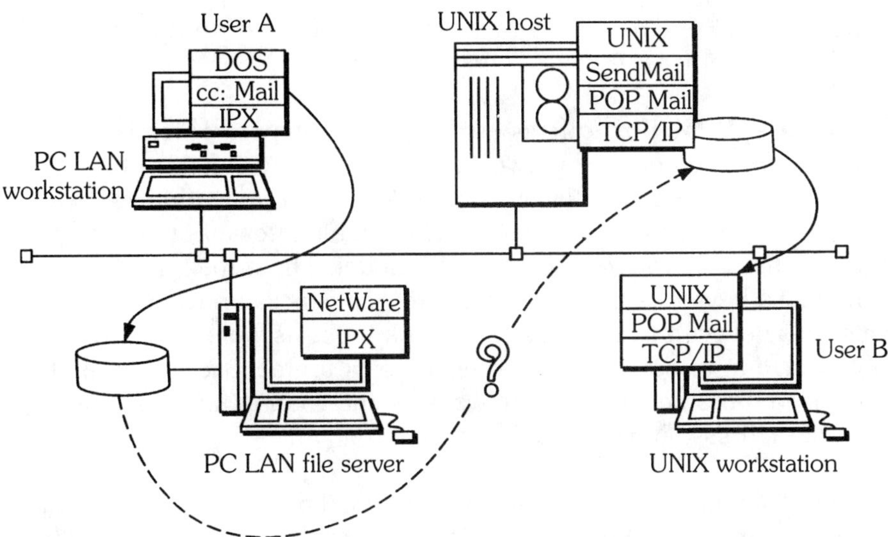

The e-mail integration problem. Learning Tree International

he doesn't have the protocol to handle image files and attachments. To carry out an integration solution in these situations, it's important to know the details of products.

✻ E-mail integration tools

There are basically two tools to integrate post offices. The first one is a router—an e-mail router, that is. It is not an OSI layer-3 device, which is a packet router, like an IP or IPX router; an e-mail router just routes e-mail messages. E-mail routers are used when the two post offices are of the same type. The other integration tool is classified as an e-mail gateway. Gateways are used when the two post offices that need to exchange mail are of different types.

E-mail routers If the post offices are of the same type, for example if they all use cc:Mail, UNIX mail, or Microsoft Mail, there is no problem in any of these situations. To connect the post offices, just install an e-mail router to move mail from one post office to another. The router solution is typically sold by the post office software vendor. Sometimes the router product comes with the post office program. For some systems, the router is an additional product and expense. In general, each router product is specific to each post

office product. Microsoft Mail routers, for example, do not work with cc:Mail post offices, and vice versa.

Gateways If the post offices are of different types, which is often the case in an enterprise network, e-mail gateway solutions are indicated. The e-mail gateway is considered an application gateway, in contrast to a protocol gateway. The e-mail gateway operates at layer 7, the very top level of the OSI model. The e-mail gateway might also have to support several protocol stacks under it if it operates between two different network types as well. It has to be able to log in to both servers or hosts, pick up mail from one, and send it to the other. E-mail gateways do the routing and the e-mail format, address, and protocol conversion.

Some companies actually specialize in e-mail gateways. One product is far and above all the others. It is made by a company called Soft-Switch. (They used to be independent but are now owned by Lotus.) They sell an e-mail gateway product that connects any e-mail environment to any other. Their software runs on many different platforms, and it is a very good technical solution, but it is expensive. It can easily cost $30,000 or more for the main program, and $5,000 for the gateway program to each e-mail environment. It can run on most platforms in the enterprise network, or Soft-Switch will sell you a dedicated platform to run the gateways for another $30,000. The total solution can easily run $70,000 to $80,000. For a large enterprise with a pressing need for integration, this is certainly a very good way to go. The cost can easily be spread over the many e-mail networks the solution will contact.

For an enterprise consisting of a few thousand nodes, and with a large computing budget, there are probably lots of different post-office types. You can then do one of two things. First, you can require everybody to use only one type of e-mail program. That is difficult to do, because you are forcing everyone to use a system that is different from the one they use for other purposes than just getting mail. If the VAX is used as the one e-mail system, then PC users have to log in to the VAX. They probably won't like to do that, and they probably won't need to, if you do the job of integration properly. This leads naturally to the second solution, which is to put in the mail system that matches each user's environment and connect them all via gateways. This is the preferred solution, and will be discussed in detail in a later section.

Components of integration

There are several components of e-mail that need to be considered when doing e-mail integration:

- E-mail product types (assuming there is more than one post office)
- Addressing schemes used
- E-mail format types being used
- Types of attachments, and whether attachments are present
- Type of post-office-to-post-office (e-mail transfer) protocol
- E-mail router products available for each post office type
- E-mail gateway products available for each post office environment
- Directory services that are employed in the network
- Delivery protocols that are being used

Keep each of these points in mind as we review the major e-mail products being used in the majority of commercial networks. This will give us a clue as to how easy it will be to integrate each post office type into the enterprise network.

Major e-mail systems

We now review the details of some of the major products presently available. Their strengths and weaknesses are reviewed from the point of view of how easy or how hard it is to integrate these with the enterprise network. We are not so concerned with whether the product is a good fit for business needs. We are assuming that the program is in place in the corporate environment for a good reason, so we concentrate on how to connect it to other post office types.

✻ The file sharing approach to e-mail

The file-sharing approach to e-mail is the most popular, because it is based on how a PC network works. Figure 4-47 describes this approach. Follow the diagram as we go through the discussion.

Figure 4-47

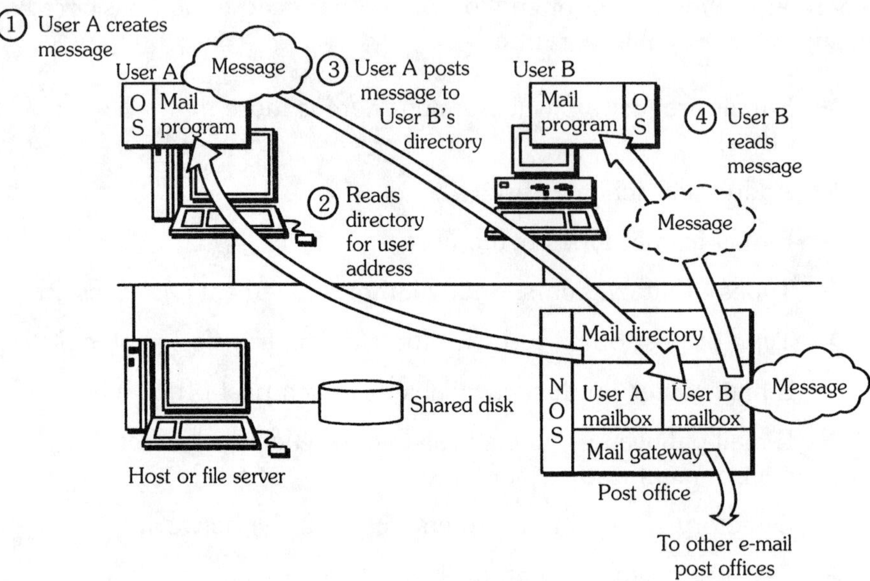

The file-sharing approach to e-mail. Note that this method has been popularized because it is based on a PC network with a file server.
Learning Tree International

File-sharing e-mail systems have a repository for messages, the post office, on the PC file server. The user runs a program on his workstation that enables him to create mail, send it to the file server, and drop it in a mailbox. Consider user A in Fig. 4-47. The first step in creating a message is to address it. User A looks up user B's address in the directory on the server. Step two is the posting of the message to user B's mailbox.

At this point user A is done. He can get off the network or go on to something else. The message will be sitting at the post office (the file server) waiting for user B to log into the network and retrieve the message.

On most PC networks a user such as B might not even have the mail program up and running in his workstation at the time user A posted the message. User B's workstation can have a small program running in the background that enables mail box monitoring. When mail comes in, a little bell goes off, meaning "You just got mail!" It's similar to the phone ringing, but better, because user B can pick up mail at any time, unlike the phone, which is so insistent. He can even finish his coffee before deciding to look at what mail came in.

✻ The Lotus solution: cc:Mail

The most popular file-sharing approach is cc:Mail. It is supported on many file server types and workstation types. It has a powerful set of APIs (Application Programming Interfaces) that allow for e-mail enabling of applications. Lotus has a sophisticated router environment, as well as many gateway types.

Since cc:Mail is the market leader, everyone tries to enable their applications to work with this program. It supports a number of API interfaces such as VIM (*Vendor-Independent Messaging*) and the Microsoft API, MAPI (*Microsoft Application Programming Interface*).

One major drawback to cc:Mail is that it is proprietary. It does not use any standard formats for messages, delivery, or routing. So anyone who writes a gateway product to connect any e-mail system to cc:Mail has to do a lot of homework and find out directly from Lotus the details of how cc:Mail works. Of course Lotus itself has produced good gateway products for cc:Mail. Soft-Switch has done a good job of creating cc:Mail gateways independently (but now they are owned by Lotus).

The reason we point this out is that it's usually better to get the gateway software from one vendor's product to the other environments directly from that vendor. Whenever faced with integrating a proprietary e-mail system, it is best to get the gateways from the folks who created the proprietary e-mail programs.

Interoperability Rating: Excellent

Are we happy with the cc:Mail program? We have to say yes, from the standpoint of interoperability. First, because it is the market leader,

and it is a good technical solution. It has routing, of course, and many other good technical features that makes it a strong enterprise-wide product. But the major reason we like it for interoperability is that it has a lot of gateways. Because it is the market leader, many other e-mail program vendors have gateways from their product to the cc:Mail world. You might decide to pick another PC network e-mail solution because it has more features than cc:Mail, but it probably has fewer gateways, and thus it will not be as easily interoperable. So stick with the market leader, especially when it has a lot of good features. The only major drawback is that it is proprietary.

✻ The Microsoft solution: Microsoft Mail

In second place behind cc:Mail is the Microsoft solution, Microsoft Mail. It's trailing behind cc:Mail just slightly. It is also a good solution for file server-based PC networks. It has lots of interfaces to lots of other e-mail environments. It supports UNIX workstations picking up mail from a Microsoft Mail post office. It also supports Windows, Windows NT, Windows for Workgroup, Macintosh, OS/2, and DOS workstations.

It uses a proprietary delivery protocol, e-mail routing protocol, and e-mail format. Microsoft has lots of gateways from Microsoft Mail to other environments. Many vendors have gateways into the Microsoft Mail world because of its market leadership.

Interoperability Rating: Excellent

Are we happy with Microsoft's Mail? Yes, for the same reasons we are happy with cc:Mail as far as interoperability is concerned. It is probably not the e-mail system to deploy corporate-wide, but if there are any Microsoft Mail post offices out there, let them stay. Just get gateways from them to the rest of the enterprise network.

✻ Enterprise-wide solutions

Our choice, and it appears to be the choice of a good many integrators also, is to use a UNIX post office as the focal point for enterprise e-mail integration. Then buy lots of gateways from that UNIX machine to all the other corporate e-mail environments for interoperability. This leads us to the second approach to e-mail: the client/server approach in general, and UNIX mail solutions in particular.

The client/server approach The client/server approach assumes that there will be several post offices in our network, and includes e-mail routing as an integral part of the system. It was geared to be an enterprise-wide solution since its inception. Figure 4-48 shows the client/server approach to e-mail. Notice that there are several post offices in the system, and that there is post-office-to-post-office e-mail transfer. Notice that the post-office server is a host computer, not just a file server. There are several programs in the host: the server program, which interacts with the user client program and delivers e-mail to the user, and the client/server programs to interact with other post offices to route mail. The users in this case also run a program on their workstations to pick up, open, read, reply to, and create mail. The HP OpenMail product and the UNIX set of programs (sendmail, POP3) are the best examples of this type of e-mail system.

The Hewlett-Packard solution: HP OpenMail The Hewlett-Packard company has taken the client/server approach in creating their e-mail system, OpenMail. It has an e-mail program to create mail on the user workstation. It also picks up mail and drops it off at the local e-mail server, the host. That e-mail server then has the

Figure 4-48

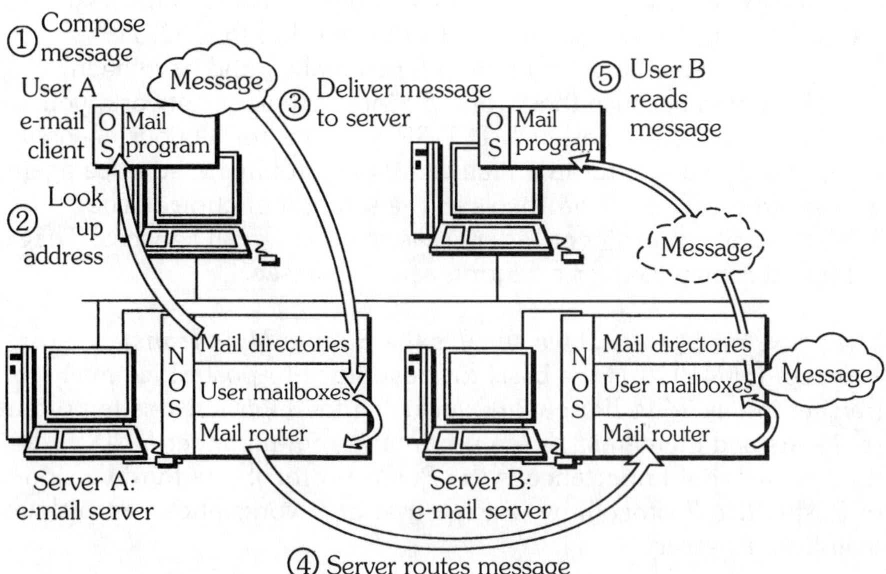

The client/server approach to e-mail delivery. Learning Tree International

responsibility of routing the message to the appropriate post office for the other user to pick up. OpenMail supports a wide variety of client workstations. The post office is meant to be set up on an HP UNIX server, but it can also run on many other types of UNIX platforms.

Interestingly enough, OpenMail uses an X.400-based message format. The HP approach is very new, and there are few gateways between the popular PC e-mail environments and the OpenMail environment. OpenMail also supports the major APIs: MAPI and VIM.

Interoperability Rating: Mixed

The client/server approach is excellent, but the OpenMail product is too new. It is acceptable if it has been already deployed in a few networks around the corporation, but it is too early to see it as a corporate e-mail backbone. It is a strong future candidate for enterprise-wide solutions.

✳ The universal solution: The UNIX approach

Looking back to other sections, you may notice that at the end of each section we suggest some TCP/IP protocol as the "universal solution." E-mail is no exception. It turns out that the TCP/IP protocols for e-mail are very rich in functionality, and have been widely adopted because they work. To integrate your network, you could do a lot worse than pick TCP/IP solutions for all your needs. It is very hard to do better in a multivendor environment, so once again, we turn to the TCP/IP protocols as the solution of choice. The TCP/IP family of protocols offers several good e-mail services. They fall into the client/server category, as we shall see.

The first e-mail protocol we study is the *Simple Mail Transfer Protocol* (SMTP). It is the basis for post-office-to-post-office mail transfer in any TCP/IP network, including the Internet. It is important to understand the details of the *sendmail* program under UNIX. The third protocol of importance is the *POP3* protocol (the third version of Post Office Protocol), by which a user at a workstation can pick up mail from a server.

✻ The UNIX and TCP/IP solution I: SMTP

SMTP is an application protocol in the TCP/IP protocol suite. The acronym means *Simple Mail Transfer Protocol*. It's not the protocol to create mail, but to move mail from one machine to another.

How does SMTP work? There are two post offices, as shown in Fig. 4-49. There is e-mail in each post office that has to go to the other. The mail was created locally on each host via a user interface. Under UNIX it would be the *maild* program. *Maild* would queue up messages for the SMTP program (probably *sendmail* if it is under UNIX) to send out to other post offices. The SMTP program embodied in a program such as *sendmail* has two components: a server portion and a client portion. It is bilateral, able to send or receive mail from another host at the same time.

If mail is sent from host A to host B, host A uses the client side and host B uses the server side. Host A will receive mail via its SMTP server component, while B will use the client side.

Figure 4-49

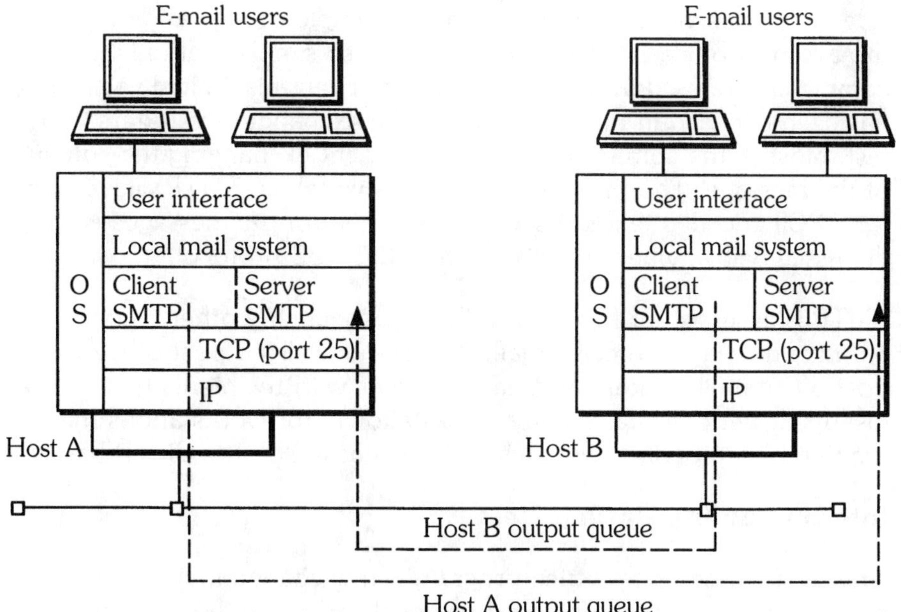

The UNIX-TCP/IP approach to e-mail delivery. Learning Tree International

When is SMTP used and by whom? It is used when a local user posts mail to go outside the local post office. And how does the local mail program know that the message is meant for another post office? By the addressee on the message. The local mail program can tell the message is meant for another machine if a fully qualified 822 address has been entered. The local mail program then puts the outgoing message into a queue for the *sendmail* program to send out. *Sendmail* periodically looks in its input queue, and sends out any mail it finds there.

The *sendmail* program (or its equivalent) on the recipient post office does not post the mail to the receiving user's mailbox directly, but passes it to the local mail program, which then posts it to the user's account.

SMTP is the protocol the two *sendmail* programs use to communicate with each other. SMTP is based on RFC 821, and has been around for a long time. (Note: the smaller the RFC number of a standard is, the farther back in antiquity the program was created.)

RFC 821 uses TCP/IP for reliable transfer. It has two parts: the client side and the server side. It defines a very simple conversation between the two post offices to exchange mail. It is so simple that only a few commands are used in the mail exchange. Even this limited vocabulary has been encoded in numbers, so that the exchange is very terse and fast. Most of the communication time is spent exchanging the content of the message. The protocol overhead is very low. SMTP allows only for ASCII encoded text messages. The format of the messages is discussed below, where we discuss the RFC 822 protocol.

SMTP is a persistent protocol in that it expects the other machine to be on and ready to receive mail. It works as well for post-office-to-post-office mail exchange. It has problems with recipients that disappear periodically, such as workstations. For workstations there is a different protocol, *POP3*, which is also based on the SMTP basics.

Interoperability Rating: Best

Any SMTP-based post office program can exchange mail with any other. This is the most universal protocol for e-mail delivery, no matter what operating system or host hardware is being used.

✻ E-mail formats

The 822 protocol The formatting of SMTP mail messages is defined by the protocol found in RFC 822. Notice this is the next protocol after SMTP (RFC 821), indicating they were defined together. When speaking of SMTP mail, most people talk about 822 formatted mail. They mean mail formatted under the specification of this RFC. So it's common to hear the expression, "Hey, that's 822 mail." What does that mean? If you're interested in doing a lot of work with e-mail, it would be worthwhile looking at the text of this protocol, since it tells you its limitations, and what is possible and not possible to send under this mail format.

Mail formatted as 822 is meant mostly for seven-bit ASCII text messages. All that is possible is seven-bit ASCII text. The RFC itself is just a guideline for programmers who wish to know how to create systems that produce properly formatted e-mail. 822-formatted e-mail messages have two components: a header and a body. The header has fields such as *recipient* and *sender*, followed (after a colon) by the value of that field. At the end of the header there is a carriage return/linefeed.

By the way, there's no specific order to the fields, but it sure is nice when the fields are in the order in which you send the message. Figure 4-50 shows a message with a simple header. As the message goes from one post office to another through e-mail routers, the header grows, gathering information as to the route it took to traverse the network. That's why the headers of e-mail messages obtained through the Internet sometimes look complex. The RFC specifies that the mail message contains within it the return paths so that the people can easily reply. (See the optional section in Fig. 4-50.) There are two required carriage returns and line feeds between the header and the message body. At the end of the message, five carriage returns and line feeds are added to signal message end.

At the very end, you get a period, which you normally don't put in there but the program inserts for you. It uses the period to tell the other machine that this is the end of the message.

There are some things 822 mail cannot communicate: attachments, multimedia, voice mail, images, and binary files. It sends nothing

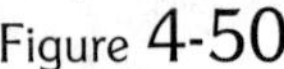

Figure 4-50

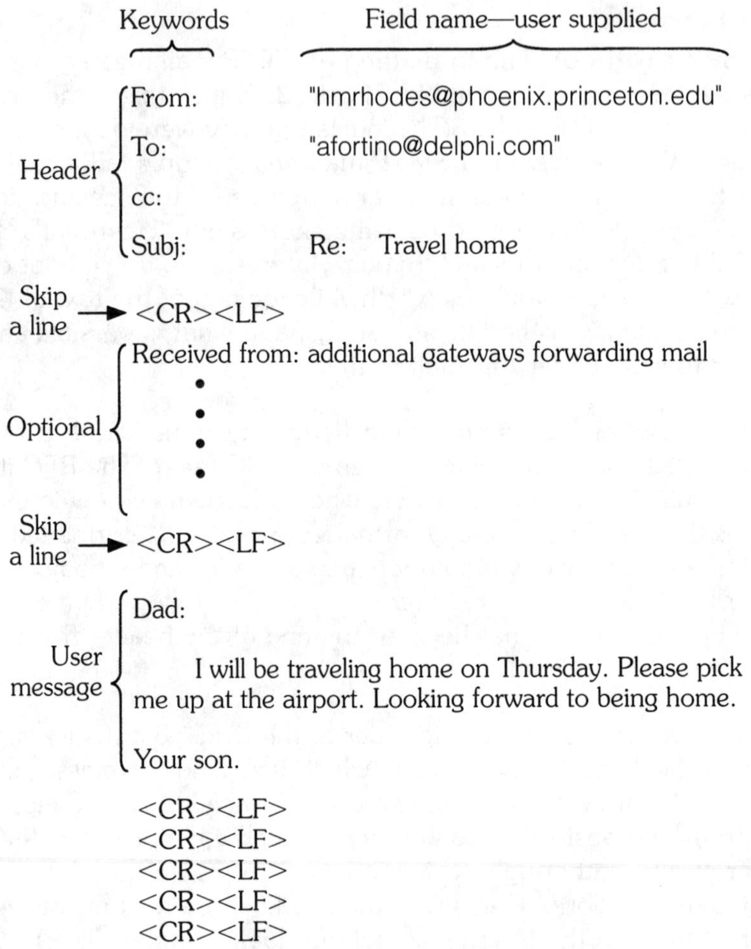

An example of an 822-encoded e-mail message. Learning Tree International

other than text. It's meant for simple mail text messages. Binary files must be converted to seven-bit ASCII text strings to send them through some networks that only accept 822-encoded mail. Programs such as BINHEX, for example, will do the conversion. When the message is received, it must be decoded by the decoding version of BINHEX to recover the original binary file.

Interoperability Rating: Mixed

This is certainly a universal protocol. It is publicly defined, has been around for a while, and works very well. Some of the largest networks in the world transmit 822 encoded messages (the Internet, for example). But the format is very simple, which means it has limited application in a communication world grown very complex.

The MIME protocol At first sight, we might be appalled by the simplicity of both SMTP and 822 messages, given our current needs for multimedia. Remember, both protocols came out of the original Internet environment, where bandwidth was at a premium and communication was simple. The Internet was put together to connect large hosts at U.S. government research facilities. The main requirements were file transfers (FTP), an occasional login to a remote machine (TELNET), and some simple messaging (e-mail, SMTP). In today's multimedia world, people want more, so the protocol has been extended. That's what gave rise to the MIME protocol (Multimedia Extension).

Mime is defined by RFC 1341. The number 1341 means it is a more recent addition to the RFC family. Until recently people were debating whether to go to X.400-based mail. X.400 is more robust. It has many other format types specified besides text. When X.400 was specified, it was a very large specification. It was meant to do everything right from the beginning. It was so complex, few vendors used it to produce products, and it just sat on the shelf as a specification.

When multimedia requirements became important, rather than produce products based on X.400, Internet engineers decided to extend SMTP rather than go to X.400 to do everything. So they came up with an extension: MIME. It extends 822 to support attachments, multipart mail, and multimedia. It can even specify to the recipient what editor to use to read the mail. It supports word-processed documents with embedded format codes, and it tells the recipient what kind of printer to use—whether a postscript printer is needed, for example. It can even specify the minimum type of terminal needed to display the message.

When looking at a post office type nowadays, the first thing to notice is that if it's SMTP-based, it will support 822-formatted mail at a minimum. The questions to ask the vendor or provider are "Do you also support MIME?" and "Which version of MIME?" In other words, can they support RFC 1341's definition of MIME, and are all of those features available?

Interoperability Rating: Mixed

MIME solves the multimedia problem of 822, and it is a *de facto* standard, but it is just too new. It needs wider adoption before becoming an important enterprise-wide solution. For that reason, we cannot give it the highest interoperability rating.

✻ The UNIX-TCP/IP solution II: The *sendmail* program

The *sendmail* program is the embodiment of the SMTP Protocol on a UNIX host. It is a daemon, which means it runs in the background. It receives mail from the local mail program, the mail daemon, makes the connection to the remote SMTP host, and sends the mail. The other SMTP host can also be an e-mail gateway. Its SMTP side receives the message and converts it, and then sends it to the Lotus cc:Mail post office from its other side.

Sendmail is managed through the creation of script files and mailing lists. Administration of a *sendmail* e-mail environment is not necessarily fun. It's a lot of hard work. For some people, e-mail administration at the corporate level is a full-time job. As with anything in UNIX and anything in TCP/IP, it requires some expert help.

Interoperability Rating: Excellent

Sendmail is very popular because, again, it comes with most UNIX systems. *Sendmail* is there, just like the rest of TCP/IP, so people use it. To configure it, to put together the lists, and to administer it is a difficult and laborious task. Many things can go wrong with it, and often do—and it takes constant maintenance. It is based on SMTP, which makes it compatible with many other programs on the corporate internetwork.

✻ The UNIX-TCP/IP solution III: The POP3 protocol

The POP3 protocol enables a user on a workstation to compose and read mail off-line. When mail is sent to the user, it arrives at a host that maintains a post office box for that user. The host in this case is called a maildrop. The user workstation does not have to be connected to the host; it doesn't even have to be on when the message arrives. This type of operation is called a maildrop system. It

is very common in a UNIX workstation environment, for example, where there are many Sun workstations. It is also common in SLIP Internet connections from workstations to Internet routers.

Figure 4-51 illustrates how the POP3 clients and servers work. When mail arrives at the post office (perhaps via SMTP and *sendmail*), the mail is held there for users. The post office host becomes a maildrop, and the user picks it up dynamically, on demand. This is similar to what happens when someone doesn't have a mailbox at home, and has to go to the post office to pick up mail.

Figure 4-51

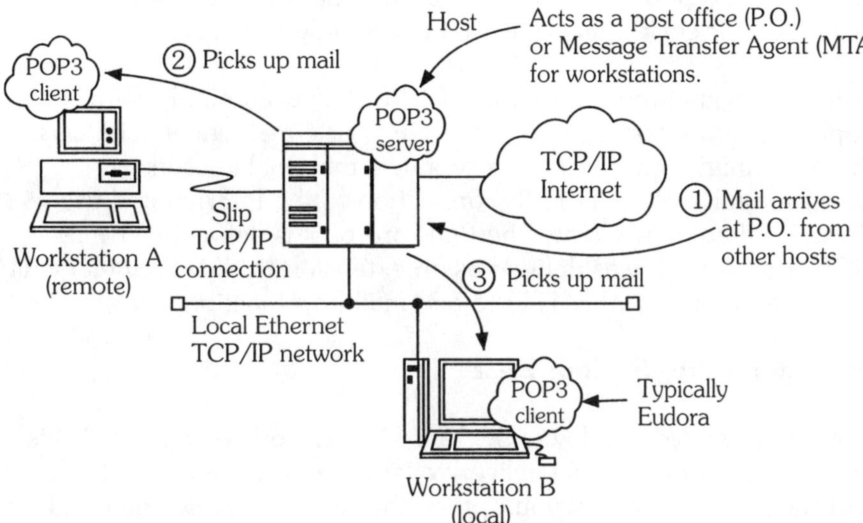

Client and server in the Post Office Protocol 3 maildrop environment. POP3 provides workstations with dynamic, on-demand access to a maildrop post office. Learning Tree International

Why would you want to use your workstations this way? For one thing, there may not be enough resources on the workstation to set up a full SMTP client/server program, such as *sendmail*. It might also not be economical for small network nodes such as workstations to stay connected to the network continuously, just in order to receive mail. Workstations dialing in over public networks to an Internet router are examples of the latter category.

It is interesting that many post offices run UNIX as the operating system. The post offices already use some form of SMTP mail to send mail to another post office. Therefore, it is natural to use SMTP to upload and download mail to workstations when they demand it. POP3 is based on SMTP mail transfers.

The post office in this case is technically considered a message transfer agent (MTA). When the user is ready to send and receive mail, he uses a program at the workstation that embodies the POP3 protocol; this is a POP3 client. The host side of the POP3 equation is considered the POP3 server. The POP3 server uses a known, fixed TCP port, port 110. The TCP port at the workstation is dynamically assigned by the user's TCP/IP environment. It's not necessarily port 110.

There are many products that use POP3. For example, Eudora is a graphics-based interface that uses POP3 for e-mail under Windows. There are many other Windows products that enable users to establish a full dial-up TCP/IP connection to the Internet and use POP3 for the e-mail client. The POP protocol is defined by RFCs 1125 and 1082. It was defined as an extension to SMTP by Marshall Rose (the originator of SMTP) to add maildrop functionality.

Interoperability Rating: Best

Whether you have Windows, OS/2, or UNIX workstations, POP3 is the best method for e-mail delivery over an enterprise network. This approach is very useful if you opt for the reductionist technique of having only one e-mail server on your network. Any server supporting the POP3 server protocol is an acceptable maildrop.

The ISO/CCITT (ITU) solutions: X.400

Another important standardized e-mail environment with de jure protocol definitions is based on the work of ISO and the CCITT (renamed ITU). ISO MHS (Mail Handling System) is an international standard specification based on the CCITT/ITU X.400 and X.500 protocols. MHS is the adaptation by ISO of the X.400 protocols suite. Although components of ISO MHS and X.400 have been

adopted by governments under the GOSIP programs, it has yet to gain widespread acceptance outside government circles.

Some people in industry are very concerned with X.400, since it still has a lot of technical problems. There are a number of good implementations of X.400. Fortunately, there are many gateways for X.400-based mail, so it's easy to move mail in and out of this environment. Some public e-mail systems, such as MCI Mail, are also based on it.

✻ The MHS architecture

The X.400 architecture is shown in Fig. 4-52. X.400 protocols provide for a host-to-host mail transfer. The hosts in the figure can be running different operating systems. The strength of the ISO protocol suite is the ability to network a variety of different host and operating system types.

Figure 4-52

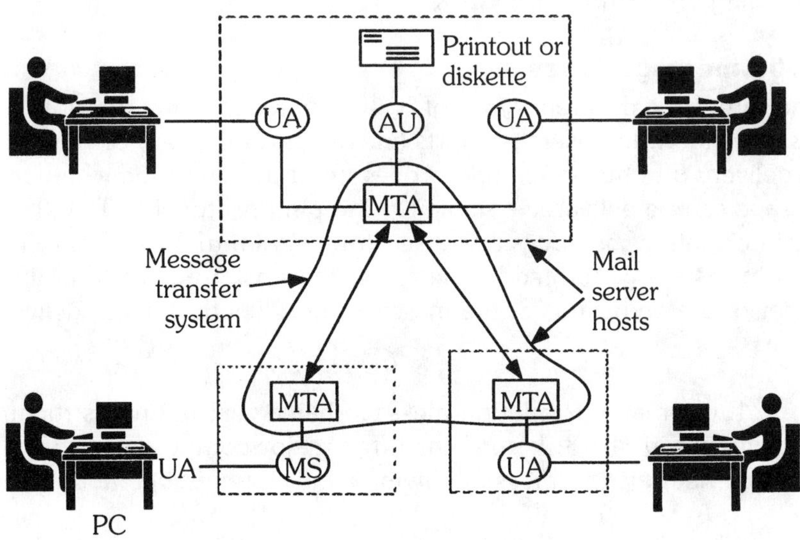

AU = Access Unit
MS = Message Store
UA = User Agent

The ISO MHS (Message Handling System Architecture) based on X.400 protocols. Learning Tree International

✻ X.400 protocol components

The X.400 protocol defines many different components. In reality it is a suite of protocols. One component is designated the *MTA*, or message-transfer agent. It is the post-office-to-post-office mail transfer protocol, similar to *sendmail* under UNIX. Another is the *User Agent* (UA), the program the user employs to create and read mail.

Additionally, the X.400 protocol suite defines a message store or a repository for messages. The message store concept and functionality was added in the 1988 version of the X.400 protocol suite. It acts similarly to the POP3 server and client environment in UNIX.

When the personal computer or user workstation is turned off, there must be a repository for mail on the network to receive incoming mail for a user. Typically, a single message store is associated with a single user agent. A message store is an optional component of X.400. Only the MTA and the UA portions and the MS portions of electronic messaging are standardized in X.400.

✻ X.400 message types

There are three different types of X.400 messages: *information messages* (between users), reports (indication of delivery or nondelivery of a message), and probes (used to determine whether a message can be delivered—similar to the ping protocol in TCP/IP). The information messages are called *InterPersonal Messages* (IPMs). The reports are generated by the message-transfer agents to notify the sender of the end status of the message, i.e., whether it was delivered or not.

The X.400 e-mail information message has a very definite structure, as shown in Fig. 4-53. It contains three major components: the *envelope header*, the *message header* and the *message body*.

The envelope header is used by the message-transfer agent to deliver the message to the end-point post office. The header has a fully qualified network address that must be used for transport of the message from end to end. The contents appear after the envelope header. The message header is created by the user agent to identify message information from the user. The third part is the body of the

Figure 4-53

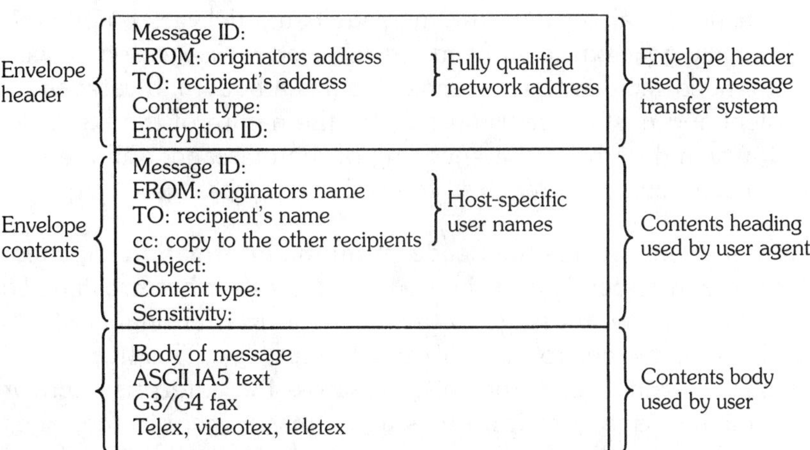

The X.400 e-mail information message. Learning Tree International

message, which can be encoded in a variety of forms. It might be ASCII, a fax, video text, or teletext. The ability to be very flexible in the type of message that it can carry, as well as the very extensive address space, is what gives X.400 such promise.

Figure 4-54 shows the message creation process, and how the various pieces of the information message are added during message

Figure 4-54

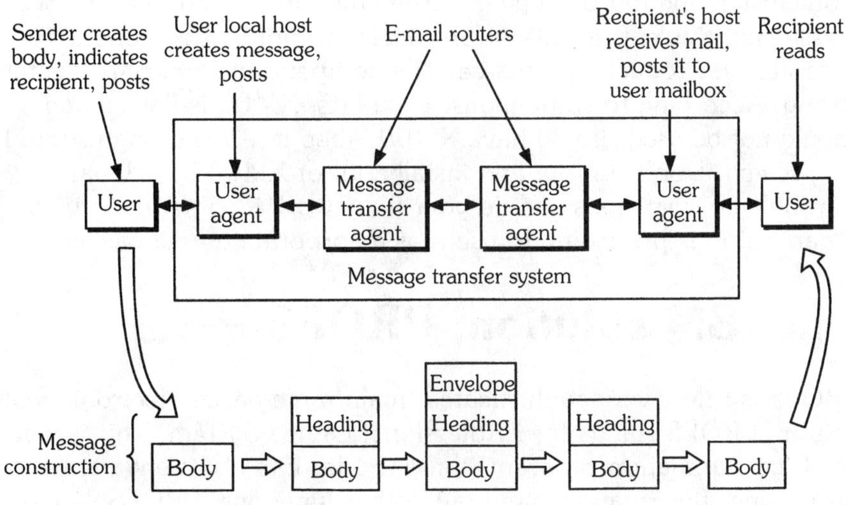

X.400 message construction. Learning Tree International

construction. It also shows the message being delivered, and how the body finally gets extracted from the received message and delivered to the intended recipient. The sender creates a body, indicates the recipient and post office it must go to, the nature of the message, the sensitivity and perhaps the encoding of the message, and the type of station characteristics required to display the message.

The user agent creates the header from the information supplied by the user, and appends it to the body that the user has created. The UA then posts the message to the message-transfer agent, which adds the envelope header to the information message. This header is required for end-to-end transmission across the enterprise network. Then the message goes from message-transfer agent to message-transfer agent until it finally arrives at the last message-transfer agent, which delivers it to the recipient's post office.

The post office is not necessarily the host the user is attached to; it may be just a mail store. In either case, the user agent will strip the envelope from the receive message and post it to the user mailbox. When the user is ready, he will open and read the contents of the message with user agent software.

Interoperability Rating: Mixed

Notwithstanding the push by governments for the adoption of ISO messaging, this entire suite of protocols has not gained general acceptance. Products are scarce. The competition from SMTP is too strong. According to some industry experts, X.400 is flawed and should not be used. If you have X.400, keep it if it is doing its job. If not, switch. Don't run out and install a lot of X.400-based mail. Certainly be suspicious and do your homework if someone in your organization is proposing to use it as a corporate e-mail backbone.

The IBM solution: PROFS

PROFS is IBM's very sophisticated, mainframe-based electronic mail system. PROFS stands for PRofessional OFfice System. This is a suite of office automation programs that includes e-mail, calendars, word processing, time management, and other functions. PROFS supports full-screen 3270 terminals, as well as 3101 line-by-line terminals.

PROFS has three message types. The first message type is a simple three-line-long message that is delivered real time, something that is called messaging in the PC world. The second is full-screen edited notes, to be delivered to an addressee once the message is created. This is the classic electronic mail. The third consists of documents created with a word processor with a redefined address, akin to the attached-file type of message delivery.

PROFS is very popular in the IBM-mainframe environment, so there are many third-party gateways between PROFS and many other electronic mail systems.

This solution is proprietary to IBM. Therefore, unless you are a pure IBM shop, it is not the best system to use as a base for enterprise-wide e-mail.

Interoperability Rating: Poor

This solution is proprietary and useful only for IBM environments.

The DEC solution: All-in-One

Digital Equipment Corporation has created a VAX-based solution called All-in-One. All-in-One is really an office automation solution (similar to PROFS). E-mail is only one of its modules. It contains word processing, database access, decision support, calendars, and many other modules.

Access to All-in-One on a VAX is supported over many different internetworking protocols. For example, it supports DECnet, AppleTalk, NetBIOS, TCP/IP and LAT. All-in-One is also supported on MS Net, LAN Manager, and many other network operating systems. The types of workstations that can access All-in-One on the VAX are DOS, OS/2, Macintosh, and VMS clients.

Interoperability Rating: Poor

This solution is proprietary and useful only for VAX environments.

Directory services e-mail addressing

To deliver e-mail, one must know the address of the recipient. Often we have the name of the user on the remote post office and the name of the host. These are put together in a conventional e-mail address. Message-transfer agents, post offices that need to forward mail, must translate this general user-friendly address into a very specific machine address to get across the network—in other words, they must use a fully qualified network address.

The problem here is that to be able to deliver mail to a user on a foreign mail system, the message-transfer agent or the post office must know how to get to the host indicated in the "To" field of the message. Then it must deliver the message to the appropriate e-mail router or gateway, or deliver it directly to the remote message-transfer agent. In general, there must be a conversion from the human-friendly username and host name e-mail address to a network compatible and understandable machine-friendly address.

✻ Directory services

Several services are available to e-mail systems to solve this problem. One of these is the Domain Name Services for TCP/IP conversion of name to network address. There is also the Network Information Services (NIS), formerly the Yellow Pages, stemming from the work of Sun Microsystems. The latter is still widely used on Sun networks.

There is a specific e-mail address format for 822-encoded e-mail. Additionally, we have the very popular and powerful X.500 directory services.

✻ TCP/IP names and addresses

Every TCP/IP network interface is identified by a 32-bit IP machine address. For example: 128.66.12.2 is a Class B TCP/IP address. Names can be assigned to any device with an IP address. Clearly names are easier for humans to remember than a four-part number, so they are more popular with people.

Name addresses and machine addresses may be used interchangeably. For example, if you were to invoke the TELNET program and type after it the number of the host, the IP address of the host you're trying to reach is TELNET 128.66.12.2. That has the same effect as if you had typed TELNET indigo.turbo.com (which is the host name corresponding to this IP address). A mapping mechanism is required for there to be true transparency of names and addresses for a user. This is where directory services such as Domain Name Services, history tables, and network information services come in.

✳ The DNS domain hierarchy

In a TCP/IP internetwork, the entire host namespace follows a convention. It is called the Domain Name hierarchy or Domain Name Service. At the very base (or top, depending on your point of view) of this hierarchy there is an unnamed root domain. You can see this structure in Fig. 4-55. Below the unnamed root domain (or just root domain), there are several top-level domains. Since the Internet and

Figure 4-55

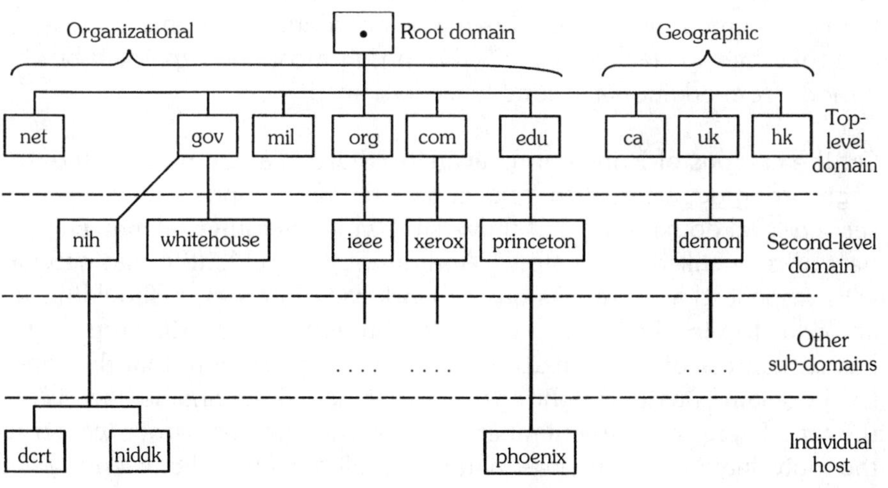

The DNS domain hierarchy. Learning Tree International

TCP/IP internetworks became popular and were developed within the United States, the top-level domain of U.S. networks identify the organization type.

For example, a "com" domain identifies hosts on a commercial network. An "edu" domain identifies hosts on educational networks, such as colleges and universities. A "mil" top-level domain identifies military networks within the United States. And "gov" top-level domains identify U.S. government agency networks.

Continuing with the top-level domains, we find that two-letter codes have been assigned to networks for countries outside the U.S. The top-level domain for Canadian networks is "ca," for example; for Hong Kong networks, it's "hk," and so on.

The fun begins below the top-level domain, where we have second-level domains plus other subdomains. At this point, the structure begins to get complex. In some cases, host names end right there at the second-level domain. For example, the computer at the White House of the United States has an Internet address of "whitehouse.gov." (Note that the root domain, not having a name, is not identified in this address, and that the top-level domain is written rightmost and separated by periods from additional second-level domains.)

On the campus of Princeton University, there is a computer in the engineering department that has a host name of "phoenix." The second-level domain for that university, which identifies all university networks, is called "princeton." And of course, to identify that network with respect to all networks on the worldwide Internet, it should have an "edu" top-level domain. Putting it all together, the Princeton host has an address of "phoenix.princeton.edu." Again, note that the top-level domain is written rightmost, and the actual host name itself is at the left. There may be a number of other subdomains before we get to the individual hosts. The host name can fall anywhere between the second-level domain and other subdomains below it.

Within an organization that is not connected to the worldwide Internet, network administrators are free to assign any combination of names to their enterprise hosts. They may or may not use the DNS hierarchy. But if the host is to be connected to the worldwide

Internet, its name must not conflict with any other name that has already been assigned. Therefore, names must be registered with an Internet authority having jurisdiction. In this case, it is the *Network Information Center* (the NIC).

Registering the name causes the name of this host and its associated IP address to propagate to all directory servers at appropriate places throughout the Internet, ensuring that mail directed to a user for this host will traverse the Internet correctly and arrive at its intended destination.

Figure 4-55 shows the DNS Domain Hierarchy. Readers familiar with the Internet will probably recognize this structure as the Domain Name structure of the Internet hosts and the Internet itself.

✳ 822 addresses

E-mail requires a fully qualified network address for delivery. The RFC822 protocol describes in detail how mail must be addressed for delivery by SMTP. The resulting address is commonly known as an 822 address. The address consists of a mailbox (typically a username on the target host) and the host DNS domain name. For example, "HMR@phoenix.princeton.edu" would be a student's mailbox at the Phoenix host in the Princeton University network of computers connected to the Internet via the "edu" domain.

As mail goes through gateways, the address information grows and becomes complex, containing the addresses of relay hosts along the way. Users are typically not concerned with this expanded address, which is only used for delivery. The SMTP host will do the conversion from the 822 *username@host-name-address* to an IP 4-byte machine address.

If the SMTP post office is not able to do the translation via local tables, it will do a look-up at a remote host table. This remote host has been set up to receive directory queries, and it is typically called a *directory server*.

✳ X.500 directory services

Under the CCITT/ITU protocol specification, there is a directory services specification called X.500. Directory services can provide

information that is needed for networking. This includes the name-to-address mapping needed for e-mail delivery. However, an X.500 directory service host may also provide authentication. The name-to-address mapping may use the name to a network-level address. Or it may use the name to an X.400 electronic-mail address.

The directory services of X.500 that are used with X.400 have two types of agents, a user agent and a server agent. A user agent is the client in the directory query. The server agent then becomes the responder to the query, and does a look-up on a more extensive table located in its database.

The entire directory information of the network is typically distributed across a number of server agents.

✻ X.400 message addressing

X.400 e-mail is delivered in what is termed *originator/recipient* (O/R) *pairs*. As we saw in Fig. 4-53, each message in the envelope contains a fully qualified address of the originator and recipient. The addresses are of the following form:

Country (e.g., gb for the United Kingdom)

Management Domain (e.g., gold400)

Organization (e.g., nottinghamuniversity)

Organizational Unit (e.g., english)

As with DNS, not all components are used in creating host addresses. The X.400 address specification is more extensible than the IP address scheme. The namespace above translates to a possible 13-octet address, compared to a DNS name translating to a four-octet IP address.

The public e-mail solution

Figure 4-56 shows a public e-mail network using the X.400 specification. Public e-mail, or a public e-mail network, can be very useful to an organization that serves as a hub connecting various

Figure 4-56

- X.400 interoperability
  — Exchanging messages between different systems

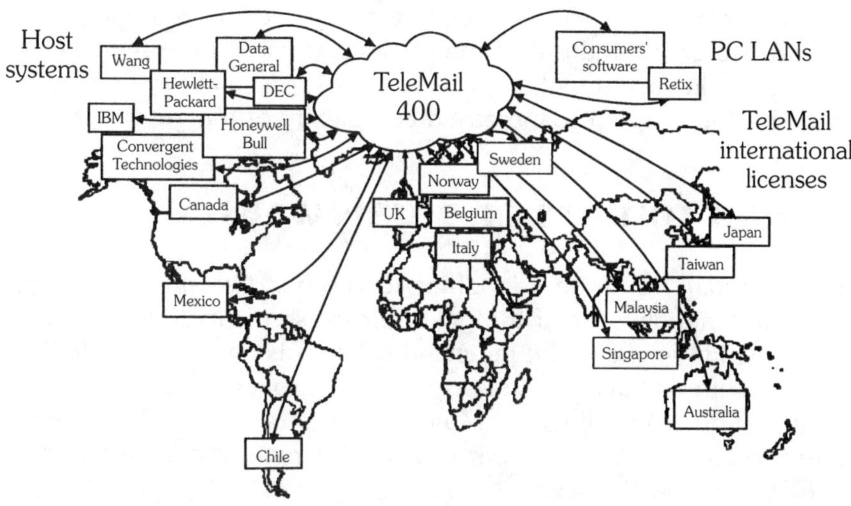

- TeleMail 400 allows the TELENET public network to serve as an e-mail hub for private networks running different systems

Using X.400 with public networked e-mail systems. Learning Tree International

geographically dispersed e-mail centers, without necessarily putting in an internetworking infrastructure to connect those centers. Services such as MCI Mail and Sprint's TeleMail are examples of public e-mail services. CompuServe also provides worldwide e-mail interconnectivity.

Figure 4-56 shows TeleMail 400, the Sprint product used for this purpose. As you can see, there are public mail e-mail hubs throughout the world that can be used to connect local enterprise post offices to the remote ones. Most of these networks are based on X.400. This can be a powerful solution for a medium-size organization that must connect all of its e-mail systems, but that lacks the resources to put in a worldwide infrastructure.

Security is one problem with public e-mail systems. If you have good solid encryption, this may not be as much of a concern to you. You should be aware that encryption is required to make the use of public facilities secure.

Interoperability Rating: Good

We consider public e-mail to be a very useful solution as a global e-mail infrastructure for medium-size companies. Gateways are required to get in and out of the public system, but the reduction in cost to put in such a global infrastructure for a small or medium organization offsets the increased complexity of the gateways.

E-mail routers and gateways

To move mail from one post office to another, two situations require two different solutions. If the post offices use compatible protocols and message formats, the indicated solution is an e-mail router. For example, if you have cc:Mail post offices throughout your organization, you will use cc:Mail routers to move mail from one post office to another. SMTP programs act as a mail router for UNIX SMTP-based post offices.

The thing to note is that router solutions are specific to the vendor whose product post office is being used. Lotus cc:Mail routers are bought from Lotus, Microsoft Mail routers are bought from Microsoft—and the two do not interoperate.

If the protocols of the two post offices that need to be connected use different e-mail and message formats, then the use of an e-mail gateway is indicated. For example, Novell has very good MHS gateway products. Lotus has very good gateways from cc:Mail to a number of other environments, and so does Microsoft for their Mail product.

✻ Routers

Using e-mail routers, we can create a very large enterprise e-mail system. But as seen above, routers can only be used for relatively homogeneous environments. For example, you must have all PC LANs or all UNIX networks or all DEC/VAX networks.

Figure 4-57 shows several cc:Mail post offices connected by a cc:Mail router. First, it's important to note that the two separate post office networks must be connected by some form of internetworking. In the

Figure 4-57

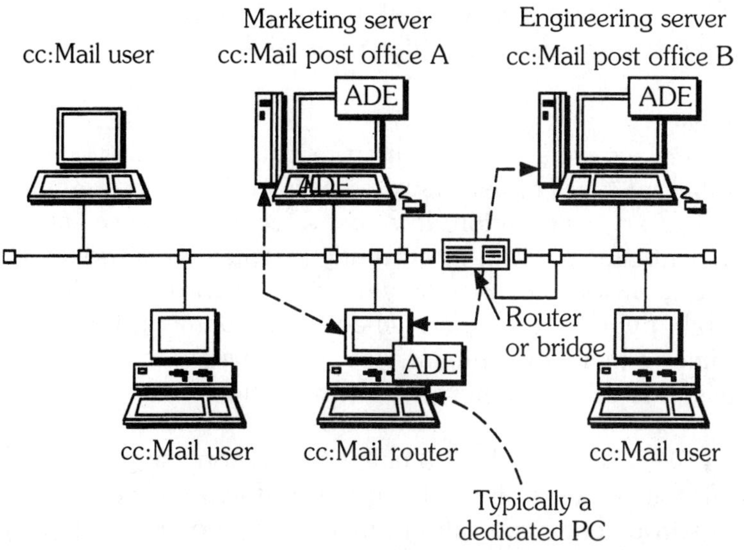

PC LAN e-mail routers. An example using a cc:Mail router.
Learning Tree International

case of Fig. 4-57, we're using either a router or a bridge to connect the two Ethernet LANs as shown. That router or bridge is not sufficient to move the mail, of course. There must be another device that will move the mail from one post office to another.

In the case of cc:Mail, the two post offices do not directly talk to each other. Why is that? It's because the post offices are LAN-based, and the file servers are only the repository of messages. There is no e-mail program running at the post offices, so there is no direct way to exchange mail between the post offices.

What is required in PC LANs is to add an additional device, a workstation or a dedicated PC, running a special program called a *router*. That router program acts as an unattended user that logs into one post office—let's say post office A on the left—looks for mail that is intended for post office B, then logs into post office B and literally transfers the mail from A and posts it to B. It also moves mail from B to A.

One additional feature available with cc:Mail and a number of other products is what's called *automatic directory exchange*. Not only

does the cc:Mail router move mail, but it can also post updates to directory lists from one post office to another.

⁂ Gateways

E-mail gateways are a costly solution, because typically you need many different types of gateways to connect the various e-mail resources of an organization. It is cost effective, because each user environment can stay with its currently installed e-mail solution while the gateway provides connectivity from one environment to another. In general, if you have a heterogeneous environment, i.e., a multivendor e-mail environment, you will gravitate toward e-mail gateways. It is the best enterprise-wide solution.

Figure 4-58 is an example of a gateway solution using an SMTP mail post office as a focal point of integration. If you have a variety of e-mail environments, you can put many one-to-one e-mail gateway solutions in place. On the other hand, if you have a great variety of post-office types, you will wind up buying many gateways under this strategy.

A solution that makes more sense is to buy a gateway product that goes from each local environment to one common environment, and let that common environment be the translator to other environments. Such a scheme is shown in Fig. 4-58. There we are basing the focal point of exchange on an SMTP-based post office.

Somewhere on the network there must be gateways from SMTP to each one of those LANs, but those gateways serve a dual purpose. They bring mail in and they move mail out. Since most large enterprises today have a fair number of UNIX machines, and are likely to have a goodly number of UNIX post offices, an SMTP-based solution is appropriate.

You must determine for yourself whether this solution fits your needs. It will depend on your environment. If you have a large number of VAXes with some PC networks, perhaps All-in-One and a gateway on a VAX might be a good focal point of integration for you. If you have a large IBM-mainframe environment, we wish you well.

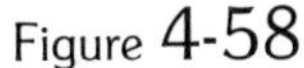

Figure 4-58

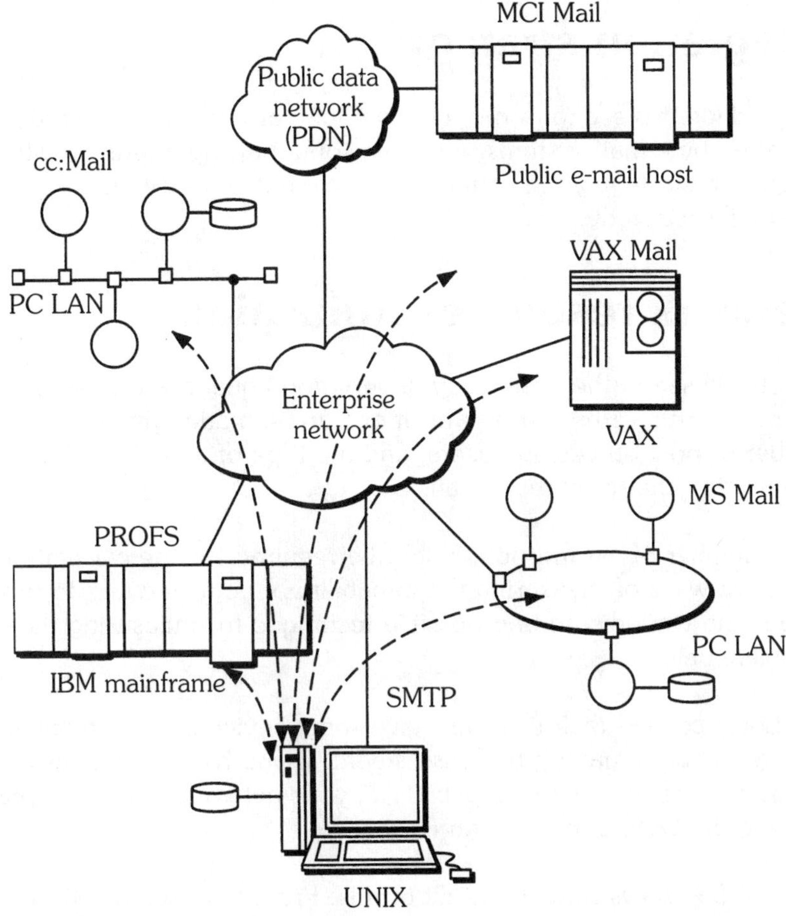

SMTP-based e-mail gateway or "e-mail server"

An example gateway solution using an SMTP mail post office as the focal point of integration. Learning Tree International

If you just have many PC networks with two or three different PC post-office types, you might not need a focal point of integration. Several gateways to move mail from one environment to the other will do the job. So you have to look at our proposal and fit it to your situation as best you can.

Integration strategy

To conclude this section on e-mail, we look at network resource utilization by e-mail systems, and give some parting words on what integration strategy you should follow. We follow this with a look at an e-mail case study.

Network resources utilization

As you will see in the case study, a very good quantitative assessment of e-mail traffic within an organization can be made, given the number of post offices and users, and the type of activity within the organization that requires e-mail.

A fair number of assumptions might be required for the calculations. There are ways of measuring the capability of your network to move e-mail, and we indicate one possible technique for measuring that capability below.

The Lotus cc:Mail technical staff give some advice in their manual on how to go about making this assessment. If you have a number of PC LANs, you can assess the capability of your network to move e-mail by using the DOS copy program.

Figure 4-59 shows how to go about this. From your workstation, using the DOS copy command, copy a large (1-megabyte) file from your workstation to your local post office file server. This will simulate sending an e-mail message with a large attached file, which is perhaps a worst-case scenario. Do some form of timing, and, if you have a protocol analyzer, put it on the segment to assess the loading impact of this transfer on your network.

Now do the same file transfer between your workstation and the file server that is running the remote post office. Measure how long this second transfer takes. These two measurements yield a worst-case estimate for e-mail transfers between remote post offices, and between user workstations and their local post offices.

Figure 4-59

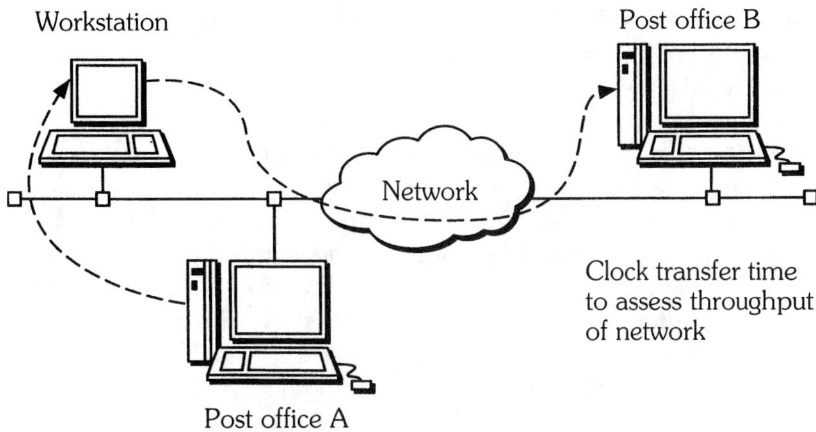

Example network for assessment of network throughput capability for e-mail delivery. Learning Tree International

UNIX networks can use FTP or the NFS *cp* command to perform a similar emulation. Generally, though, e-mail does not place a heavy burden on network resources. It tends to be bursty, and due to small file sizes, does not load down the network as NOS file services do. For large enterprise e-mail systems, there may be sizable steady e-mail traffic on the backbone. This might typically account for up to 30% of total network capacity, so although the e-mail messages are small and the file transfers that occur are relatively light, the volume of e-mail traffic might easily load down the network.

Integration strategy

By now, you know what our advice will be for a heterogeneous environment. Take the reductionist approach. It is desirable to minimize the various types of e-mail systems being used. It's not as crucial an issue as it might be with other interoperability situations tackled in this chapter, because there are so many products that are available to integrate e-mail systems. But in general, the reductionist approach makes good business sense.

By and large, your strategy should be to allow each user community to use whatever local e-mail is convenient. Some components might already be in place. Otherwise, if you are buying new, buy an e-mail

system that shares the characteristics of the local environment: cc:Mail or MS Mail for PC networks, SMTP mail for UNIX users, All-in-One for VAX users, etc.

Follow the principle of minimizing the variety of products used. Obtain solutions that are based on generic protocols such as SMTP or X.400. (We tend to favor SMTP as a more universal solution.) Provide routers within one product family to connect homogeneous post office types. When that is not possible, move to gateway products to connect the heterogeneous e-mail networks. When choosing gateways, make sure you base your decision as much as possible on standard and universal protocols such as SMTP.

One last issue that you must provide for is some form of compatible directory services on your enterprise backbone. X.500 is quickly becoming the favorite to provide this service, although in most networks, DNS still does yeoman service. You might in some cases have a heterogeneous set of directory servers.

Network e-mail case study

To exemplify all of the principles just discussed concerning integration of e-mail systems, let's now tackle the analysis and design of an e-mail integration project for a medium-size company.

Background

A medium-size organization ($50 million to $100 million dollars per year in gross sales) might have in place an e-mail network similar to the one shown in Fig. 4-60. Notice the wide dispersion of e-mail resources. We are looking at many geographic locations—six cities throughout the world. And we are looking at a number of e-mail systems that need to be integrated.

The requirement here is that any user within the organization must be able to exchange e-mail with colleagues anywhere on the enterprise-wide network. They must also be able to exchange e-mail with their colleagues in other organizations via the Internet connection.

Before we begin the analysis and put resources in place for a solution, we assume that these post offices start out isolated. The

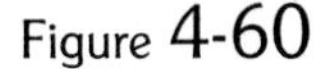
Figure 4-60

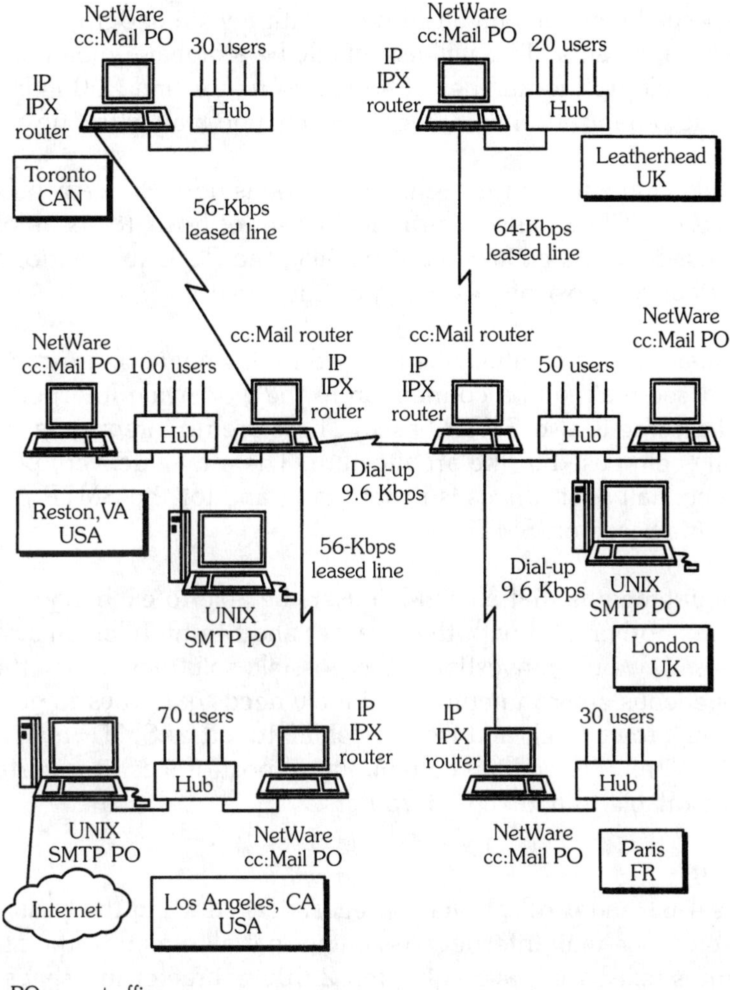

Example e-mail network. Learning Tree International

post-office types are cc:Mail in the PC networks, SMTP mail in some of the UNIX networks, and one Internet connection currently available at one location.

✻ Analysis

As an analysis exercise, let's review the components and possible integration issues for this system. The first thing to note is that there

are six e-mail locations involved here, with a wide geographic distribution of users. This will essentially be a global e-mail network. When we add up the number of e-mail users, we find 200 individuals within the organization who need to exchange mail with one another.

We would characterize this e-mail network as both PC LAN-based and UNIX-based. There is no central host or set of hosts to speak of; it is not a host-centered environment. Finally, the PC LAN solutions that are in place are post offices based on file servers.

The e-mail products initially being used in this network are cc:Mail, SMTP-based mail, and a connection to the worldwide Internet, which we will assume is also SMTP-based. There are no integration tools currently being used in the architecture. The e-mail delivery protocol for the cc:Mail post offices is proprietary, and for the SMTP post offices, it's based on SMTP.

The requirement is that all cc:Mail users be able to exchange e-mail with one another, and with the Internet and Sprint TeleMail system. To this end, we now investigate the possible solutions. Since this is a heterogeneous environment, it definitely needs gateways to go from SMTP mail to cc:Mail, and from TeleMail to either SMTP mail or cc:Mail. Within the cc:Mail community of post offices, it is certainly very reasonable to add cc:Mail routers.

✻ Solution

Figures 4-61 and 4-62 provide spreadsheets showing the analysis of traffic on this e-mail Internet. Assuming that all e-mail traffic at the moment is text-based, averaging ten 2,000-character messages per hour for each user, the percentage of LAN and WAN bandwidth used by e-mail is given in the charts.

On the LAN side, we are using a very small fraction of the bandwidth, less than 0.03% in Toronto, 0.10% in Reston, 0.02% in Leatherhead, and so on, as you can see from the upper chart in Figure 4-62.

The bandwidth on the WAN links, of course, is much smaller. The installed bandwidth between sites is 56 Kbps in most locations. The link between Europe and America is 9.6 Kbps. This narrow

$$\text{Each user} = 10\ \frac{\text{msgs}}{\text{hour}} \times 2{,}000\ \text{chars} \times \frac{8\ \text{bits}}{\text{char}} \times \frac{1\ \text{hour}}{3{,}600\ \text{sec}} = \frac{45\ \text{bits}}{\text{sec}}$$

Figure 4-61

Assume 10% overhead ≈ 5 bps, total = 50 bps

PC Mail Traffic

	Number of users	bps/user	Total traffic	LAN %
Toronto	30 in + 30 out	50	3,000 bps	.03 @ 10 Mbps
Reston	100 in + 100 out	50	10,000 bps	.10 @ 10 Mbps
Leatherhead	20 in + 20 out	50	2,000 bps	.02 @ 10 Mbps
Los Angeles	70 in + 70 out	50	7,000 bps	.05 @ 10 Mbps
London	50 in + 50 out	50	5,000 bps	.05 @ 10 Mbps
Paris	30 in + 30 out	50	3,000 bps	.03 @ 10 Mbps

WAN Traffic

Assume that half the messages are for remote post office; the other half stay local.

	Messages/ hour	Total traffic	% BW WAN
Toronto-Reston	15	750 bps	1.8% @ 56 Kbps
L.A.-Reston	35	1,750 bps	4.1% @ 56 Kbps
Paris-London	15	750 bps	10.4% @ 9.6 Kbps
Leatherhead-London	10	500 bps	1.04% @ 64 Kbps
Reston-London	100 (worst case)	5,000 bps	69.4% @ 9.6 Kbps 11.9% @ 56 Kbps

Solution worksheet for e-mail case study. Learning Tree International

bandwidth can intuitively be seen as the weak link. The numbers in Fig. 4-62 bear this out.

The Toronto-Reston link, with roughly 50 messages per hour, consumes roughly 2% of the WAN bandwidth. The Los Angeles-Reston traffic consumes 4% of available bandwidth. The Paris-London traffic consumes 10%, and the Leatherhead-London consumes 1%. When the bandwidth is much smaller, of course, we consume a greater portion of it with roughly the same number of messages.

Since the Reston and London post offices are the key connection between the Western hemisphere and Europe, we have a great number

Figure 4-62

Additional Internet e-mail

$$\text{Each user} = \frac{2\text{ msgs}}{\text{hour}} \times 2{,}000\text{ chars} \times \frac{8\text{ bits}}{\text{char}} \times \frac{1\text{ hour}}{3{,}600\text{ sec}} = \frac{9\text{ bits}}{\text{sec}}$$

Assume 50% overhead = 13.5 bps total

All the traffic is seen in Los Angeles LAN.

	Number of users	bps/user	Total	% BW
Los Angeles LAN	300 in + 230 out	13.5	7,155	.07 @ 10 Mbps
L. A.-Reston WAN	300 − 70 = 230	13.5	3,105	7.4 @ 56 Kbps
Reston LAN	230 in + 130 out	13.5	4,860	.05 @ 10 Mbps
Reston-Toronto	30	13.5	405	.01 @ 56 Kbps
Reston-London	100	13.5	1,350	24 @ 9.6 Kbps
				3.2 @ 56 Kbps

Solution worksheet for adding internet mail to existing e-mail case study network. Learning Tree International

of messages per hour crossing the Atlantic, with a great deal of traffic consuming almost all of the bandwidth if the link remains at 9.6 Kbps.

The recommendation is to keep the WAN links in all but the Reston-London connection at the same bandwidth. For the Reston-London connection, it would be wise to increase the link from 9.6 Kbps to at least 56 Kbps. With that increase, we see that the percentage of bandwidth consumed by e-mail will go down to roughly 12% from the previous 70% consumption.

In our analysis, we have assumed that every user is just as likely to send an e-mail message to every other user, and that's how we come up with the number of messages per hour. Adding up the 10 messages per hour, each 2,000 characters long, with 8 bits per character and 3600 seconds in an hour, we get a data rate of 45 bps consumed by each message. Adding a 10% overhead for each message for the message header, protocol, Ethernet, TCP/IP, and other ancillary protocols that are involved, brings it up to about 50 bps per message. This is the base figure used for the analysis.

We have also assumed that WAN links have an additional 25% inefficiency for the time required to set up the link and the retransmissions that are required to maintain the link error free. The WAN links are therefore only 74% efficient. The Ethernet LAN, on the other hand, will be assumed to be 100% efficient, with the transmission protocols added into the per-message overhead.

Let's add to the PC LAN mail routing the additional fact that we want to connect each user to colleagues outside the organization via the Internet. Assume that each user receives two Internet messages per hour, each 2,000 characters long. Now, this is a very conservative estimate. Certainly there might be more messages per hour for some who are very active on the Internet, while most users do not receive any messages. Typically the messages are not a whole screen's worth (2,000 characters) of information. By and large, this is a realistic estimate and a good assumption.

What would the additional loading on the LAN and WAN links be due to this traffic? Assume we were going to connect all of the parts of the organization to the Internet for receiving and sending e-mail. Figure 4-63 shows the required additional analysis. Assuming two messages per hour, 2,000 characters long, 8 bits per character, and 3600 seconds per hour, each of those messages in and out of the Internet consumes 9 bps of bandwidth with a 50% overhead. The messages have a great deal more protocol overhead than they do in the cc:Mail case. We thus have roughly 13.5 bps of bandwidth consumed by each Internet e-mail message.

Since the Internet connection is made via the Los Angeles network on a UNIX SMTP post office, all of the Internet traffic in and out of the organization will filter through that connection. That traffic will then disperse throughout the organizational network, loading the LAN and WAN links unevenly. The Los Angeles LAN, of course, will see all of the Internet messages meant for people in and out of the organization. We are looking at roughly 300 Internet messages per hour. The relative consumption of bandwidth on the LAN side, again, is very small. This results from the fact that the Ethernet has a very high bandwidth, 10 Mbps.

The LA-Reston WAN link sees roughly all but the Internet traffic meant for LA, and that link carries a fair amount of e-mail traffic, 7.4%. This

Figure 4-63

Characteristic	SMTP	MHS
Acronym	Simple Mail Transfer Protocol	Message-Handling Standard
Principal protocol suite	TCP/IP	OSI
Promoter/developer	*De facto* Internet	ISO
Major program implementing it	Send Mail POP	MHS/FTAM
Based on protocol	RFC RFC 822	X.400
Typical type of message	822.MIME	X.400
Addressing scheme defined by	822	X.400
Directory services	DNS.NIS	X.500
Message structure defined by	822.MIME	X.400
Interoperability	+	0

Summary of the characteristics of two of the more popular e-mail environments. Learning Tree International

adds 4% internal traffic to the LA-Reston WAN, so we are getting close to 12% bandwidth consumption between the cc:Mail-routed traffic and the Internet-routed traffic. The Reston LAN, which sees all of the traffic meant for the rest of the organization, is again not loaded very heavily, because of the high bandwidth capacity of Ethernet.

The Toronto-Reston WAN link has little loading. The Reston-London link is heavily loaded with this Internet traffic. When you add the

Internet traffic to the already existing cc:Mail LAN traffic, we consume almost the entire bandwidth available on that link, if it is kept at 9.6 Kbps. If we increase the link to 56 Kbps, as we are proposing, the consumption of bandwidth by both types of traffic adds up roughly to 15%.

We should be fairly happy with that. It's a very high load, but it's not nearly the consumption of a third of the bandwidth that is typical of a very heavily loaded e-mail net. So we haven't reached the high-water mark we might have, and there is no need to jump to the next higher WAN-link bandwidth, T1, which might be prohibitively expensive.

✳ Additional issues

E-mail is often used to say "Hey, we're having a party on Friday night. You're invited!" with a carbon copy to everyone in the organization. There is a lot of personal use of e-mail within an organization. Different organizations take different approaches to the use of their e-mail systems. Some allow this, some don't, and don't even recognize that this is one use of their e-mail system.

✳ Privacy

Privacy is an important e-mail issue. Is your employer allowed to look at your e-mail to see what you're doing with your time, and then decide to fire you because you're running a lot of frivolous e-mail that has nothing to do with work? If you make sexist remarks on the network via e-mail, or make some racist remarks, should you be brought up on disciplinary charges? Is your e-mail really private, or does it belong to the organization? These are ethical and business-law issues that are just now being decided in the courts.

✳ Untraceable mail

Some e-mail programs actually erase all traces of where the e-mail came from, producing anonymous e-mail. This can give rise to all kinds of incidents, like hate mail that is hard to trace to its source being sent through an organization's mail system. There are actually some routers that will erase where a certain piece of mail originated. Managers are very concerned about this. There can be all manner of inappropriate anonymous mail being sent. Keep an eye on this before it becomes a major issue on your network.

⁕ Housekeeping

My very favorite cartoon is having two people sitting across a desk from each other. The wall behind them is full of awards. And there is one particular award that one of the individuals is very proud of. The caption reads: "That award I got for filling up a 100-megabyte disk with 1,000-character e-mail messages."

That's what happens sometimes. Your server disks become full of "Hey, come to our party tonight," and other six-month-old messages. One of the biggest problems is cleaning up proliferating e-mail files. In some environments, when you carbon copy to a long distribution list, you actually send copies of the file to everyone on the list, and that really fills the server's disk fast. So an e-mail administrator has his job cut out for him: he's got to be really fast on his feet to keep the place clean.

Many e-mail systems have rules-based administration. E-mail should be held for so many days before deleting, for example. Some rules-based systems will sort e-mail for the user. They allow the display of mail on a priority basis: your boss first, for example. And this helps keep the system clean of junk mail. User education is of paramount importance. Teach users to clean up their mailboxes. A good rule is to delete any mail that has been read after seven days, and delete any unread mail after thirty days from the date of the posting.

Summary

The chart in Fig. 4-64 summarize the major points of the e-mail section.

Interoperable database access

Accessing a database from an application on a standalone computer is a relatively simple task. There are many products that help us do this today. A more challenging task is for a local application to open a database on a remote machine, and read or write records over a high-performance network.

Figure 4-64

Program	Protocol suite	Protocol	Major function	Typical platforms	Promoter/ developer
SendMail	TCP/IP	SMTP	P.O.	UNIX	*De facto* Internet
POP3	TCP/IP	SMTP	Pick up mail, M.S.	DOS, Mac Windows, OS/2	*De facto* Internet
MHS gateway	OSI	X.400	Gateway	Varies	Novell ISO Gov't GOSIP
cc:Mail	PC LANs IPX NetBIOS	Lotus prop.	P.O.	PC server DOS, Win. Mac, OS/2	Lotus
OpenMail	TCP/IP	SMTP	P.O.	UNIX	HP
Microsoft Mail	PC LANs IPX NetBIOS	Microsoft prop.	P.O.	PC server DOS, Win. Mac, OS/2	Microsoft
PROFS	SNA	IBM prop.	P.O.	IBM MVS host	IBM
VAX Mail	DECnet	DEC prop.	P.O.	DEC VMS host	DEC
DNS	TCP/IP	DNS	Direct serv.	UNIX	*De facto* Internet

Comparison of the characteristics of major e-mail delivery systems. Learning Tree International

If the database files and the access programs are homogeneous on all the remote hosts, as well as on the local machine (for example, we might be using Oracle on all platforms), this problem is also technically unchallenging. The most challenging problem is to access a remote database from a workstation application, with the remote database being of a different type than the one we're using on the workstation. See Fig. 4-65.

Take, for example, a typical database scenario. We have written an application on our personal computer, perhaps an MS-DOS machine,

Figure 4-65

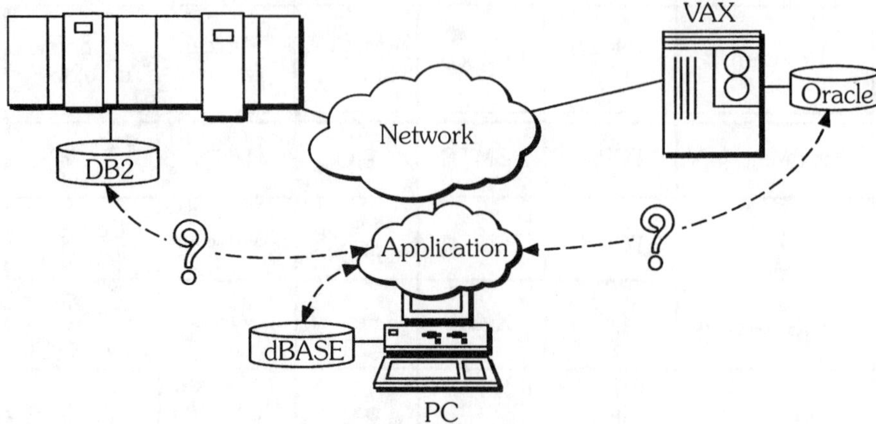

The challenge of a heterogeneous database environment: a PC application uses dBASE on the local machine and must simultaneously access Oracle and DB2 databases on remote machines. Learning Tree International

which uses dBASE as a database manager. That same application might also need to access records on a VAX, which may be using the Oracle database access product, and simultaneously may need to open database files on an IBM mainframe that is using the DB2 product. That is the nature of the interoperable problem tackled in this section. The problem is sometimes called interoperable database access.

The classic solution to the interoperability problem is to standardize to the most popular database access environment. This environment is ANSI SQL, which is a standard for database organization and access. The problem with this solution is that vendors, in their pursuit of market share, have enhanced the basic functionality of ANSI SQL to the point where it is difficult to interoperate among the various vendor products. In other words, not all SQLs are the same, even though they are all based on the same standard.

The reason for all the extensions is that the minimum standard SQL provides interoperability, but it has poor performance. Vendors have therefore tended to create SQL implementations that are based on the standard but that perform much better. Performance enhancements bring with them incompatibilities among the various

vendor products. Extended SQL (proprietary) provides good performance, but lacks interoperability.

The solution is to have a common SQL on the requester, that is, the workstation, and convert to extended SQL on the servers. We have three approaches available for interoperability:

- Client driven—Microsoft ODBC and Borland IDAPI
- Gateway driven—Sybase Open Server
- Database driven—IBM DRDA or ISO RDA

The client-driven solution

There are two popular client-driven techniques. The first is the Microsoft *Open Database Connectivity* (ODBC) specification. The second is the Borland *Integrated Database Application Programming Interface* (IDAPI).

Of the two, the Microsoft specification appears to be gaining market share, and will probably be on the market longer. The Microsoft solution provides a common set of SQL statements at the client for access to heterogeneous local and remote databases. It has a common syntax that is translated at the client, so normally there are drivers provided by the DBMS vendors that convert from specific extended SQL vendor products into the Microsoft ODBC, and it is done at the client.

Currently, ODBC makes approximately 50 API calls available. Unfortunately, not all drivers from all vendors support all the ODBC calls, so in a limited number of cases, programmers might have to use proprietary calls. The new version of ODBC API, version 2.0, which appeared late in 1994, helped to solve some of these problems.

The Borland Integrated Database Application Programming Interface (IDAPI) is an emerging specification produced under a joint effort between Borland, IBM, and about 64 other organizations. It is still in draft form, and the final version was due out in late 1994. It is meant

to compete directly with Microsoft ODBC, providing the same functionality: a gateway for SQL on the client.

IDAPI has a larger call set than ODBC (approximately 80 calls). There has been some criticism about the number of Borland-specific calls in this set. It appears at this point that the Borland specification may be too late to gain enough market share over the Microsoft specification. That, coupled with the fact that Microsoft, at the moment, owns the desktop due to the proliferation of Windows and Windows NT, makes ODBC a clear winner for this approach.

The gateway-driven solution

The gateway-driven solution requires having a computer on the network that converts the calls from specific workstation SQL implementations to the appropriate server SQL language. We consider this a gateway because it does protocol conversion.

One very good interoperable solution is provided by Sybase. Another very good solution of this type is provided by Oracle. Both of these require a computer on the network to run the gateway software. It might be one of the Oracle or Sybase database servers themselves that is conversant with other vendors' specifications, and that receives calls from the workstations that are written in the common SQL language. See Fig. 4-66.

This solution requires that the gateway vendor provide a large number of client interfaces, as well as server interfaces, on the gateway computer. Both Oracle and Sybase provide a very good solution in this respect.

Interoperability Rating: Best

At the moment, this solution appears to be the most mature of the three integration approaches. Both Sybase and the Oracle products provide excellent interoperability to all other major vendor products. Of course, because of the conversion that takes place within a gateway, this solution tends to have poorer performance than the other two. But its maturity and the number of choices available today to implement this gateway solution make this an appropriate best choice at the moment.

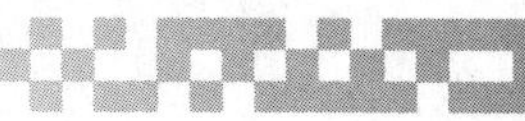

Figure 4-66

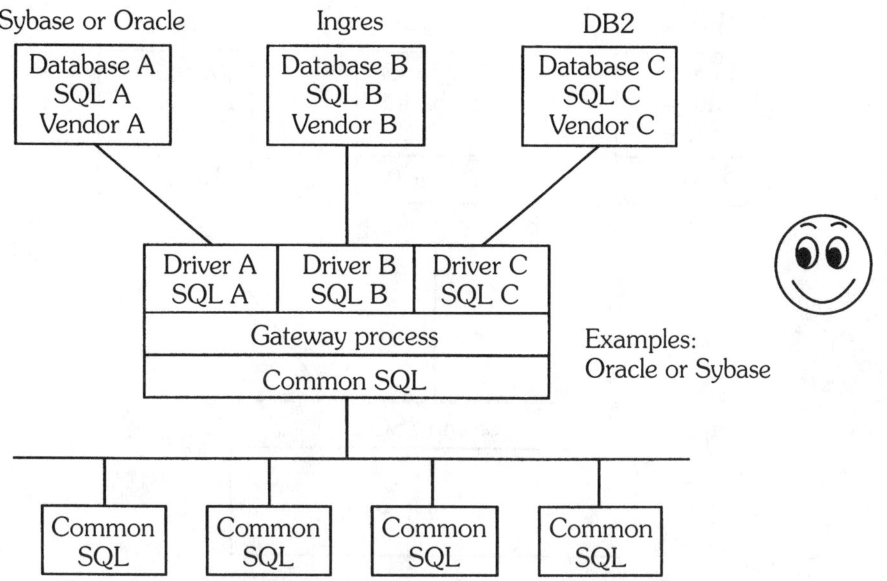

The gateway-driven solution to the interoperable databases problem. Learning Tree International

The database-driven solution

In the database-driven solution, a common SQL front end is installed at the client, and it makes common calls to the SQL databases. Each vendor's product on the server has a converter from the common SQL to the vendor-specific extended SQL environment. See Fig. 4-67.

For example, there might be a Windows NT or DOS or UNIX client workstation that uses a common SQL specification from IBM or the ISO specifications. It makes calls to Sybase, Oracle, and DB2 database servers, all of which contain drivers that convert the common SQL calls into Sybase, Oracle, and DB2 calls. The two competing specifications for the common SQL calls are the IBM *Distributed Relational Database Architecture* (DRDA) and the ISO *Remote Data Access* (RDA).

IBM's DRDA provides access to IBM and non-IBM databases through a common protocol. IBM claims to have support from up to 74

Figure 4-67

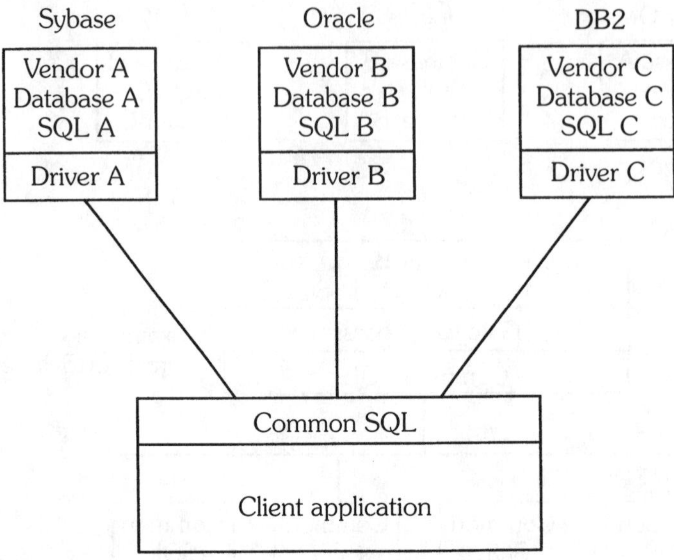

The database-driven solution. Note that the conversion in this case is performed at the server itself. Learning Tree International

companies that have committed themselves to implementing this architecture. At the moment, IBM does not support the Microsoft ODBC initiative. In other words, there are no drivers to support Microsoft ODBC, which is a drawback of this approach.

Also, this approach only specifies SQL access for client/server environments, since it only allows access to one database at a time, which makes it difficult to implement distributed transactions.

RDA is a competing specification created by ISO. It is an emerging standard for multisite transaction processing, and it is still in the draft stage. Unlike the IBM DRDA, it will provide concurrent access to multiple databases. See Fig. 4-68.

Interoperability Rating: Poor

This third approach is too new, with no proven track record, and very few products on the market. Therefore, at this point, it is not a viable alternative for those who need to develop applications right away.

Figure 4-68

IBM's DRDA	ISO RDA
Performance centered	Portability centered
Based on SNA	Based on OSI
Each server uses native SQL	Only a common Subset of SQL supported
Maximize support for existing applications	Minimize effort to port tools to different servers
Focus is on IBM interoperability	Focus is on multivendor, heterogeneous, database interoperability
IBM, plus? others	Every major DBMS vendor (except IBM)

Comparison of the IBM DRDA approach with the ISO RDA approach. Learning Tree International

Interoperability strategy

This chapter covers the seven major problems and solutions to interoperability:

- Remote terminal access
- Remote procedure execution
- File transfers
- Distributed file systems
- PC network services
- Electronic mail
- Distributed database access

There is an eighth problem, and that is the more complex subject of client/server applications. It encompasses all of the above situations, plus new sets of application programming interfaces. We cover the special requirements of client/server in Chapter 6, and leave the discussion of its interoperable needs at that point.

Figure 4-69 compares the functionality of the major protocol families in support of each of these interoperable problems. The TCP/IP, OSI, Microsoft, Novell NetWare, and IBM SNA protocol suites are considered.

Figure 4-69

Function	TCP/IP	OSI	Microsoft Windows NT (native)	Novell NetWare (native)	IBM SNA
Typical end-to-end protocols	IP TCP UDP	ISO CONS ISO IP TP0–TP4	NetBIOS NetBEUI	IPX SPX	APPN
DFS	NFS	FTAM	SMB	NCP	—
File transfer	FTP TFTP	FTAM	copy	ncopy	IND$FILE
Remote terminal	TELNET rlogin	VT	—	—	3270 gateways TN3270
Remote procedure at the server	Procedure (server daemon)	Application	Server program	NetWare Loadable Module (NLM)	Program
Remote procedure protocol	RPC sockets	ROSE	Windows sockets	Sockets	APPC for transactions CPI-C
Typical e-mail components	SMTP SendMail POPmail DNS, NIS	ISO MHS X.400 X.500	MS Mail All PC mail programs	All PC mail programs MHS gateway	PROFS OfficeVision
Typical SQL database	Oracle Ingres	Oracle	SQL Base Sybase	NetWare SQL Sybase	DB2

Comparison of the solutions to interoperability in each of the major protocol suites. Learning Tree International

If we look at the integration tasks, especially from the standpoint of interoperability, we see that there is a progression from easy integration tasks to much harder integration tasks. This progression begins with a single host or a monolithic system in which all the applications and services are on that single host. In that case, our integration task is almost nonexistent.

The next step is to move the users and their terminals across the network. The applications still run on the host, and the users now require remote terminal access. There might be multiple-host networks at this point. We progress from the simple task of remote terminal access to needing to do some simple file transfers from one host to another. We might have multiple-host networks in this case, and applications still run on the user hosts. The task is one of moving entire files from one host on the network to another.

Distributed files add a significant step in complexity at this point. We are still working with multiple-host networks. Applications are now on the user host or the user workstation. The data to be used with the application is typically across the network on another host or server. To make this task easier for PC clients, we introduce PC network operating systems at this point. PC NOS is a special case of a distributed file environment.

Lastly, the most complex task is probably distributed database access. Here we are working with multiple-host networks. Applications are typically on user hosts or on user workstations. The data is now organized in SQL records and files that are accessed via queries across the network. The data can be distributed over many hosts that might need to be accessed simultaneously.

The ultimate complex task of all is implementing the new area we call client/server. This falls under the umbrella of distributed application execution, where we make use of very sophisticated techniques such as remote procedure calls—splitting an application into client and server components that are acting together across

the network. We may also have some very sophisticated transaction-based remote applications as a part of this environment.

The farther you go in this progression, the greater the need for interoperability, and the harder the job of integration becomes. In some cases, integration might not even be possible.

We show the scale of complexity so that an integrator can judge how easy or difficult a particular integration problem will be to solve.

As a strategy, one should be sure to support and have solutions ready for the simpler tasks at the base of this progression, and carefully consider what to do for the more complex tasks. See Fig. 4-70.

Concepts review

You can use the chart in Fig. 4-71 to test your understanding of some of the concepts presented in this chapter. At a minimum, you should be able to identify and classify the acronyms. The answers can be found in Appendix C.

Figure 4-70

Scale of difficulty in providing interoperable solutions in network integration. Learning Tree International

Figure 4-71

	Transport/ network protocol	Remote program execution	File transfer	Database access	Remote terminal access	DFS	PC NOS	E-mail	Network management	Directory services
TCP										
NetWare										
FTP										
NFS										
X.500										
X.25										
TELNET										
ROSE										
DNS										
TP4										
APPC										
DB2										
TFTP										
NCP										
SMB										
NetBIOS										
Advanced Server										
X.400										
Sockets										
UDP										
IPX										
VT										
SNMP										
CMIP										
SMTP										
POP3										
PROFS										

Identify the acronyms and classify them according to function. Learning Tree International

5

Case studies in integration

CHAPTER 5

IN Chapter 4, we discovered the approach to integration from the server side. In this chapter, we take up the study of how client machines on a network can access servers, wherever those servers might be. In particular, you are going to discover the techniques necessary for a client machine to access several servers simultaneously: the multiprotocol stack situation.

First, there is a survey of all the various client types that require access to servers, and all of the server types that provide those services in today's corporate networks.

One problem of particular interest is the management of the user namespace. By that we mean how users cope when there is a proliferation of user names and passwords, i.e., as the networks get more and more complex and the users need to access more and more servers of different types. As you will see, sophisticated solutions to these problems are just coming on the market.

The rest of the chapter is devoted to cataloging, by workstation type, the specific ways a workstation accesses the various servers. In particular, we look at the necessary protocol stacks, transport and networking protocols, and the various application protocols that are typically used in each situation.

At the end of each workstation section, there is a summary chart outlining how a particular type of workstation might have the network services, discussed in Chapter 4, satisfied by the types of servers in today's corporate environment. In particular, we are interested in the file transfer protocols, the type of distributed file system being used, whether there is multiprotocol support, and if it is needed at all, how printing is done, the use of remote terminal emulation and the appropriate type for each situation, diskless workstation support for that environment, and what type of e-mail is recommended for that workstation type.

We conclude the chapter with some real-life scenarios of networks requiring integration. In particular, we look at four case studies, each one exemplifying a different situation. The first case study is the integration of PC, Macintosh, and UNIX workstation networks. The

second involves integrating large hosts such as VAXes and IBM mainframes. The third case study is the integration of workstations in large-host environments. The last one is purely an application-level integration for database work. We hope that by the end of these four case studies, the reader will have a good understanding of the various techniques and the possible scenarios that require integration in today's corporate networks.

We provide two exercises to conclude the chapter for the reader to try his hand at the procedure we develop for analyzing and integrating multivendor network situations. The solutions to these two exercises are found in Appendix C.

The client/server problem

What problem are we trying to solve in this situation? The problem is that of a user on a workstation connected to a high-performance network with multiple servers, and the user needing to have access to all those servers simultaneously. This would not be a big problem if the servers were all of the same type. Let's say they are all Novell NetWare, or all UNIX machines. In today's networks, the opposite is generally true: a user needs to log into a variety of servers.

Each server type requires different types of accounts set up for each user. Each server type might require a different type of internetworking protocol, such as TCP/IP or IPX or NetBIOS for access. The protocol family might be connection-oriented, as in SNA, rather than connectionless like TCP/IP.

We have an added problem. The types of workstations we see in a corporate environment vary greatly across the corporation. Some users use UNIX workstations. Some users use DOS or Windows. We are also seeing the appearance of many more Windows NT workstations on our networks these days. There are still a considerable number of Macintoshes. Some users might be using terminals on UNIX hosts to reach other servers through that UNIX host across the network. Figure 5-1 exemplifies this situation.

Figure 5-1

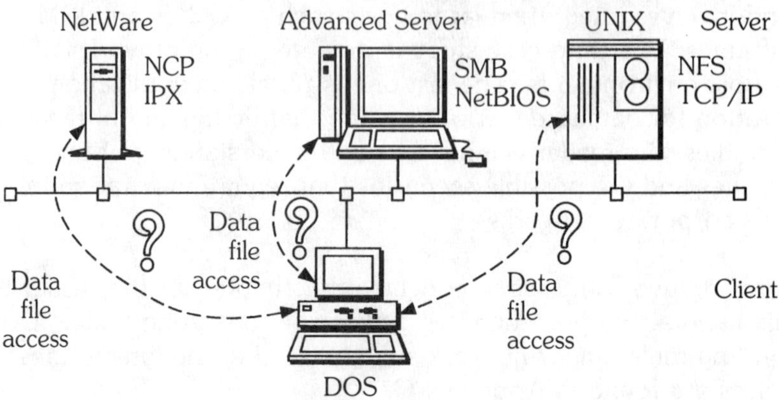

The client/server problem of a workstation having to access different servers simultaneously for many different services. Learning Tree

If we were to make a list of all the various clients found in today's corporate environments, the list would be quite long. Figure 5-2 is a reasonable list of client machines common on networks today as clients needing access. What do these workstations need to access?

Figure 5-2

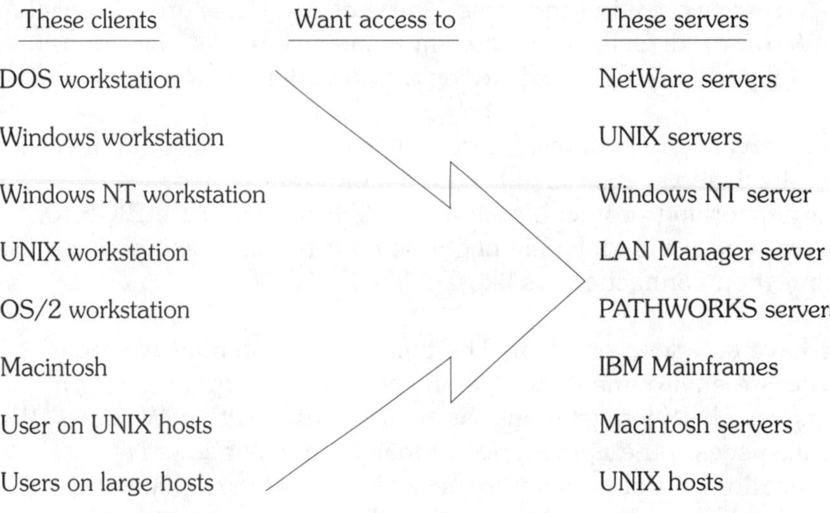

- Clients and servers typically are attached to a high-performance network, a WAN or LAN

List of client and server types. Typically the clients need access to several of the servers simultaneously. Learning Tree International

On the right-hand side of this list, we see the possible kinds of servers to be accessed: we have DOS workstations, Windows workstations, Windows NT workstations, UNIX, OS/2, Macintosh workstations, users on UNIX hosts, and users on large hosts such as IBM mainframes. These need access to file servers, print servers, application servers, electronic mail post offices, procedure servers and document servers. These platforms might use various kinds of network operating systems. The types of services also vary.

These clients and servers are typically connected to one another by a high-performance network, usually a wide area network or a local area network. In a few cases, some workstations might be coming in through a dial-up, direct-access line such as a SLIP or PPP connection into a network. This case is typically seen on the Internet, but for the most part, we are not going to discuss large public networks. We are going to focus on corporate internetworks where the workstation connection is via a local area network, and the various networks around the corporation might be connected by wide area network links. We leave the case of an Internet direct dial-up workstation connection as a special case.

Typical services

What types of services do these clients require from our servers around the network? Chapter 4 dealt extensively with the types of services that servers provide. Typically, the workstations require access for simultaneous operation of many tasks, with data coming from multiple servers. The workstations also need to run applications on the remote host via either remote terminal access or, more recently, via client/server remote procedure executions.

In general, most workstations are going to run some form of application on the workstation and access data files on remote servers. In this case, the servers are providing multivendor, multiprotocol distributed file access. Together with file access, as we saw in Chapter 4, the servers provide printing services.

Another important issue is the ability to transfer a file, either from one server to another through the workstation directed by the user

sitting at that workstation, or from a server to the workstation, or from a workstation to the server. In other words, the issue is the ability to upload and download files, as well as the ability to initiate server-to-server file transfer.

We are also interested in how each workstation type is supported for electronic mail—in other words, the types of post offices that each workstation can access on the various servers. As we go through one workstation after another, we enumerate as far as we are able how these services are provided for that workstation type with respect to each server type.

Another service that is required by a workstation is access to SQL databases. This was taken up in Section 8 of the last chapter.

The multiprotocol stack problem

For a workstation to access multiple servers simultaneously, it must often run two or more protocol stacks simultaneously. The problem that crops up in this situation is how these multiple protocol stacks on one workstation share the use of one LAN card in that workstation.

This problem shows up most severely in the DOS workstation. We have essentially three solutions for the DOS workstation: two have been promoted by vendors, and the third is from the academic world.

The first solution is the NDIS environment from Microsoft. The second solution is ODI from Novell. The third was created at Clarkson University, and it's called the Clarkson Packet Drivers. All cases require a special program that is embodied in each of the three solutions, to allow either the network protocol or, in the case of NetBIOS, the session protocol, to access a single network card.

Single-stack solutions can be executed on a workstation simultaneously. But if each stack is unaware that others are there, they can conflict with one another, hanging up the machine. In other words, in a DOS workstation, you can load IPX and NETx, both DOS programs that execute under DOS, to allow the workstation to reach a Novell server. You can also load TCP/IP to allow that workstation to reach UNIX servers at the same time, and you can do the loading

of one program after the other. First you load IPX.COM. Then you load NETX.COM and then TCPIP.EXE. All three execute and are operational. But as soon as two of them, IPX and TCP/IP, for example, try to use the network card for their purposes, there is a high likelihood that they will conflict, and cause the machine to hang.

So there must be a protocol that mediates between IPX and TCP/IP for the use of the network card without causing the machine to hang, which is why these three solutions have been implemented.

Figure 5-3 shows two situations. One is a DOS workstation running multiple protocol stacks (IPX and TCP/IP) using the ODI solution to resolve contention for the network card. The second workstation, on the right, shows a Windows NT workstation running the TCP/IP protocol stack simultaneously with the NetBIOS protocol stack using the Microsoft NDIS solution for resolution of the conflict between the two protocol stacks.

Figure 5-3

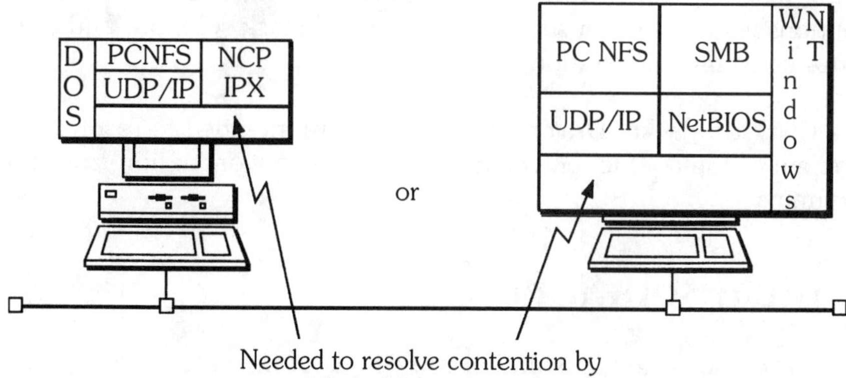

- NDIS is the Microsoft solution
- ODI is the Novell solution
- The Clarkson packet drivers are the TCP/IP solution

Two access servers running different protocol stacks: the workstation is often forced to load and run different protocol stacks as well. The text discusses the problem and several solutions. Here is an example of two workstations employing two different methods to solve the problem: NDIS and ODI. Learning Tree International

The multiprotocol stack solution works very well for DOS, as we mentioned, as well as for DOS machines that run Windows and Windows NT. It is also useful in the Macintosh environment. There, of course, the Macintosh operating system handles the use of multiple protocol stacks without having to bring in external services.

There would not be a conflict if you were to run multiple protocol stacks on a UNIX machine. In general, we tend not to run anything other than TCP/IP on UNIX workstation or UNIX servers.

Servers, of course, are a different story. Many servers, whether they are UNIX, NetWare, LAN manager file server, or an application server, typically run multiple protocol stacks. Vendors handle any conflict between the multiple protocol stacks with their own proprietary solution. For example, the ODI solution created for the Novell server to run multiple protocol stacks migrated as a very good solution to run multiple protocol stacks on DOS workstations. The same thing happened with the NDIS solution from Microsoft. These solutions began life as a server multiprotocol stack solution, and then migrated to the workstation.

The Clarkson Packet Drivers, of course, were intended to resolve the problem of the conflict between IPX and IP at a workstation from the beginning.

Typical solutions

Let's take a look now at some of the details of how the three major DOS workstation multiprotocol stack support programs work. We'll take up the case of Novell first, move from there to NDIS, and complete the discussion with the Clarkson Packet Drivers. We also comment at the end of each section on how useful that particular protocol is for interoperability.

Remember our comment concerning interoperability criteria. We're not so interested in how good the solution is technically, or how popular it is, but whether it is a good solution to consider deploying network-wide for integration.

Novell ODI

The Novell solution, called ODI (*Open Data-link Interface*) is shown in Fig. 5-4. Novell enables a workstation to have up to two different network cards. That is why we show two cards as NIC A and NIC B. Multiple drivers in the case shown in the diagram are the NE 2000 driver from Novell and the SMCPLUS driver from Standard Microsystems.

Figure 5-4

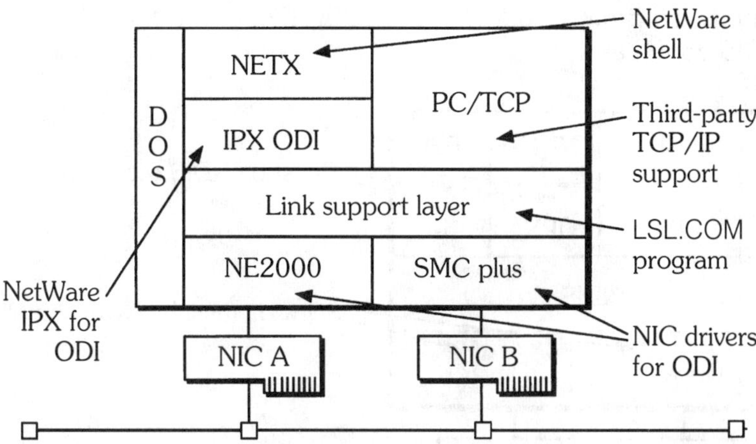

The Novell ODI solution to the workstation multiple protocol problem. Learning Tree International

A program called the *Link Support Layer* (LSL.COM) resolves contention between IPX and TCP/IP in this case. This is the typical client set-up under DOS. Notice that the version of IPX running in this case is not the same version you would run for the workstation to access the Novell server with IPX only, and there is no need to run multiple protocols. So an IPX.EXE program other than IPXODI.EXE is needed here. The shell here (NETX.EXE) would be the same in both cases.

Microsoft NDIS

Microsoft has put together a similar protocol environment to resolve contention on DOS workstations. That environment is called NDIS (*Network Device Interface Specification*). The solution has been widely adopted by many other vendors, especially those who were

OEMs of the Microsoft LAN Manager program several years ago. This version of NDIS works not only under DOS, but also under OS/2 and Windows NT. This approach to the multiprotocol problem has wider appeal because it is applicable to a wider range of workstation types.

NDIS also requires NDIS-compliant components: the drivers for the network cards and network protocol. Figure 5-5 shows a DOS workstation running NDIS with two possible network cards and two drivers (which must be NDIS-compliant). In that case we are also running three protocol stacks: NetBIOS, IPX, and IP, which must all be NDIS-compliant.

Figure 5-5

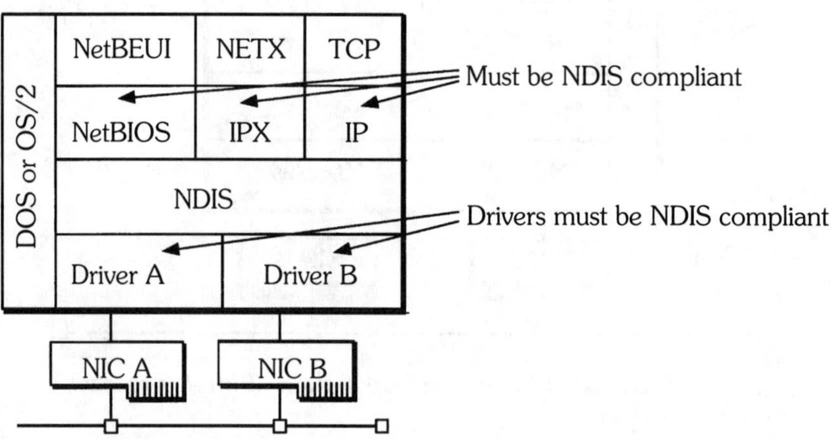

The Microsoft NDIS solution to the workstation multiple protocol problem. Learning Tree International

Typically, the NDIS-compliant components are shipped with any network operating program, utility, or operating system sold by Microsoft. In other words, Windows for Workgroups comes with an NDIS solution out of the box. Windows NT comes with an NDIS solution out of the box, and Microsoft makes the NDIS environment—drivers, protocols, and the NDIS program itself—available to users through their bulletin boards, their Internet server, and in many other ways, including all of their resource guides.

For that reason, we consider the NDIS solution, which has wider acceptance and wider applicability, to be better than the ODI solution. They are both technically excellent. They both work with a

wide range of protocols. But because Microsoft has a larger installed base in the workstation world, the NDIS solution will prevail over the ODI solution.

✻ The Clarkson Packet Drivers

The third solution, *Clarkson Packet Drivers*, is a public domain program, or group of programs, that came out of a Clarkson University project. This was an early solution to the multiprotocol problems on DOS clients. Its main function, as it is with the other two solutions, is to resolve contention for use of the network card by protocols (IPX, NetBIOS, and IP) that might be operating in a workstation simultaneously. Again, there is a program between those protocols and the network card, and it is the packet driver. There is a specific driver for the network card that must be compliant with the packet driver, and all the protocols—IPX, NetBIOS and IP—must be written to be compliant with the packet driver protocol itself.

Since this solution is in the public domain, many TCP/IP vendors use it as the basis of their offering to ensure interoperability of their workstations with multiple servers. The packet drivers are in the public domain, and for that reason we consider its interoperability very good. On the other hand, because it is in the public domain, the documentation for it is not commercial-grade, although it is very extensive and very well documented. NIC drivers for the network cards you might be interested in are not always available. So not all network cards are supported, and it is sometimes difficult to get drivers of all kinds to support the packet driver environment. For that reason, it is probably not as useful a solution for integration as ODI and NDIS.

Of course, the vendors are going to focus on the DOS workstation, so they will come out with a DOS program. They might make one that is also Windows compatible, in case there are some differences between how a program will work with Windows and DOS. Then they have some choices. Do they do ODI? Do they do NDIS? Or do they do a packet drive-compatible driver?

The largest installed base is probably with NDIS, because Microsoft owns the desktop. That's why we consider NDIS to be the favorite for interoperability. ODI is the next most favorable solution. The packet driver solution comes last. But in this problem and in this situation,

we'll take anything we can get. All three will work perfectly well if you can get yourself a complete set of drivers and protocols that are compatible for the workstation in question.

Figure 5-6 shows the packet driver solution to the multiprotocol problem.

Figure 5-6

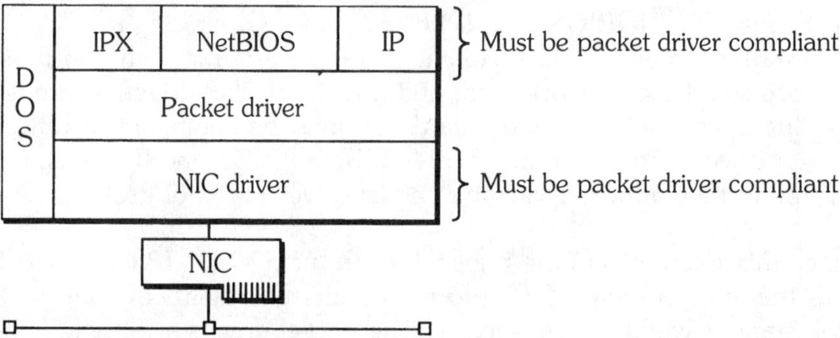

The Clarkson University Packet Drivers solution to the workstation multiple protocol problem. Learning Tree International

✻ Management of the user namespace

Consider the situation depicted in Fig. 5-7. A user at a workstation accesses a number of servers: the CompuServe on-line host, an Internet remote host, a UNIX server within their own corporate network, and perhaps multiple NetWare servers. In each situation, the user has to deal with a username and a user password. This is not so farfetched. Most corporate workers today have to deal with at least two, if not more, different user names and passwords to reach the corporate resources and external business resources that they need for their work.

The problem is how a user can deal with so many different passwords and names. How does a user keep track of names and passwords? What's a user to do? Write them down? Do they get embedded in script files, which can become a security breach if others have access to the workstation and can see what's in those script files? Do they keep asking the system administrator every time they forget their

Figure 5-7

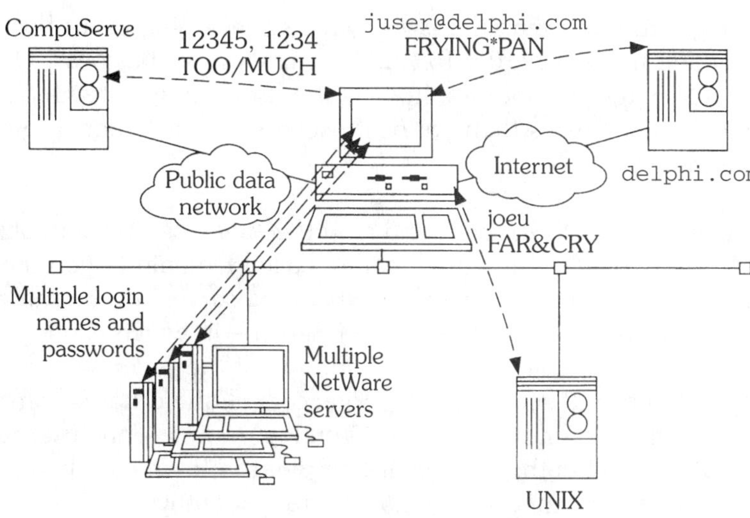

When too many user names and passwords become an unmanageable situation and a probable security risk, it is time to consider some solutions. Learning Tree International

username and password? This happens very frequently in some of the networks that we deal with.

So users write them down. Where do they put them? In their address book? On little 3M Post-it notes on their monitor or under their blotter? Do the user names and passwords become so simple as to allow the user to remember them all? These approaches can become security breaches.

Strike a balance between security and convenience. This is truly the razor's edge of our very complex networks these days.

There are a number of solutions vendors are beginning to offer to help with this problem. The problem might be termed the *user namespace* problem. Of course, the very first solution that comes to mind is our reductionist approach. The reductionist approach goes something like this: reduce the number of user names and passwords that a user has to remember.

How do you do that? Reduce the number of hosts a user needs to directly log into. You can restrict access to the number of hosts that

he can log into by forcing the issue, which might restrict his business solutions and make it harder for him to carry on his business. Or you can provide gateways and one host as the user's point of entry to the network. The gateways then connect users to all the other resources transparently.

In PC networks, LAN Manager and now the Advanced Server solutions from Microsoft have a very good feature called Domain Logon that allows you to do just that. It works well among servers of the same type—a LAN Manager or an Advanced Server—in which you log into one server, automatically logging you into all of the other servers, giving you rights and setting you up to access all the other servers on the network that you need to access. You log in once—one username and password—get authenticated at one point in the network, and from there have all the connections you need transparently.

Now we're seeing Novell following suit with their NetWare 4.0 directory services, where there is a function for multiple server login. Banyan VINES and Banyan in general provide a service called ENS, *Enterprise Network Services*, which is basically an umbrella service that allows the user to log into one server and access all the multiple servers through that one login.

The interesting thing about the Banyan solution is that it works not just for their own VINES server, but it extends to multivendor servers: LAN Manager, Advanced Server, LAN Server, and Novell NetWare servers. For interoperability, this looks like a winner in the PC network environment.

For UNIX hosts, there is the trusted host utility, in which logging into one host and having that equivalency as a user in other hosts automatically authenticates you to use the resources of those other remote hosts. This, of course, can be a security problem in the UNIX environment, so it has to be used with discrimination and care.

In the client/server world, a security server or service in which the user is authenticated on one server, enabling him to have access to other services around the network as allowed by a network administrator, is becoming popular.

Certainly, the final word in this particular area has not been written. There is definitely a problem with the proliferation of user names and passwords. There have been some good starts at comprehensive solutions. We hope to see more in the future.

✻ Concepts review

To review the concepts we just covered in this section, please answer the following questions. The answers can be found in Appendix C.

1. What are some of the most popular client platforms, by operating system type?
2. What server and host types do these workstations typically connect to?
3. When dealing with a workstation requirement to employ multiple protocols simultaneously, what are the most popular interoperable solutions, and the vendors or organizations that promote them?

   __________________, ________________________, ____________________

   __________________, ________________________, ____________________

4. What are some solutions to the problem of proliferation of user names and passwords as the degree of connectivity of a workstation increases?

Client platforms

Let's take a look at how various types of workstations connect to servers and hosts over a network. We are looking at the client side of the multivendor integration equation. The criteria for the technical details will be which protocol stack and which services need to be active and operating under that workstation's operating system to connect to the different types of hosts and services over the network. In some cases, we might even indicate the required multiprotocol stack solution—for example, in the case of DOS and Windows. At the end of each section, there will be a drawing summarizing the characteristics of connectivity for that type of workstation, server by server and service by service.

The UNIX workstation

The UNIX workstation is fairly straightforward. It's a user machine running any one of a number of different UNIX operating systems. The principal protocol stack at the workstation level is TCP/IP. This is the classic solution for a UNIX workstation to connect to a network. The typical network card for this type of workstation is Ethernet. Some vendors make other network cards available, but for the most part, the Ethernet NIC is the most popular. There is really no other choice for a UNIX workstation.

All of the services that TCP/IP makes available (NFS, DCE, POP3 for Mail, FTP, and TELNET) are used most often in this type of workstation. This says that in UNIX workstation environments, the servers should have TCP/IP running on them for interoperability. A UNIX workstation can reach an IBM SNA host and interoperate with it by using the TELNET 3270 solution.

Figure 5-8 demonstrates the typical protocol stack and services required by a UNIX workstation.

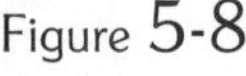

Figure 5-8

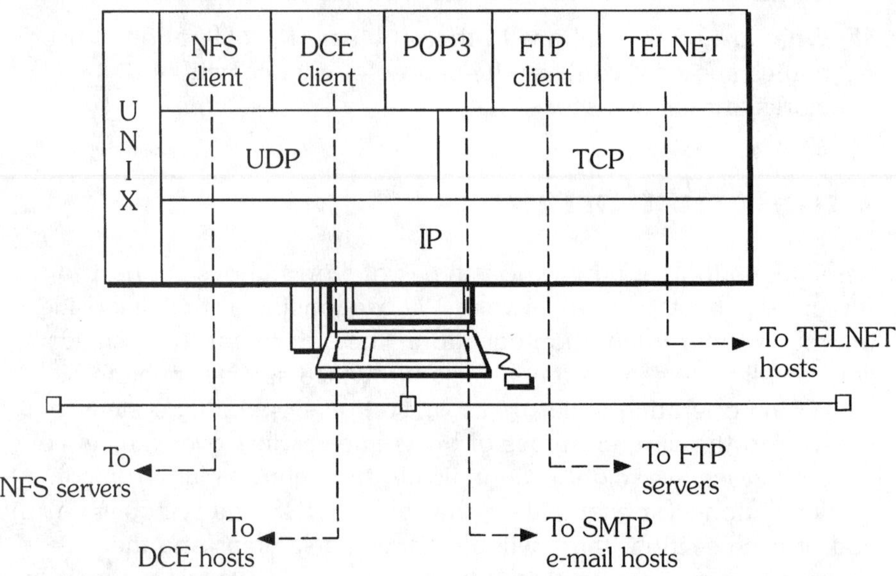

Techniques to access servers from a UNIX workstation. Learning Tree International

✻ The UNIX client

There is another UNIX client situation besides a user on a UNIX workstation, namely a user at a terminal on a multiuser host that is running UNIX. Through that host, this UNIX client or UNIX user can reach the same services that the UNIX workstation reaches. Those terminals, of course, might be character based, such as a VT-100 connected to a DEC VAX, or they might be sophisticated X Windows terminals connected to the UNIX host via a local area network. In either case, the user is on a terminal. The services are provided by the multiuser host to reach other hosts and servers. Again, as in a UNIX workstation, all of the connectivity is provided through TCP/IP on the host, and all services are TCP/IP application services in OSI layer 7.

Figure 5-9 shows techniques to access servers from a UNIX client running on a large UNIX host.

Figure 5-10 is a summary of the connectivity characteristics of the UNIX workstation.

The NFS protocol is the appropriate distributed file system protocol for a UNIX workstation. There is really no need for multiprotocol

Figure 5-9

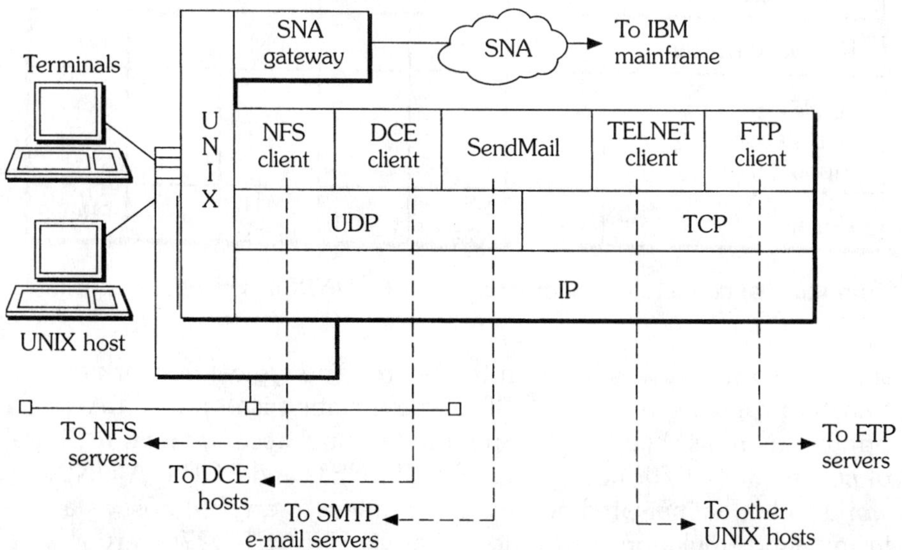

Techniques to access servers from a UNIX client running on a large UNIX host. Learning Tree International

Figure 5-10

Characteristic	Novell NetWare 3.x server	Microsoft Windows NT Advanced Server	DEC PATHWORKS host	UNIX server/host	IBM LAN server	IBM host
Network/ transport protocols	TCP/IP	TCP/IP	—	TCP/IP	—	TCP/IP
DFS protocol	NFS	NFS	—	NFS	—	—
Multiprotocol support	—	—	—	—	—	—
Printing	LPD	LPD	—	LPD	—	—
File transfer	FTP	FTP	—	FTP	—	FTP
Remote terminal	—	—	—	TELNET	—	TN3270
Diskless workstation support	—	—	—	BOOTP	—	—
E-mail	cc:Mail	MS Mail cc:Mail	—	SMTP POP3	—	PROFS

Summary of connectivity characteristics of a UNIX workstation. Learning Tree International

support, so that row is blank in the figure. The typical network or transfer protocols are all TCP/IP-based. Printing is done via LAN printer daemons. File transfer, remote terminal access, and e-mail are of course all TCP/IP-based, via FTP, TELNET, and POP3. A UNIX workstation, as indicated before, can connect to an IBM host as a terminal by emulation, and it does that via TELNET 3270 very readily. To use the IBM host for mail, you would access PROFS on the mainframe through the TELNET 3270 client on the workstation.

This chart applies to both UNIX workstations and a UNIX client on a multiuser host.

The DOS and Windows client

We are going to lump the DOS and Windows client into one category. True, they present radically different user interfaces, one being character-based and command-line driven, and the other being graphical, but as far as connectivity goes, they are roughly the same. To run Windows, which is only a user interface on top of DOS, you must still have DOS underneath. Since any of the connectivity solutions, protocols, and multiprotocol stack solutions must still run under DOS, they are essentially DOS solutions.

Figure 5-11 shows how a DOS client accesses servers.

Figure 5-11

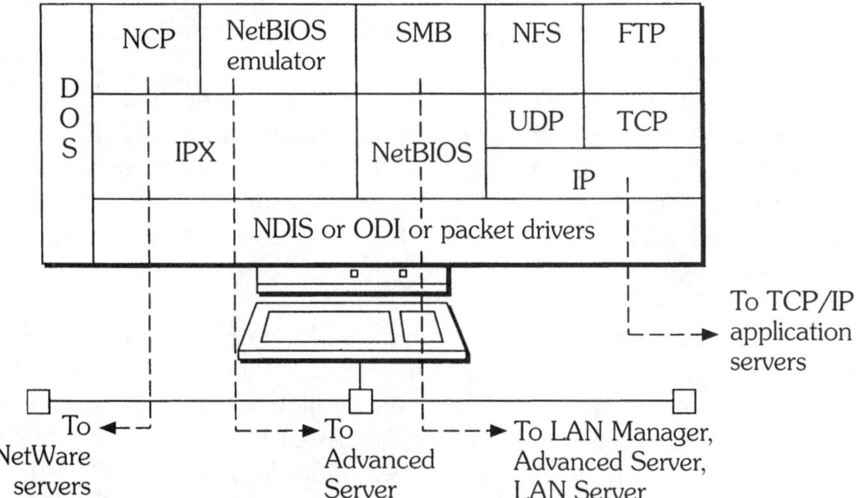

Techniques to access servers from a DOS-based workstation. Learning Tree International

A PC with DOS probably has the greatest range of connectivity options. It is probably the most popular type of workstation on the market today, so there have been many products that provide network services for it. There is great market pressure on vendors to produce inexpensive software and hardware for this type of workstation, again, because it has become a commodity market. So there are many

different drivers available under DOS for different kinds of network cards, and DOS supports probably the widest range of protocol families, enabling this workstation to connect to a wide range of servers.

DOS (with Windows) is the platform of choice for most users in today's corporate networks. It supports NDIS, ODI, and the Packet Drivers for multiprotocol support. It connects to almost any type of host as a terminal, and to every type of server as a client.

We see from the figure the typical protocols that can run at the same time under DOS. IPX connects to NetWare servers and the Windows NT Advanced Server. You can also have NetBIOS connect to the LAN Manager, Advanced Server, or IBM LAN Server, and you can have TCP/IP to connect to UNIX hosts. You can also use TCP/IP to connect to NetWare servers, Windows NT Advanced Server, or NFS UNIX servers, and under all three protocol stacks, we have, of course, the NDIS or ODI or Packet Drivers for multiprotocol support.

Shown above the Network and Transport layer protocols are the application services. For the IPX protocol stack we have NCP. For NetBIOS we have SMB, and for TCP/IP we show NFS and FTP, although, of course, you might also have NCP on top of TCP/IP to connect to a NetWare server, the latter being a recent development by Novell.

Figure 5-12 shows the techniques that can be employed on a DOS workstation to connect to hosts as a terminal. You can use IPX and the Novell protocol stack, and LAN Workplace for DOS, to enable a DOS workstation to run a 3270 session on an IBM host via a Novell gateway that is running IPX on the LAN side. You can use TCP/IP and TELNET to connect to a TELNET host. You can use TELNET 3270 over TCP/IP for the DOS workstation to connect to an IBM host that is running TCP/IP, or via a TCP/IP gateway into the SNA world. You can use terminal emulation over NetBIOS to an IBM host. You can add a special board, called SDLC hardware, to enable the DOS workstation to directly emulate and connect to an SNA network. This special hardware turns the DOS machine into a virtual 3270 terminal, enabling it to connect directly to a cluster controller. Lastly, you can run LAT over an Ethernet board and provide VT-100 connectivity for the DOS machine to a VAX host.

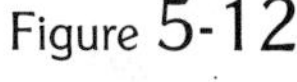

Figure 5-12

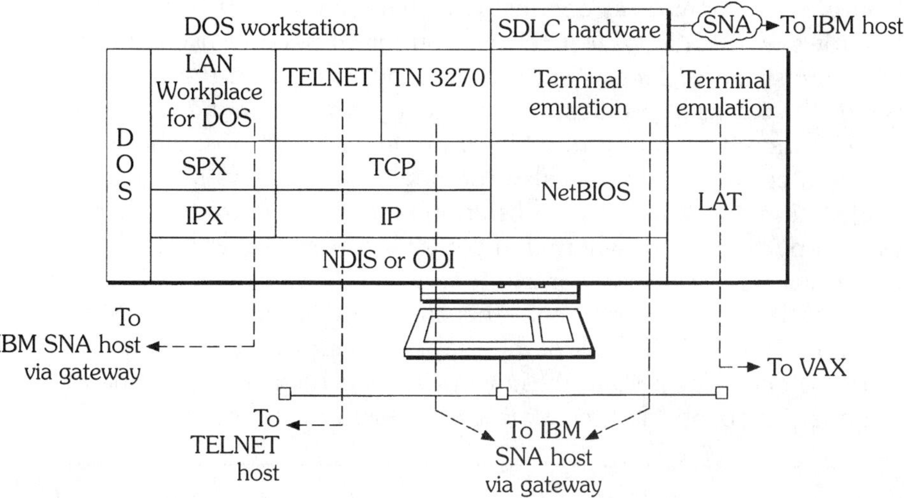

Techniques to access a large host from a DOS-based workstation acting as a terminal. Learning Tree International

The Windows client typically uses DOS under Windows in the workstation. Windows is the graphical user interface. All the interoperability solutions available for DOS work with Windows, but with some caveats. You must load the internetworking protocols under DOS before you load Windows, so that they are available from all Windows sessions. Although the application protocol written for DOS can be run in DOS sessions under Windows, it is better to use new application protocols written for Windows directly. For example, with e-mail interfaces, you can run a DOS e-mail interface in a DOS window under Windows, but it is far better to use the graphical e-mail interface. The e-mail program cc:Mail has both a DOS and a Windows interface in its e-mail program. There are some interfaces that are written only for Windows, such as NetManage from Chameleon, which has TCP/IP, POP3 client software. Other solutions are strictly Windows-based, such as the Mosaic client, which does not run under DOS at all.

There have been a number of application protocol interfaces written directly to work with the Windows socket. The Windows socket is one capability under Windows that is not available under DOS. It makes DOS with Windows a very viable client platform for writing client/server software specifically because Windows sockets, UNIX RPCs, and DCE RPCs are all available to produce this new type of software.

Figure 5-13 summarizes connectivity characteristics of DOS and Windows workstations. Notice that you might need to use a different type of distributed file system protocol, depending on the type of server the workstation must reach. Most printing is done via queues, except when connecting to a UNIX server or UNIX host where the line printer daemon (lpd) is also running. File transfers in the DOS workstation and Windows workstation environment are usually done via a copy command, where you copy files from the workstation to the server, server to the workstation, or server to server, rather than a separate file transfer protocol. This capability is found under NFS for TCP/IP. The TCP/IP environment also provides for the use of the FTP protocol. For e-mail, these workstations have a wide range of options: Microsoft Mail, cc:Mail, and a host of others.

Figure 5-13

Characteristic	Novell NetWare 3.x server	Microsoft Windows NT Advanced Server	DEC PATHWORKS host	UNIX server/host	IBM LAN server	IBM host
Network/ transport protocols	IPX	NetBIOS	DECnet	TCP/IP	NetBIOS	SDLC
DFS protocol	NCP	SMB	SMB	NFS	SMB	–
Multiprotocol support	ODI NDIS PD	ODI NDIS PD	–	ODI NDIS PD	ODI NDIS PD	–
Printing	Queue	Queue	Queue	LPD	Queue	–
File transfer	NCOPY	COPY	COPY	FTP	COPY	IND$FILE
Remote terminal	–	–	–	TELNET	–	3270 E.M.
E-mail	MS Mail cc:Mail	MS Mail cc:Mail	MS Mail cc:Mail	POP3 SMTP	MS Mail cc:Mail	PROFS

Summary of connectivity characteristics of a DOS or Windows workstation. Learning Tree International

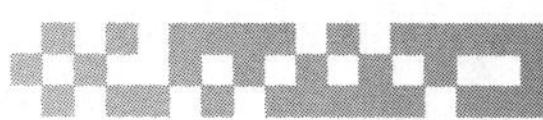

✻ The Windows NT workstation

Windows NT is Microsoft's latest workstation operating system. As an operating system, it runs on many different platforms, including a wide range of Intel-based machines. It runs on the DEC Alpha workstations. Windows NT, because it is a Microsoft solution, uses NDIS for multiprotocol support. NDIS supports a variety of network (layer 3) protocols on the client, such as IP, IPX, and NETBUI/NetBIOS connectivity (layer 5). These enable a Windows NT client to interact with a variety of hosts and server platforms, principally Microsoft servers such as LAN Manager and Windows NT Server. It also works well with the IBM LAN Server and TCP/IP hosts, and with NFS servers and Novell NetWare servers.

IBM mainframe connectivity is supported via a protocol from Microsoft called the *Data Link Control* (DLC) protocol, which attaches the Windows NT workstation directly to an IBM 3745 front-end processor. Figure 5-14 shows the Windows NT workstation connectivity solutions.

It is of interest to note that Microsoft has produced a whole new interface between the network protocols and the application

Figure 5-14

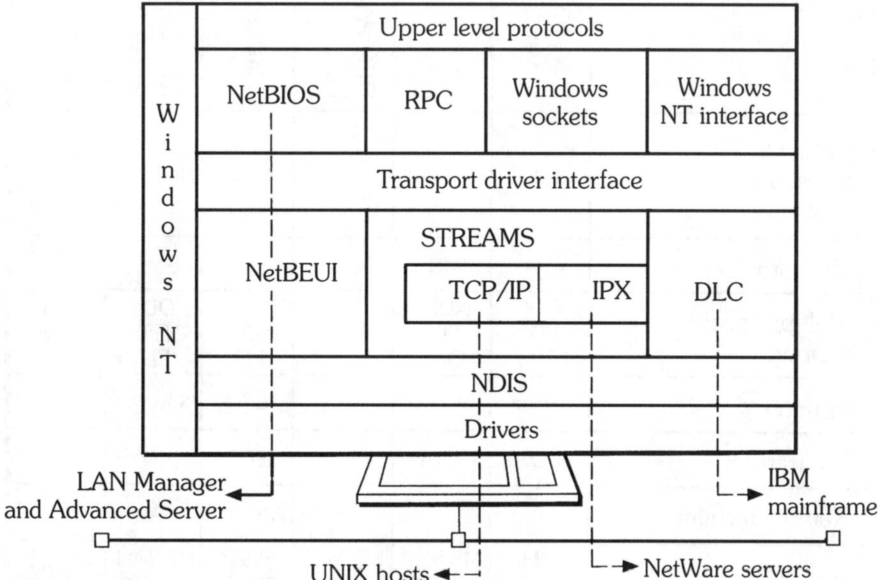

Techniques to access servers from a Windows NT workstation. Learning Tree International

protocols. Certainly this is great theory, but how much of it is working at this point is debatable. Can you run TCP/IP over NETBEUI and NetBIOS? The answers to these questions are not clear. Can you run Windows Sockets over IPX? Certainly some of the vertical stacks are completely working, but probably not all. This situation is very fluid—as time goes on and Microsoft continues to do more work, more and more of these combinations of vertical stacks or profiles will be made to work.

Figure 5-15 summarizes the connectivity characteristics of a Windows NT workstation. Notice that this is very similar to the DOS and

Figure 5-15

Characteristic	Novell NetWare 3.x server	Microsoft Windows NT Advanced Server	DEC PATHWORKS host	UNIX server/host	IBM LAN server	IBM host
Network/ transport protocols	IPX	NetBIOS	–	TCP/IP	NetBIOS	SDLC TCP/IP
DFS protocol	NCP	SMB	–	NFS	SMB	–
Multiprotocol support	ODI PD NDIS	NDIS PD ODI	–	ODI NDIS PD	ODI NDIS PD	ODI NDIS PD
Printing	Queue	Queue	–	LPD	Queue	–
File transfer	NCOPY	COPY	–	FTP	COPY	FTP
Remote terminal	–	–	–	TELNET	–	3270 E.M.
E-mail	MS Mail cc:Mail	MS Mail cc:Mail	MS Mail cc:Mail	POP3 SMTP	MS Mail cc:Mail	PROFS

Summary of connectivity characteristics of a Windows NT workstation.
Learning Tree International

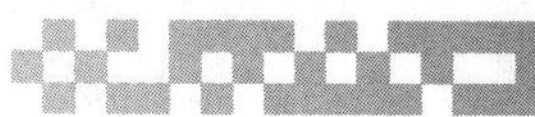

Windows environment—as it should be. This solution is an extension of what Microsoft has done with Windows and DOS. Windows NT is not a radical departure in terms of connectivity. The ODI solution shown under Novell NetWare 3.x server multiprotocol support, of course, comes from Novell, and Novell makes it a point to provide the ODI solution for a Windows NT workstation.

We strongly suggest that you stick to Microsoft protocol products on the Windows NT workstation to reach all the different kinds of servers (since these will probably be working best). The Microsoft solution might not be the most technically complete or highest-performing solution, but it is certainly the safe solution that will work in most cases.

✳ The OS/2 workstation

The OS/2 workstation has a growing installed base. It is a strong competitor to Windows NT. It has a graphical user interface; like Windows NT, and unlike the DOS Windows solution, it is an integrated operating system and GUI all rolled into one. In an IBM environment, OS/2 is the platform of choice. It has many connectivity tools to connect to an IBM mainframe, as it should. This is strictly an IBM solution. We are working now with the IBM OS/2 solution, not the Microsoft OS/2 program. The latter has essentially been abandoned by Microsoft.

An OS/2 workstation connects to most hosts as a terminal, and it can connect to many servers as a client. But it probably has fewer connectivity options than a DOS or Window workstation, and probably just about as many as a Windows NT workstation, although the Windows NT workstation probably has more options overall.

The OS/2 workstation best supports NetBIOS; then IPX, which should really come from Novell; and then TCP/IP. These are the three main protocol stacks supported on the OS/2 workstation. The NDIS multiprotocol support program is highly recommended as the solution here. The solution from Novell, which has ODI for multiprotocol support, is acceptable, and it comes with the Novell product called LAN Workplace for OS/2. If you are using the NDIS solution and the IPXODI driver, you should obtain the ODISUP or the ODI2NDIS driver to connect the client and server.

Figure 5-16 shows an OS/2 workstation with all the options to connect to the different types of servers and hosts. In some cases, we indicate where a particular protocol or solution should be procured. We strongly recommend, for example, that you look at the TCP/IP for OS/2 solution from IBM if you plan to run TCP/IP connectivity from this type of workstation, rather than looking for a third-party solution.

Figure 5-16

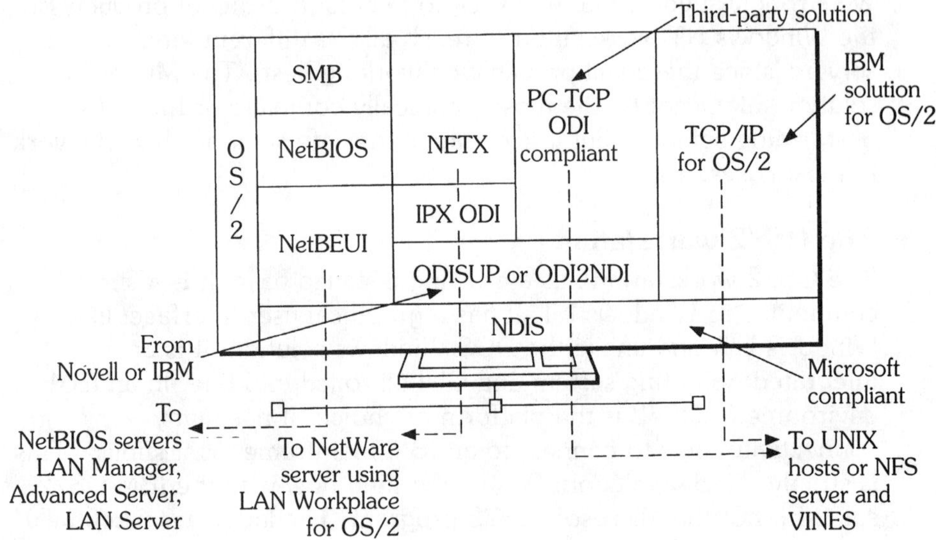

Techniques to access servers from an OS/2 workstation. Learning Tree International

Figure 5-17 summarizes the connectivity characteristics of an OS/2 workstation. It is very similar in nature to DOS and Windows NT.

✳ The Macintosh workstation

The last type of popular workstation you might encounter in corporate networks these days is the Macintosh workstation, which has a fair amount of popularity. The number of connectivity options here is limited. The Macintosh workstation primarily works, if there is going to be any network at all, as a client to an AppleTalk server. Macintosh workstations are typically found in small workgroups that share graphics and data files. It typically uses AppleTalk to connect to most PC file servers. TCP/IP might be installed on some of the workstations to connect NFS servers.

Figure 5-17

Characteristic	Novell NetWare 3.x server	Microsoft Windows NT Advanced Server	DEC PATHWORKS host	UNIX server/host	IBM LAN server	IBM host
Network/ transport protocols	IPX	NetBIOS	NetBIOS	TCP/IP	NetBIOS	NetBIOS CS/2
DFS protocol	NCP	SMB	SMB	NFS	SMB	–
Multiprotocol support	NDIS ODI	NDIS ODI	NDIS	NDIS ODI	NDIS	–
Printing	Queue	Queue	Queue	LPD	Queue	–
File transfer	NCOPY	COPY	COPY	FTP	COPY	–
Remote terminal	–	–	–	TELNET	–	CS/2
E-mail	cc:Mail MS Mail	cc:Mail MS Mail	cc:Mail MS Mail	POP SMTP	cc:Mail MS Mail	PROFS

Summary of connectivity characteristics of an OS/2 workstation. Learning Tree International

The Macintosh environment assumes that most network operating system vendors either support AppleTalk on the server to provide Mac connectivity, or provide some form of TCP/IP and NFS support. By and large, these are the two protocols that are most popular on an Apple workstation, with the AppleTalk profile being the more widespread of the two.

A Macintosh workstation, as we saw in the last chapter, can be set up as a file server using AppleShare and AppleFile print services.

Figure 5-18 shows these two types of connectivity on a Mac workstation. Typically, the Mac workstation is running System 7 as the operating system, and AppleTalk is the protocol stack of choice. These both come with the Mac workstation.

Figure 5-18

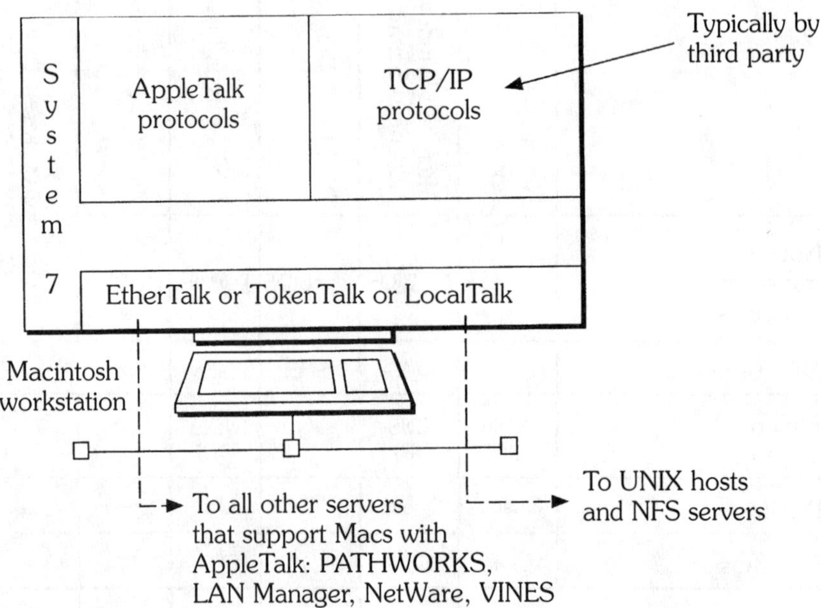

Techniques to access servers from a Macintosh workstation. Learning Tree International

Under AppleTalk, there are three choices for data link services: EtherTalk, TokenTalk, and LocalTalk. TCP/IP, as mentioned before, is supported on an Apple workstation, and it is typically supplied by a third party. It allows the Mac workstation to connect to a UNIX host as a terminal over the network, or to access NFS servers.

Figure 5-19 summarizes the Macintosh workstation. Notice that it again has very similar characteristics to the other types of workstations that we have seen so far. It prints to network queues, uses the *copy* command to most servers for file transfers, and has the same typical e-mail options.

This ends the review of workstation connectivity characteristics—what we have termed the client side of the multivendor integration

Characteristic	Novell NetWare 3.x server	Microsoft Windows NT Advanced Server	DEC PATHWORKS host	UNIX server/host	IBM LAN server	IBM host
Network/ transport protocols	Apple-Talk	Apple-Talk	Apple-Talk	TCP/IP	–	SDLC
DFS protocol	AFP	AFP	AFP	NFS	–	–
Multiprotocol support	–	–	–	–	–	–
Printing	Queue	Queue	Queue	LPD	–	–
File transfer	COPY	COPY	COPY	FTP	–	–
Remote terminal	–	–	–	TELNET	–	3270 E.M.
E-mail	cc:Mail MS Mail	cc:Mail MS Mail	cc:Mail MS Mail	POP3 SMTP	–	PROFS

Summary of connectivity characteristics of a Macintosh workstation. Learning Tree International

equation. We are going to put it all together, next, with a look at a number of case studies of integration.

Case studies

In this last section, we take up the subject of how to approach an integration problem. We go through the various steps that must be considered and the tasks that must be carried out. We favor a top-down design approach, in which you first gather information and then you apply your skills with that information to resolve the

problem. The problems to be resolved first are those of interoperability, followed by those posed by internetworking.

We apply this approach to four case studies so that you can see how the methodology works. We hesitate to call it a methodology, because it's not that formal—it is more of a tried-and-true method we employ when faced with a multivendor integration task. This is a very common-sense approach. There is nothing mysterious about it, and you do not have to learn a new theory before you can apply it. You are probably unconsciously doing this already, not only in the area of multivendor networks, but when faced with other computer or business problems.

The first case study tackles the integration of a number of workstation environments connected by LANs. The second case study addresses the other extreme, a number of large hosts that need connectivity. The third case study integrates the micro world to the mainframe world. The last deals strictly with the upper layers, where we tackle a database integration problem. We do not have a case study on e-mail integration, because that was thoroughly covered in Chapter 4.

We have two exercises in which we present the problem and let you work out the solutions for yourself. We propose some solutions, and you can see these in Appendix C after you have had an opportunity to try them out for yourself.

The approach to integration

In tackling a multivendor integration problem, we propose that you use the following approach.

1. Always focus on the most important question: What is the user trying to do?
2. Perform a thorough inventory of the assets on hand.
3. Determine the interoperable situation that needs to be resolved. Classify it according to the classes of interoperability found in Chapter 4.

4. Select the best interoperable solutions to satisfy all situations found in 3.
5. Then, and only then, consider what internetworking changes or additions must be made to support the interoperable solutions being proposed.

Let's look at these action items in more detail.

Step 1. Ask the user the most important question: What are you trying to do? You have to extract from the user, in the user's own words, exactly what problem they are having with the network that they would like to have fixed or changed. Get down to specifics. In their own words, are they trying to get a file from a computer that they are not connected to? Do they need access to a server of some kind? Try to get from them as much of the requirements and detail as possible. Focus on requirements. There is time enough for specifications in Step 4.

We strongly urge you not to use technical language with users. Speak in the user's language. Do not come to them with solutions in mind already, especially with solutions that have labels, such as acronyms. They probably won't understand you. Get from them the parameters of the task and the scope of work that needs to be done. Use the active listening technique of repeating back to the user your understanding of what needs to happen. Again, use the user's vocabulary. You will eventually be translating the problem into the technical jargon that you and your colleagues are most familiar with, but, with the user, keep it simple. The KISS principle is best.

Step 2. At this point, do not try to classify the solution. Do not even attempt to find a solution. First perform a thorough inventory of the environment that the user is operating in, so that you know what is available. Document the user applications that require network access, the client and server platforms involved in this problem, and any network operating systems that are involved. Document any internetworking protocols. For example, are the users running IPX, or are they running NetBIOS? Document all types of network interface cards in servers and clients. Document any connectivity

devices that are already in place, such as bridges, routers, or gateways, and document any application protocols that might already exist on servers and hosts and on the client workstation, even if they are not currently being used (but might be of use to you).

Throughout the inventory, keep your eyes open for any tools already in place that might be used to solve the problem without further purchases or installations.

Step 3. The next step is to classify the solution approach. Determine what interoperable situation needs to be resolved. Classify the situation according to the eight classes of interoperability discussed in Chapter 4. We repeat them here for the benefit of the discussion.

1. Remote terminal access
2. Remote procedure execution
3. File transfer
4. Distributed file system
5. PC file server integration
6. Electronic mail access or integration
7. Distributed databases
8. Client/server applications

Classifying the type of integration problem helps you look for solutions to the problem. From the user's statement of the problem, you might extract more than one of these as the scope of work. That's perfectly acceptable. As you will see, we normally have more than one interoperable problem to solve at any time. Some of them go hand in hand, so you might wind up with a list of two, three, even four interoperable situations that need solutions.

Many of these are classified according to the pair of devices that need the connectivity. You might have one interoperable problem to solve from one type of client to a particular type of server. Then, you might have to solve a different interoperable problem from another type of client to another kind of server. All of this might be one integration problem that you are tackling. Classify the problem according to the

interoperable situations that need a solution, and which client/server pairs require that particular solution.

Step 4. Design from the top down. We are finished with the requirements—now we are looking for the specifications of a solution.

To design from the top down, look at the highest levels of the protocol stack and work toward the lowest levels. It makes sense, then, according to the way we have broken up network integration, to first look at interoperability, and then look at internetworking.

We strongly urge you to stay away from internetworking. Don't think of the bridges, routers, or gateways that are needed to solve the problem before you understand whether it's a file-transfer problem and what solutions are available, because the nature of the interoperable solution will modify the required internetworking infrastructure that supports it.

Look, then, for the interoperable protocols, programs, or network operating systems that provide the most general solution. Strive for generality. Find a solution that is as universal as possible. That's why we have labeled TCP/IP a universal solution in most of our interoperable situations in Chapter 4. It has the widest possible applicability. Look for a solution that not only provides an answer to the problem that you are tackling, but also addresses the future and the growth of your organization. Your present solution might provide additional solutions in the future if needed.

Sometimes you are forced to use a solution that has a limited future or potential for growth. If there is no other solution that solves the current problem, then by all means go ahead and put it in. A poor solution is better than no solution at all. But strive for generality if you can.

Once you have classified the interoperable problem, look through the solutions offered in Chapter 4, or any others you have learned along the way. This is where, as a network integrator, it helps to accumulate a bag of tricks that you can dip into whenever an interoperable problem comes up.

Also, if there are a number of interoperability situations that you have identified that require solution, do not move on to integration unless

you have come up with a plan to provide for interoperability in all of the required situations. In other words, it's best to find solutions for all of your interoperable problems at the level of interoperability before moving on to internetworking.

Step 5. Once this is done, and you have an idea of what interoperable solutions are required, you can look at the internetworking infrastructure already in place and determine what new internetworking to add to support that interoperability. At this point, you have to ask what must be added or changed in the network infrastructure to make your interoperability solutions work. You might wind up having to replace all the network cards in a particular set of workstations in a certain local area network just to make it work with the new protocol that you are going to introduce to solve your interoperability problem. That's perfectly acceptable, but you won't know that until you have decided upon the interoperable solution.

You might introduce additional protocol stacks at certain workstations—in other words, do a client-side solution in some cases, for which you will require a multiprotocol stack protocol solution. Turn to the earlier part of this chapter to find out whether you are going to use NDIS, ODI, or the Packet Drivers, for example.

In some cases, you will find that the users' requirements force internetworking integration in your network. This might be the tail that wags the entire network dog. As we saw with e-mail integration, as soon as users begin requesting that the entire organization share e-mail, you begin considering the corporate-wide infrastructure of connectivity. In other words, you begin introducing bridges and routers to connect everyone together, if they have not already been connected.

Don't forget to consider all the other issues involved in solving this problem. Three major ones are corporate direction, future growth, and corporate culture and politics. Whatever solution you come up with should be in line with corporate direction, in order to maximize the chances of being subscribed to by upper management. You might be going off in a new direction, which you in your heart of hearts believe would be the right one for the corporation. If this solution happens to be the first one—the tip of the spear—to head off in that direction, then you have a big selling job ahead of you. Be prepared for that.

Also, when looking at interoperable and internetworking answers, be sure to consider future growth of the corporate network. This means proposing something that will still be around five to ten years from now, and that is capable of providing other solutions that you know people will need in the future.

If your solution is a radical departure from the corporate culture and there is no way to move the corporate culture in your direction, then you might have to go back and rethink the technical solution along the lines of what's already in place.

This brings us to one last point, which is to consider corporate culture and corporate politics. Be careful that you do not run afoul of managers or vested interests within the corporation, that the lines are crossed properly, that you do not encroach on people's turf without their permission, and that everyone buys into this integration. Otherwise, you will have a hard time implementing what might appear to be a perfectly good technical solution—but one that makes absolutely no political sense to those who have the power to approve or disapprove of it.

We now have enough information to tackle a number of case studies, and you will see this approach applied to these situations.

✳ Case study 1: NetWare/UNIX/Macintosh integration

In this first case study, we take up integration of a NetWare/UNIX/Macintosh environment. Please refer to Fig. 5-20, which shows the network environment in question. In each of the case studies, we will follow a certain plan of attack. First, we will look at the requirements—in other words, we have asked the user what he is trying to do, and from his response boiled it down to a few sentences describing the task at hand. Then we are going to go through the analysis—essentially, step number two—to find out what is in place, the inventory of all the network protocols, servers, operating system platform types, and so on. Then, to follow the procedure we outlined earlier, we are going to look at the interoperable solutions, and then we will see what internetworking changes must be made to provide for the new interoperability.

Step 1: Requirements definition In this case, the marketing PCs in Network C, Building 1; the engineering Sun workstations in Network

Figure 5-20

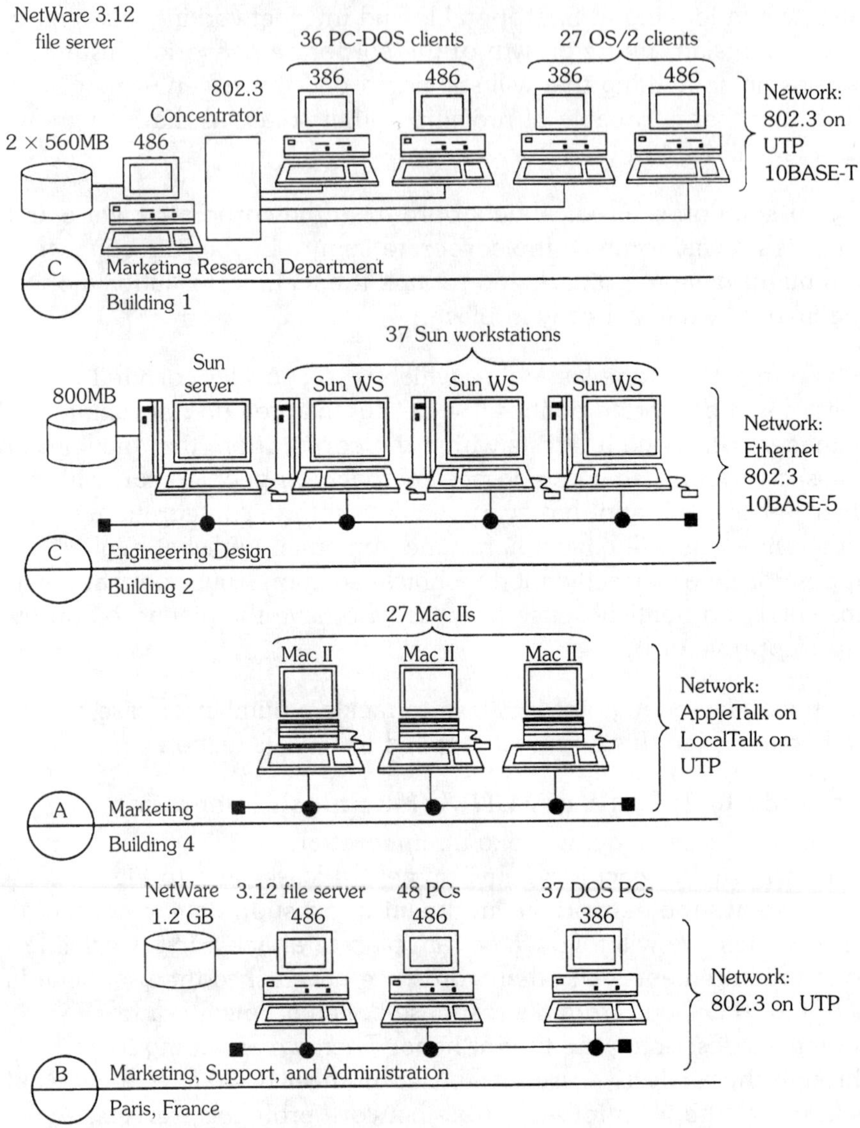

NetWare 3.12 file server
36 PC-DOS clients
27 OS/2 clients
386
486
386
486
802.3 Concentrator
2 × 560MB 486
Network: 802.3 on UTP 10BASE-T
C
Marketing Research Department
Building 1

37 Sun workstations
Sun server
800MB
Sun WS
Sun WS
Sun WS
Network: Ethernet 802.3 10BASE-5
C
Engineering Design
Building 2

27 Mac IIs
Mac II
Mac II
Mac II
Network: AppleTalk on LocalTalk on UTP
A
Marketing
Building 4

NetWare 3.12 file server
1.2 GB
486
48 PCs
486
37 DOS PCs
386
Network: 802.3 on UTP
B
Marketing, Support, and Administration
Paris, France

The networks to be integrated in Case Study 1. Learning Tree International

C, Building 2; and the marketing Macintosh workstations in Network A, Building 4, must exchange files and print to one another's printing environments. In addition, we must integrate whatever e-mail systems are in place, or add new e-mail systems, to achieve electronic mail connectivity among all users on any of these networks.

We should add that the marketing personal computers in Network B—the Paris, France, branch of this company—must also take part in any interoperability we want to provide to the rest of the workstations just described. All four local buildings are located in San Jose, California.

We assume that all of these workstations reside within one company (we are talking here of a corporate network), and that you have been given the responsibility to make a proposal and report directly to the appropriate level of corporate management, perhaps the CIO, who can direct the integration of these networks. Needless to say, you must work hand-in-hand at a peer level with the department heads and the users in each department to be sure that they agree with whatever solution you come up with. But the decision for integration is going to be approved or disapproved by your boss in upper management, perhaps a member of the Board of Directors of the organization.

Step 2: Analysis In this step, you must take an inventory of the protocols, operating systems, network operating systems, platform types, applications, and so on, that bear upon the problem. By looking at the network, you discover that Network C, Building 1, has a NetWare server. There are DOS machines and OS/2 machines of the Intel 386/486 variety. They are wired with 802.3, 10Base-T, and although it is not stated, more than likely they are running IPX and NETX as the network protocols to talk to their NetWare server.

On Network C, Building 2, there is a group of Sun workstations, with a Sun server, running TCP/IP and the entire range of TCP/IP application protocols: FTP, TELNET, NFS, etc. They are wired with Ethernet 10Base-5. The network protocol in this case is IP.

Network A, Building 4, the marketing network, is a group of Macintoshes. They are Mac IIs running AppleTalk, sharing files in a peer-to-peer networking environment. They are running AppleTalk over LocalTalk.

Network B, in Paris, is a group of DOS PCs connected with 802.3 10Base-T. The Paris PCs are connected to a NetWare file server. They are using IPX and NETX as the protocols for connectivity.

Let's enumerate the operating systems. On the workstations we have DOS, OS/2, Sun UNIX, Macintosh System 7. For file servers we have NetWare 3.12 and Sun UNIX. The protocols we are dealing with are Novell IPX, TCP/IP, and AppleTalk. The network card types installed (i.e., the data link protocols employed) are Ethernet and LocalTalk.

Step 3: Identification of interoperable problems The next step is to identify the interoperable situations we are dealing with here. Since the first requirement is file exchange, this is a file-transfer interoperable situation. The second situation is printing. Typically, we consider printing part of the network operating system connectivity, so it is a PC network integration problem.

The third situation is e-mail integration. cc:Mail post offices are currently installed in Network C, Building 1, Network A, Building 4, and Network B in Paris, and the Sun workstation network has an SMTP mail post office on the Sun server with a Sun workstation using some form of POP3 client protocol.

Step 4: Proposed interoperable solution There are many ways to tackle this problem and satisfy user requirements. What follows is one possible approach, and it is the approach that the authors are familiar and comfortable with. You might have a different background and different experience, so you might see other ways of solving this problem. You should do what is familiar to you, as long as it is reasonable and it follows the basic principles outlined here.

One basic principle to follow is to implement a solution that allows for future growth. Another is to develop solutions that are as universal as possible. Be sure the solution is efficient. This solution is not necessarily optimal, and is not the most technically correct solution. It is definitely not an elegant solution. But it's a solution that will work. It is a solution that costs the least, and will make the users reasonably happy with the performance.

For file transfer, we propose connecting all the Novell file servers, so that users move files from one server to another. For the PCs, we do not propose peer-to-peer networking, but actually use the file server as a focal point for transfer. To facilitate the Sun workstations communicating with the DOS machines, we install NetWare NFS on the

Building 1, Network C, Novell server. That will allow both file transfers and print services between the two environments (DOS and UNIX).

To connect the Macintosh environment to the rest of the networks, we propose putting NetWare for Macintosh on that same file server. This approach will provide file transfer capabilities from any machine to any machine via file servers. It also solves the second requirement of printing anywhere on the network by connecting the print queues in Novell, UNIX, and the Macintosh environments to one another via these two products (NetWare NFS and Network for Macintosh).

For the third requirement of e-mail integration, we tie all of the electronic mail post offices to one another with a cc:Mail router, possibly installed on Network C, Building 1, and provide a cc:Mail SMTP gateway, perhaps on the same computer, to connect the Sun post office to the rest of the enterprise.

Figure 5-21 is an overlay of our solutions on the network diagram, showing file sharing, print sharing, and e-mail exchange solutions. What remains is to determine the additional internetworking changes that must be made to satisfy these interoperable solutions.

Step 5: Necessary changes to internetworking Notice immediately that these networks are not connected to one another. There must be some form of router or bridge connection among the various networks. Since we only have two protocols to deal with, IP and IPX, and they are both eminently routable, we propose a routing solution. We assume that the Building 1, Building 2, and Building 4 networks are close enough to install one router. As a matter of fact, to make the solution inexpensive, we propose that the Building 1, Network C file server become the router for IP and IPX, and from it provide a wide area network connection to the NetWare server in Paris. This connection only needs to route IPX.

The connection between the Sun server and the NetWare server in Building 1 is going to route IP, and the connection between the Macintosh network and the NetWare server in Building 1 has to route AppleTalk. You can put a LocalTalk adapter in the Building 1 NetWare server, or switch the Macintoshes to Ethernet and put an Ethernet adapter in the server in Building 1 for the Mac segments.

This completes the analysis and solution for Case Study 1.

Figure 5-21

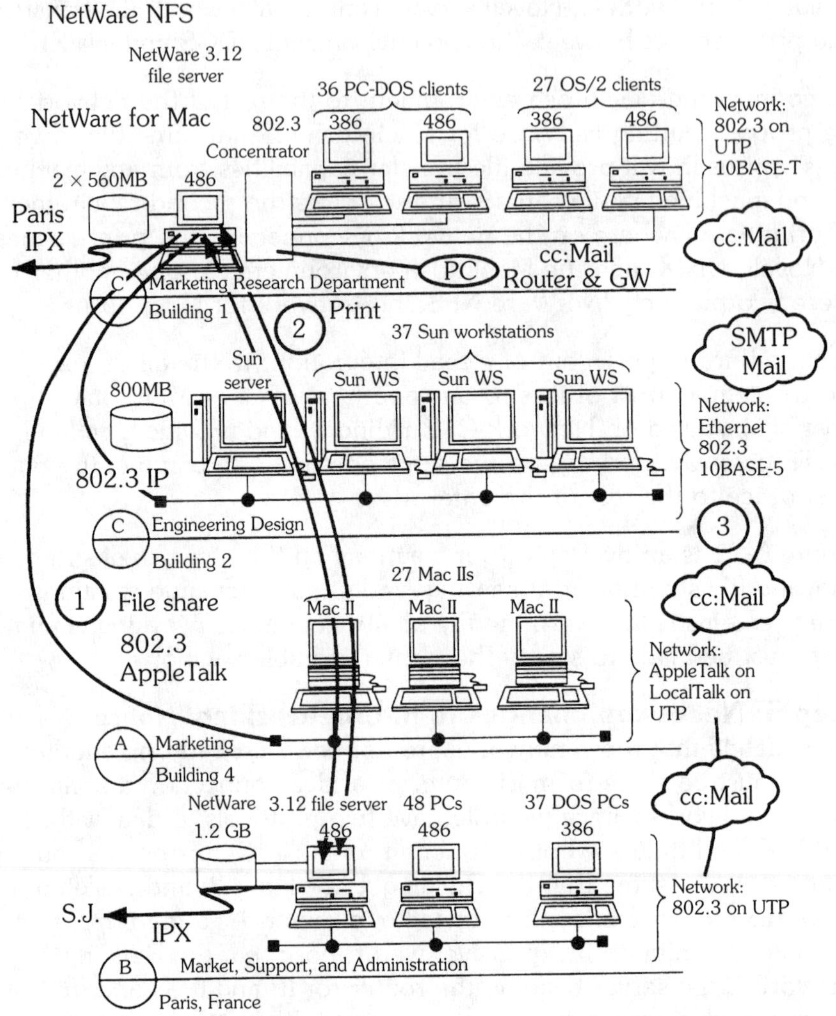

One possible solution to the problem presented in Case Study 1.
Learning Tree International

✻ Case study 2: VAX/IBM/large host integration

This environment has three large multiuser computers: a DEC VAX, an IBM mainframe, and an HP 9000 minicomputer. Figure 5-22 shows three multiuser hosts and their terminal environments. The three machines are collocated on the same campus in three buildings. These three buildings are in San Jose, on the west coast of the United States, and need the following connectivity.

Figure 5-22

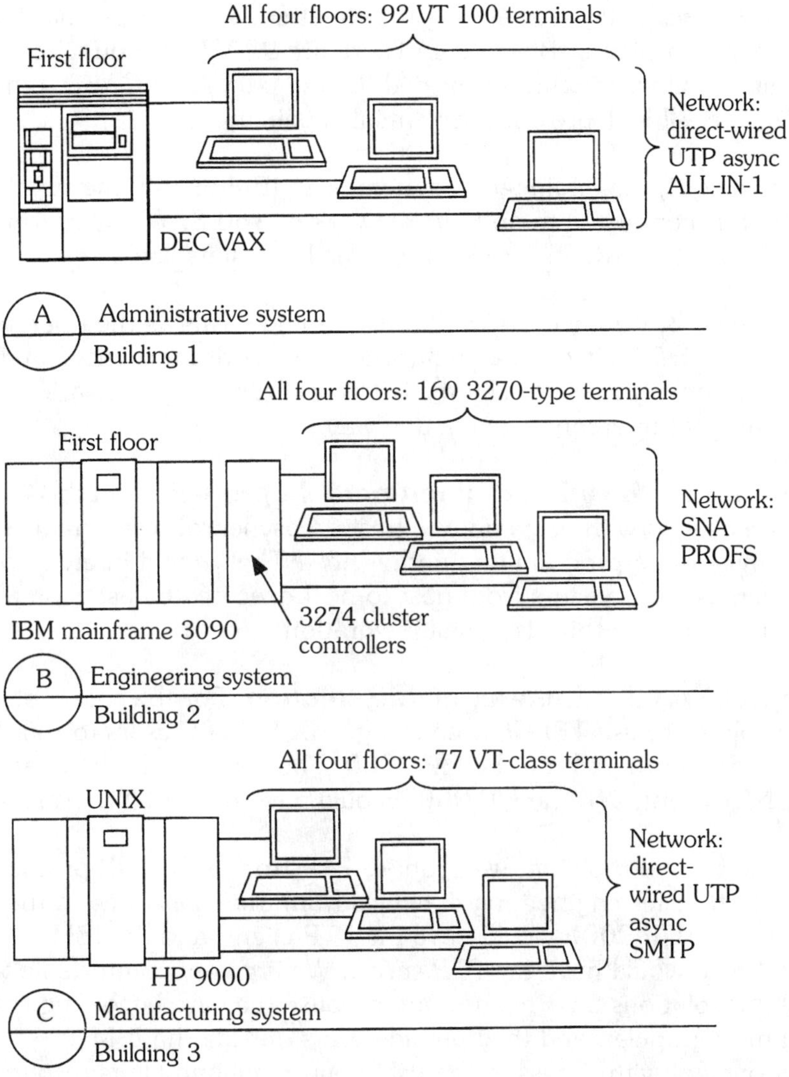

The networks to be integrated in Case Study 2. Learning Tree International

Step 1: Requirements definition The first requirement is that the VAX users have access to a job-tracking system on the manufacturing HP 9000 computer. The second is that manufacturing personnel using the HP 9000 must download engineering drawings produced by engineers on the IBM mainframe. Lastly, we must integrate the overall e-mail system.

Step 2: Analysis In analyzing the situation, we discover that there is a multiuser host in Building 1. There are 92 VT-100 direct-connect terminals, all wired with unshielded twisted pair. A DEC VAX running VMS with All-In-1 provides the e-mail capability.

In Building 2, the engineering system is an IBM mainframe running MVS. It is connected to an SNA network of 160 3270-type terminals, and is running PROFS for electronic mail for all its users.

In Building 3, we have the HP 9000 running UNIX as the operating system, with 77 VT-class asynchronous direct-wire terminals, and it is running some form of SMTP mail. These three computers are currently not interconnected in any way.

Step 3: Identification of interoperable problems In this case, we are dealing with users on one host on asynchronous terminals who must run a program on another host. The second interoperable problem is file transfers from host to host executed by users on a terminal, and the third is e-mail integration.

Step 4: Proposed interoperable solution For the first problem, we propose to use TELNET, enabling VAX VT-100 users to run the job-tracking program on the HP 9000. The VAX would then have the TELNET client, and the HP 9000 would have the TELNET server.

For the second problem, we propose FTP from a TCP/IP protocol suite to transfer engineering drawings from the mainframe to the HP 9000. The HP 9000 would have the FTP client, and the IBM mainframe would have the FTP server. We are very comfortable with TCP/IP solutions on a mainframe, because they are becoming more and more popular, and they provide users outside the IBM environment with access to the IBM mainframe and all its resources. It is very common to run TCP/IP and provide server services on the IBM mainframe from a TCP/IP protocol suite.

For the third requirement, we plan to find a gateway spanning the three e-mail systems. We follow the principle of keeping the users on their existing individual e-mail systems, and provide background e-mail exchange among the hosts via some other type of computer that hosts the e-mail software. For this solution, we can use either a product that

runs on one of the hosts, or a separate computer that does the exchange with the one-to-one gateways between All-In-1, PROFS, and SMTP mail. (See Fig. 5-23.)

Step 5: Necessary changes to internetworking Figure 5-23 shows these solutions in place, with the additional requirement of

Figure 5-23

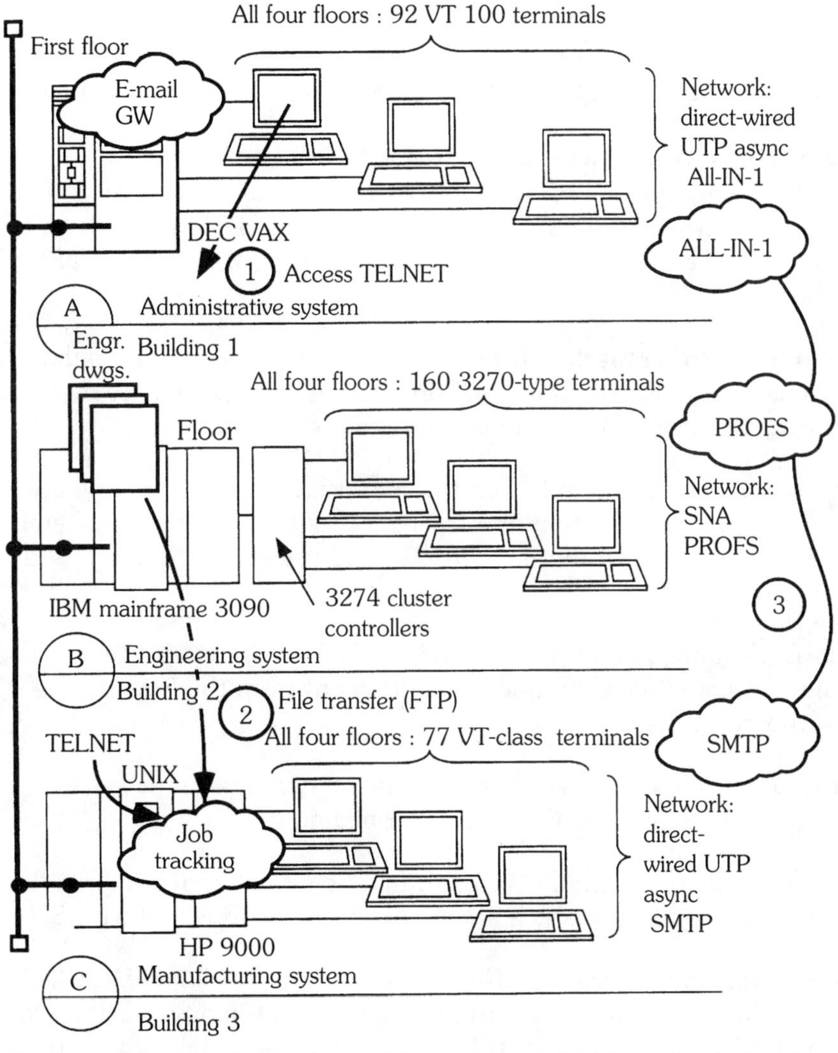

One possible solution to the problem presented in Case Study 2. Learning Tree International

internetworking among the three large hosts. We propose an Ethernet network with three Ethernet adapters, one for the VAX, one for the IBM mainframe, and one for the HP 9000, with TCP/IP as the protocol suite in each machine for connectivity. If the buildings are close enough to one another, one Ethernet segment might be sufficient. Otherwise, you might need several segments with repeaters. In the worst case, routers might be needed—some with wide area network connections—to put this connectivity in place.

We are actually going to run the e-mail gateway as shown in Fig. 5-23 on the VAX, since there are many VAX solutions for e-mail gateways.

❊ Case study 3: PC workstation/large host integration

In this environment, we have the IBM mainframe in Building 2 with 160 engineering terminals, and a Novell NetWare Network B, also in Building 2. The engineering network in Building 2 is a group of Sun workstations.

Step 1: Requirements definition The first requirement is that the documentation users and engineering design users be able to download drawing files from the IBM mainframe to their workstations.

The second requirement is that the Sun users and PC users be able to run a drawing preparation and tracking program on the mainframe. This program was written in COBOL, and runs under CICS.

The third requirement is that the engineering design personnel on the Sun workstations be able to upload marked-up drawings to the mainframe, so this is an upload requirement from the Suns to the mainframe.

Step 2: Analysis Figure 5-24 shows the network in question. Notice that all three networks are in Building 2.

The first network is the IBM mainframe network, Network A, with 160 3270-type terminals. It's a multiuser host connected with SNA.

The documentation system, Network B, has a Novell NetWare file server. The workstations are running DOS and Windows. They are connected to the server via Token Ring on twisted pair. They are running IPX and NETX to connect to the Novell server.

Figure 5-24

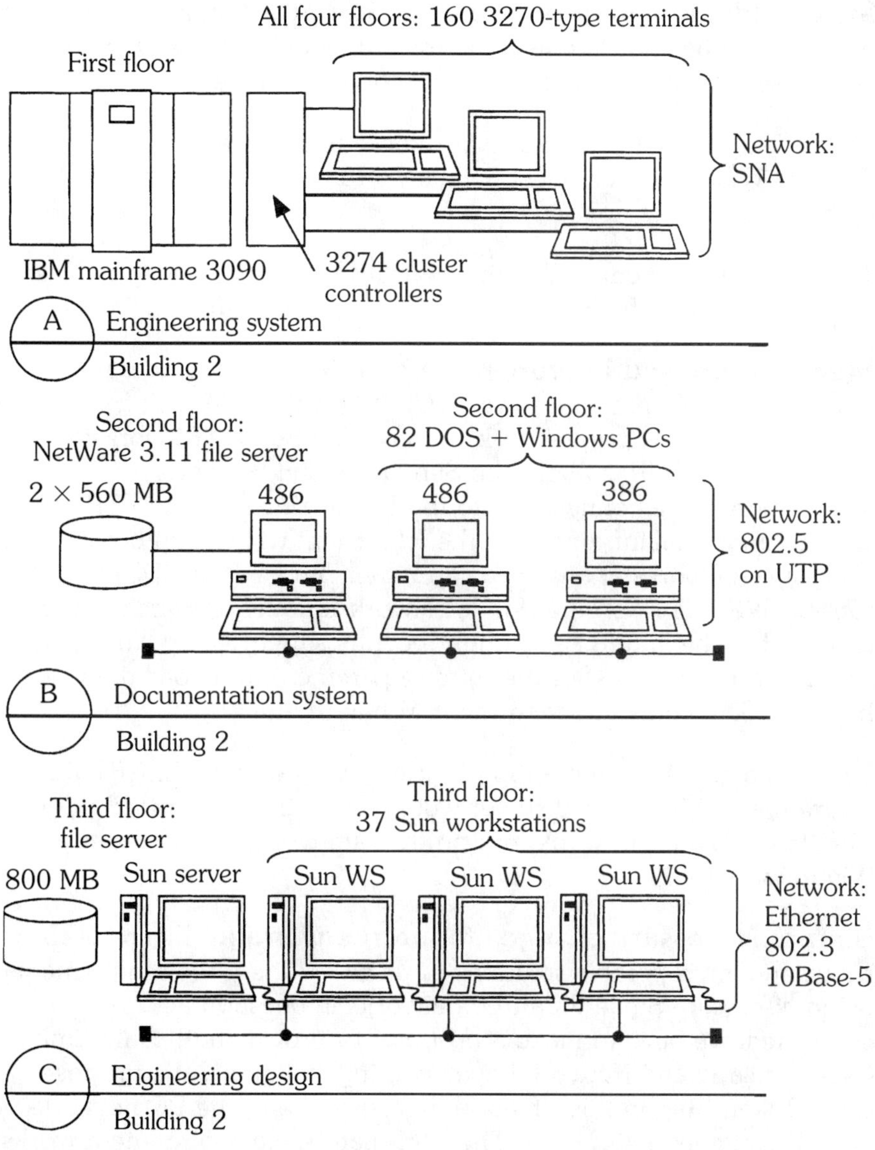

All four floors: 160 3270-type terminals
First floor
Network: SNA
IBM mainframe 3090
3274 cluster controllers
A Engineering system
Building 2

Second floor: NetWare 3.11 file server
Second floor: 82 DOS + Windows PCs
2 × 560 MB
486
486
386
Network: 802.5 on UTP
B Documentation system
Building 2

Third floor: file server
Third floor: 37 Sun workstations
800 MB
Sun server
Sun WS
Sun WS
Sun WS
Network: Ethernet 802.3 10Base-5
C Engineering design
Building 2

The networks to be integrated in Case Study 3. Learning Tree International

The engineering design system, Network C in Building 2, is a Sun UNIX network with several workstations running UNIX, and TCP/IP as the protocol stack. They are all connected via Ethernet 10Base-5 wiring.

Step 3: Identification of interoperable problems The first problem can be classified as a file transfer problem. The DOS and UNIX users must download files from the mainframe.

The second requirement is one of remote terminal access. Again, the Sun and Windows users must be able to run a program on an IBM mainframe, a large host.

The third requirement is also for file transfer, in that drawings must be uploaded from the Sun workstations to the mainframe.

Step 4: Proposed interoperable solution To satisfy user requirements, we propose to run TCP/IP on the IBM mainframe, and IPX on the DOS Windows workstations. The Sun UNIX workstations already run TCP/IP, so both the Sun UNIX and Windows workstations have connectivity to the IBM mainframe. An FTP server will run on the mainframe, and the FTP client will run on both categories of workstations. You can either run NetWare ODI to support IPX and IP on the Windows workstations, or switch entirely to TCP/IP. The latter is a solution recently supported by Novell. The FTP solution also satisfies the third requirement to upload drawings from the Sun workstations to the IBM mainframe.

To run the drawings operation program, we propose TELNET 3270, where the TELNET 3270 server would be on the mainframe and the TELNET 3270 client would run on the Sun workstations and the Windows PCs.

Step 5: Necessary changes to internetworking Figure 5-25 shows the interoperable solutions in place for the three requirements being discussed. Since the three networks in question are not connected, we have to make a proposal to provide both Data Link (network card) and Network Layer (routing) solutions. We propose using Token Ring in this situation, because one of the PC networks is already running Token Ring. The IBM mainframe environment works best with Token Ring solutions, rather than Ethernet, and it is easy to bring the group of Sun workstations into a Token Ring network via a Token Ring/Ethernet bridge.

Our proposal, then, is to put in a 16-Mbps Token-Ring backbone with the mainframe directly connected to it, and to connect the NetWare

Figure 5-25

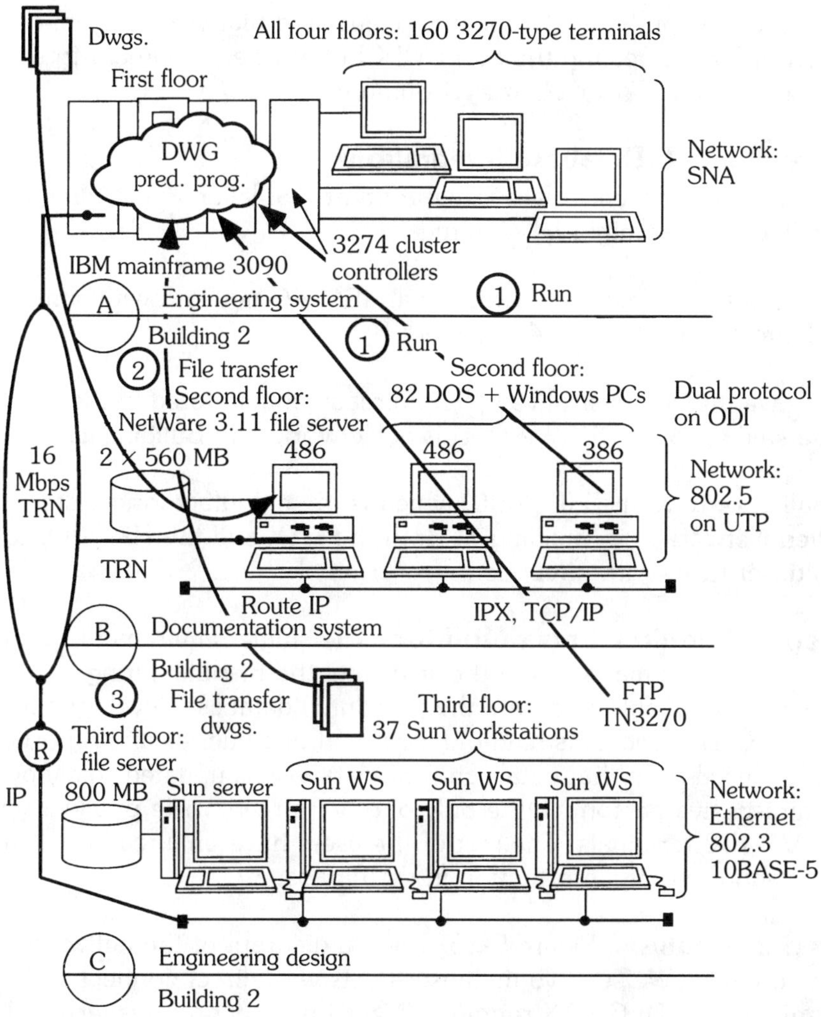

Dwgs.
All four floors: 160 3270-type terminals
First floor
DWG pred. prog.
Network: SNA
3274 cluster controllers
IBM mainframe 3090
Engineering system
1 Run
A
Building 2
1 Run
2 File transfer
Second floor: NetWare 3.11 file server
2 × 560 MB
Second floor: 82 DOS + Windows PCs
Dual protocol on ODI
486
486
386
16 Mbps TRN
Network: 802.5 on UTP
TRN
Route IP
IPX, TCP/IP
B
Documentation system
Building 2
3 File transfer dwgs.
Third floor: 37 Sun workstations
FTP TN3270
R
Third floor: file server
800 MB
IP
Sun server
Sun WS
Sun WS
Sun WS
Network: Ethernet 802.3 10BASE-5
C
Engineering design
Building 2

One possible solution to the problem presented in Case Study 3.
Learning Tree International

file server in the Building 2 documentation system to the 16-Mbps backbone via an additional 16-Mbps NIC in the Novell NetWare 3.11 file server. The file server in this case would be acting as an IP router, enabling the IP file transfer and IP terminal access protocols to run through it to the mainframe.

For the Sun workstation environment, we propose an IP router connecting Ethernet on one side to the 16-Mbps Token Ring on the

other. This is probably the most efficient, and higher-performance, solution for connecting the Sun UNIX Ethernet environment to the IBM mainframe Token-Ring environment.

✳ Case study 4: Database integration

The last case study involves integration at OSI layer 7, including any middleware that might be required.

We have a sales network, which is an IBM PC network with some Macintoshes in Building 4, Network B.

We have the administrative system already encountered in the other case studies, on a DEC VAX. This is Network A in Building 1.

Finally, we have the HP 9000, which is the manufacturing system called Network A in Building 3. Both the VAX and the HP 9000 are multiuser hosts with direct-connect terminals.

Step 1: Requirements definition The single requirement in this case is that the managers in the sales and the manufacturing department be able to access the personnel database developed for the VAX. This access is read-only, with ad hoc queries and reports to be produced from that database. The database is updated solely by administrative personnel (the only ones with read/write privileges) via the VT-100 terminals. The rest of the users have only read privileges, and cannot modify information in the database.

Step 2: Analysis Figure 5-26 shows a diagram of the network environment. We see two multiuser hosts with direct-connect terminals: the DEC VAX running VMS with asynchronous terminals, and the HP 9000 running UNIX, also with asynchronous direct-connect terminals.

The Building 4 sales network, with a NetWare 3.11 file server, has a mixed population of DOS and DOS+Windows workstations, and some Apple Macintoshes. The Building 4 sales network is running the 802.3 Data Link protocol on 10Base-2 wiring.

It is interesting to note the database types used in all three environments. The manufacturing system has an Oracle database, but

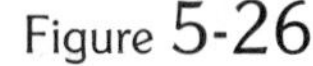
Figure 5-26

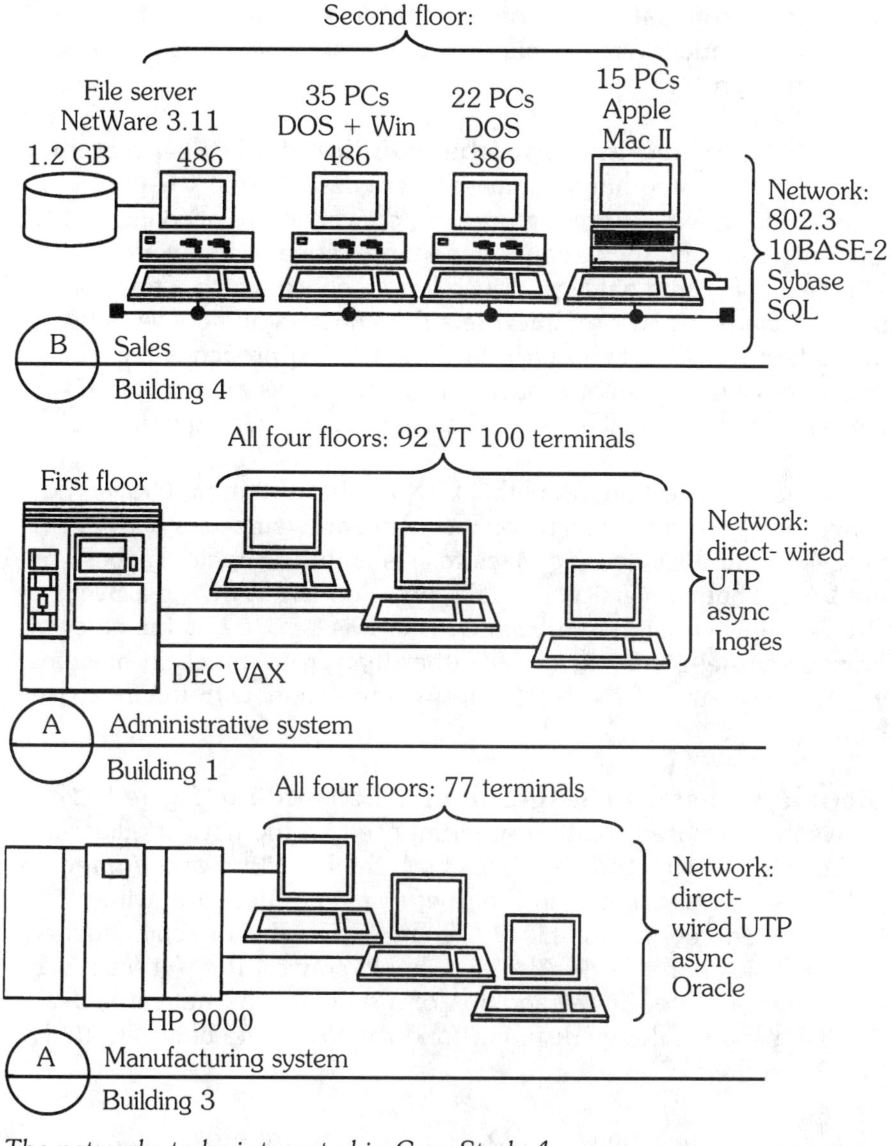

The networks to be integrated in Case Study 4. Learning Tree International

this is unrelated to the problem that we are trying to solve here. The DEC VAX is running Oracle (not shown), and the NetWare people on the Building 4 sales network are running on a Sybase SQL server.

Step 3: Identification of interoperable problems In this case, there is only one interoperable problem to be solved: the database integration problem.

Step 4: Proposed interoperable solution Remember that three approaches to integrating databases were described in Chapter 4. The first is user-driven, where the protocol conversion to account for SQL differences among the various vendors was done at the workstation. The next is called a gateway-driven approach, in which a third-party program and/or platform interprets SQL queries and converts them to the target SQL database format. The third approach is the database server-driven approach, in which queries are converted to the target SQL language by the database server being queried.

Since the HP 9000 and the DEC VAX are both running Oracle, no conversion is necessary. The only conversion we need is between Sybase on the PCs and the Macintoshes and the Oracle database on the VAX. That conversion can take place on the VAX, via a Sybase Oracle gateway purchased from Oracle. It is best to buy the gateway from the target machine vendor rather than from the client machine vendor to be sure of having the best compatibility with the database being queried.

Step 5: Necessary changes to internetworking Figure 5-27 shows the proposed solution superimposed on the network diagram. To be able to reach the VAX from both the HP 9000 and from the PC network, we propose an Ethernet local area network, with Ethernet adapters for the HP 9000 and the VAXes directly attached to that Ethernet network. Also, we should convert the NetWare 3.11 server into a router by the addition of a second Ethernet card. Use TCP/IP between the workstations and the VAX, and between the HP 9000 and the VAX.

We propose connecting the VAX and the NetWare server, and use a dual protocol stack on the Macintosh by running AppleTalk on one side and TCP/IP on the other. TCP/IP is needed to access the VAX server.

The Sybase SQL gateway, in this case, is attached to the backbone. Alternatively, we might have run a database gateway between Sybase and Oracle on the VAX machine.

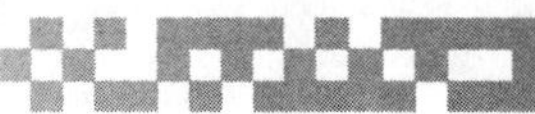

Figure 5-27

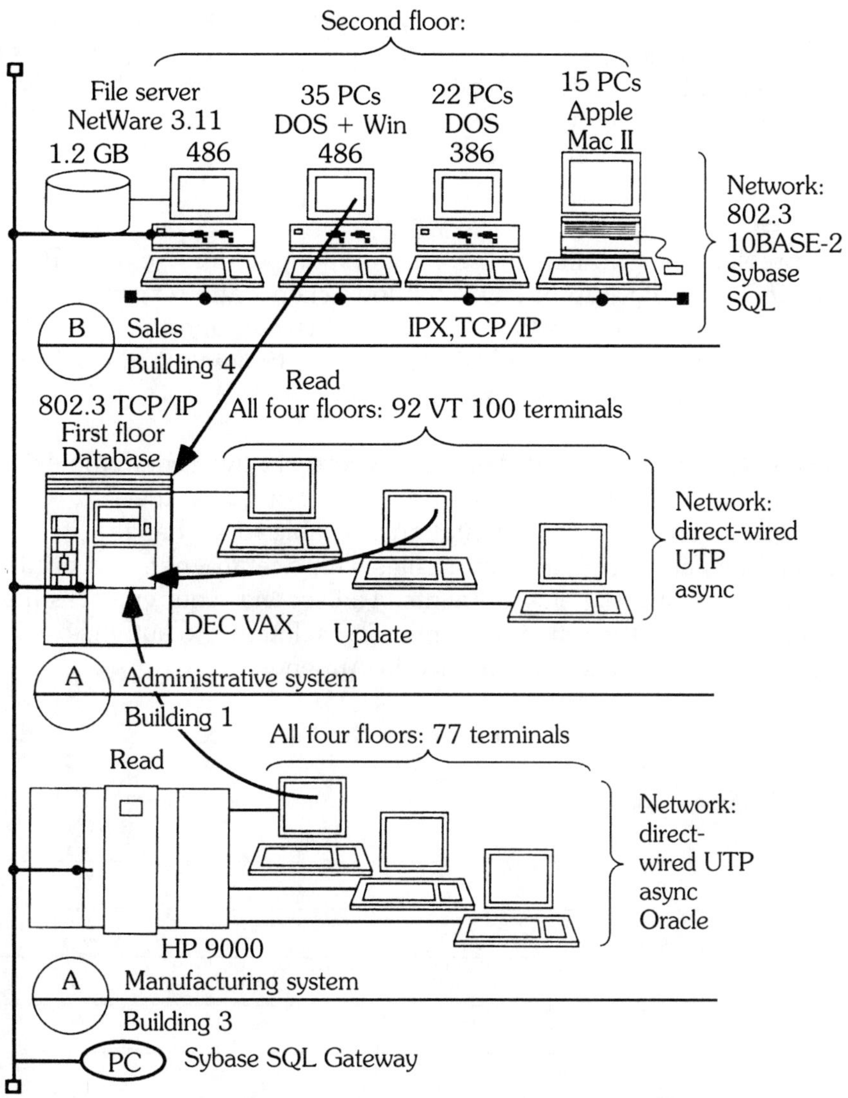

One possible solution to the problem presented in Case Study 4. Learning Tree International

Now that you have seen four integration situations, we would like to challenge you to try your hand at integrating two network environments for the same company. Again, assume that you have been given responsibility to investigate the problem at hand, find out

the users' requirements, and return your proposed solution to the appropriate management. We urge you to use our five-step process of investigation, being careful to take care of interoperability first, and then to supply internetworking changes to support your interoperability solutions. Have fun.

* Exercise 1

This problem has two networks: the marketing design Network B in Building 1 with a group of DOS Windows clients and Macintosh clients running Novell NetWare 3.11 on Ethernet, and the engineering design department Network A in Building 2 running Ethernet through a Sun server.

Step 1: Requirements definition There are two requirements in this exercise. The first is that the personal computers, the Macintoshes and the Sun workstations, be able to exchange spreadsheets and word processing files with one another, and print to one another's printing environments. The second requirement is the full integration of the e-mail systems. The solution the authors propose for this problem is outlined in Appendix C.

Figure 5-28 shows the two networks to be integrated.

Step 2: Analysis

____

____

____

____

____

Step 3: Identification of interoperable problems

____

____

____

____

____

Figure 5-28

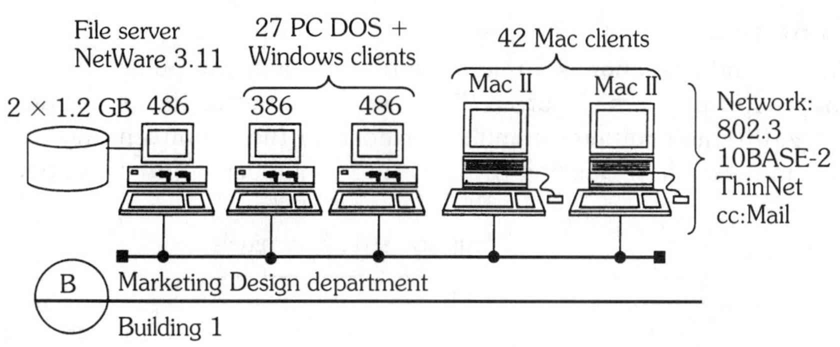

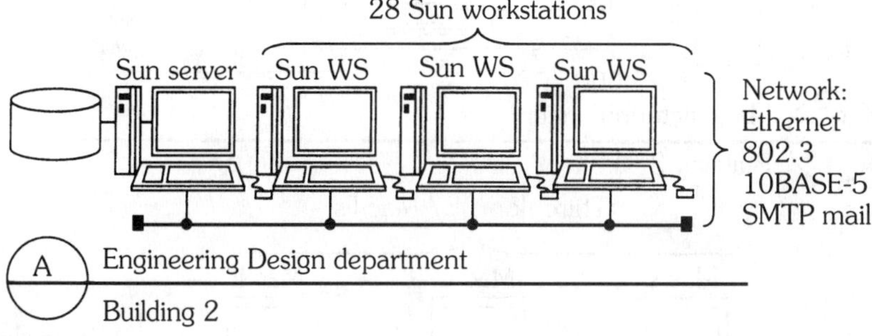

The networks to be integrated in Exercise 1. Learning Tree International

Step 4: Proposed interoperable solution

Step 5: Necessary changes to internetworking

✻ Exercise 2

The second challenge is a microcomputer-to-minicomputer integration problem. Figure 5-29 shows three networks. One is an HP 9000 minicomputer, a multiuser host for the manufacturing system network in Building 3, which has a number of direct-connect

Figure 5-29

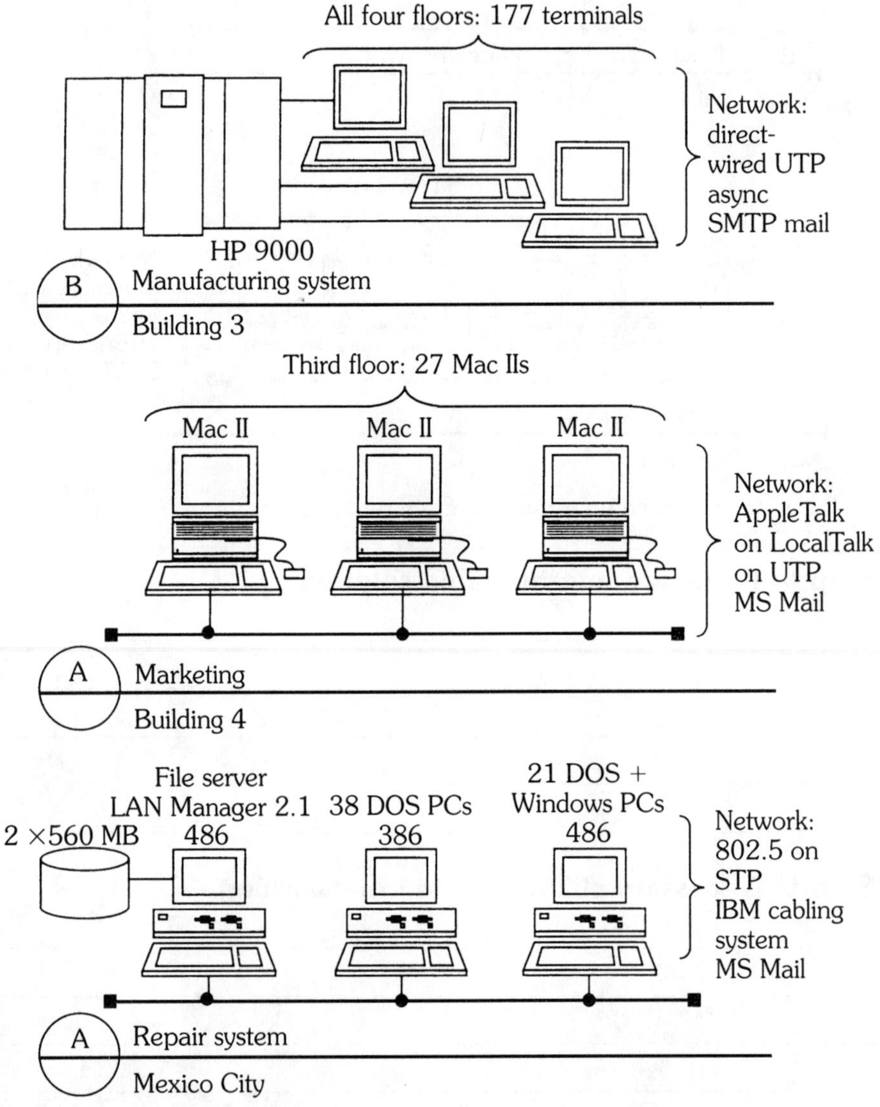

The networks to be integrated in Exercise 2. Learning Tree International

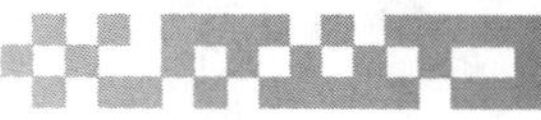

asynchronous terminals. There is a marketing network of 27 Macintoshes using AppleTalk on LocalTalk in Building 4. Finally, there is a remote network in Mexico City, consisting of over 50 personal computers, some running DOS, some running Windows connected with Token Ring to a LAN Manager file server. Remember that Buildings 3 and 4 are located in San Jose, California, with the third network in Mexico City. Keep in mind the distances between the two locations.

Step 1: Requirements definition There are three requirements in this case. The first is that the Macintoshes and the PCs be able to log into the HP 9000 host and run programs on the HP 9000.

Step 2: Analysis

____

____

____

____

____

Step 3: Identification of interoperable problems

____

____

____

____

____

Step 4: Proposed interoperable solution

____

____

____

____

____

Step 5: Necessary changes to internetworking

OSF/DCE, client/server, and downsizing

CHAPTER 6

WE have seen that there are a number of approaches that can be taken when attempting to build integrated networks. Almost all of these approaches favor the use of some form of standards. This raises additional problems, since selection of a standard is challenging in today's volatile marketplace. Ideally, we would like to have a standard that is vendor-independent and provides maximum flexibility in integrating all network components. The standard has to meet the needs of a variety of individuals, including application developers, network managers, and end users. How can this be accomplished?

During the days of monolithic environments, it was possible to perform all integration at the central computer site. By relying on a single environment, organizations could easily integrate programs, data, and networks. This is no longer possible, for a number of reasons. Given the large processing requirements of most organizations, it has become impossible to purchase a single machine that satisfies the total demand. This is especially true if you consider the cycles needed to maintain a graphical user interface at a single desktop station. In addition, reliance on a single processor platform approach poses a serious danger. Failure of the central site can severely impact the organization's ability to operate. For years, the solution has been sophisticated disaster recovery and contingency planning. While this has generally been successful, the cost of providing the contingency has been quite high. Some better way is required.

When you factor in the high cost of owning and operating a large central computer, it becomes evident that an alternative is required. So we have to consider approaches that provide a solid foundation for future systems while meeting current requirements. Many techniques have been suggested; we examine three of the most common philosophies.

We begin by examining the Open Software Foundation's (OSI) Distributed Computing Environment (DCE). We follow this with a brief overview of the client/server movement and the impact that it is likely to have on building applications. Finally, we conclude with a short section on the various sizing options that are available and how they can be applied.

Open Software Foundation/Distributed Computing Environment (OSF/DCE)

A number of years ago, the major UNIX manufacturers (with the exception of Sun and AT&T) formed a consortium known as the Open Software Foundation. The intent of this organization was to provide a set of software tools that would integrate the technology being provided by the manufacturers. In order to do this, the OSF examined the technology offered by each member organization, selected what they felt was the best, and then integrated the components. The resulting software was then licensed back to the member organizations. Each organization was then responsible for implementing the product in its own environment.

The power of this approach derives from the fact that all OSF implementations are essentially the same code running on different hardware platforms. This means that you can buy a directory implementation from any of the member companies with a fair degree of assurance that it will interoperate with products from other OSF members.

There were four principal areas for which OSF provided products: a common user interface (Motif), a common operating system (OSF/1), an integrated set of network services (DCE), and a set of tools to manage the environment (DME). To date, the user interface and the network services have been the most widely adopted and have received the most attention. We focus our attention on the Distributed Computing Environment, since it is of great interest when considering how to build an integrated network.

OSF/DCE architecture

Figure 6-1 shows the basic architecture of DCE. The intention of DCE is to provide a series of distributed services that create an

Figure 6-1

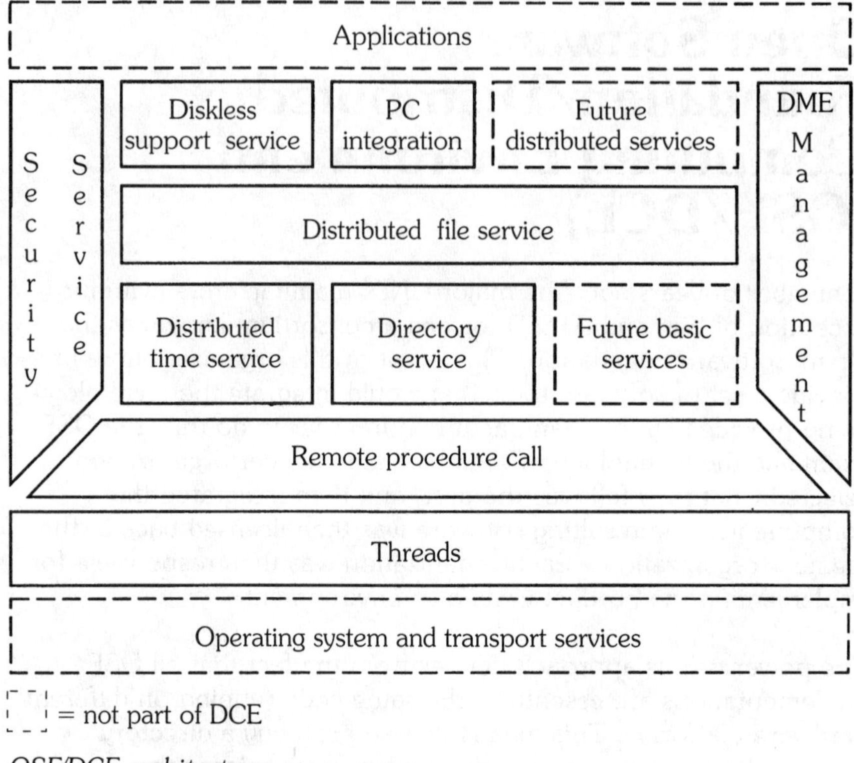

OSF/DCE architecture. Learning Tree International

integrated environment for applications and users. A variety of services are available, including directory, time, security, and file access. One way of characterizing DCE is that it is middleware that sits between the operating system/network and the application programs. Ideally, the application developer gains maximum flexibility and portability, since code should be easily transportable among different vendors' environments.

In addition, to the benefits mentioned above, DCE also provides an excellent environment for the development of distributed applications. Its use of a Remote Procedure Call enables the programmer to use resources as appropriate. Rather than trying to do everything in one location, DCE enables processing tasks not only to be distributed but also to run in parallel. This functionality is further extended by the use of threads. Threads provide the capability for applications to perform multiple tasks concurrently. In essence, threads enable programmers to write multitasking applications. While more complex than "linear" coding, the use of threads makes it possible to use multiple resources (both local and distributed) concurrently. This is a fundamental requirement of the well-integrated network.

DCE also provides a more robust environment. In the past, the failure of a single processor generally implied that users could not access programs or data that resided at the failed location. With DCE, not only are the applications and data distributed, but it is also relatively simple to implement replication techniques that provide a seamless transfer to backup copies of programs and/or data.

A key feature of DCE is its file system, known as the *Distributed File System* (DFS). While a variety of distributed file systems have been previously discussed (Novell NetWare, Microsoft Advanced Server, Sun Network File System), the DFS goes further: it provides a single-system image onto all files, regardless of their location or local storage format. DFS is highly automated, and provides the administrator with a rich set of functions.

DCE can track data and programs as they move through the organization. Pointers can be maintained that provide links to the current data locations. Programs and users do not have to know (or care) about the actual, physical location where the data is stored. It is sufficient to reach a location that knows where the data is stored. DCE provides this functionality through its directory service and a service known as the *Fileset Location Server*.

The OSF/DCE cell

The basic unit of administration and management in DCE is called a *cell*. In order to implement a cell, a minimal set of DCE services must be available. These include the DCE client, security, time, and directory services. In addition, the DCE run-time library is also required. Note that the file server is not part of this base set. (See Fig. 6-2.)

DCE provides two directory services. The first is a local directory called *Cell Directory Services* (CDS). In order to enable cells to

Figure 6-2

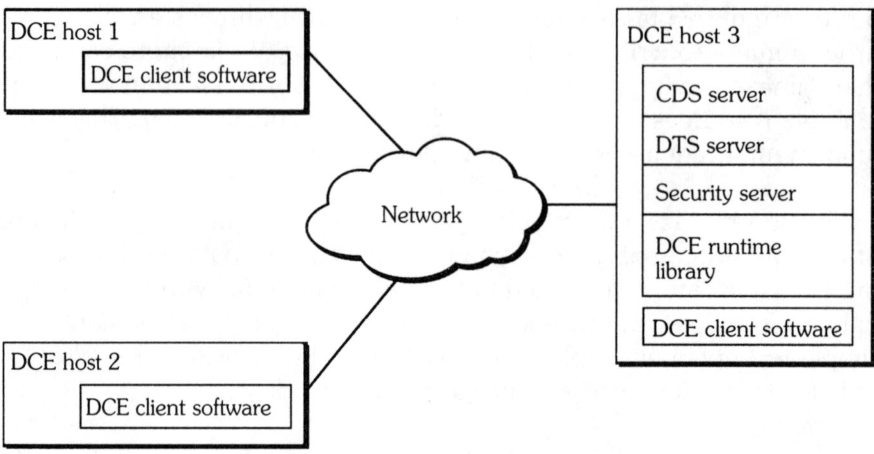

A simple OSF/DCE cell. Learning Tree International

discover resources external to themselves, DCE also provides a *Global Directory Service* (GDS). GDS is an implementation of the OSI X.500 directory service. Support for the Internet *Domain Name Services* (the DNS) is also provided. This enables the cells to use either naming convention to reference resources at a global level.

DCE implementation issues and concerns

A detailed treatment of DCE and its components is beyond the scope of this book, but there are a number of implementation issues and concerns that should be mentioned. Prior to beginning an implementation, ensure that selected vendors will support DCE in their software environment. In addition, new administrative agencies within the organization must be created. A central authority should be created to allocate cell names, since they must be unique. Within each cell, a number of configuration issues must be addressed. In particular, the location of the core services (either centralized or distributed) must be determined. Issues involving security, replication, and all administration will also have to be handled.

There are a number of concerns still outstanding with respect to OSF/DCE. This is still a relatively new technology, and we do not yet have enough empirical information (via case studies) to understand the performance/operation of a DCE environment. Interoperability issues have yet to be fully addressed among the various vendors. While the theory is sound, this does not necessarily translate into a well-integrated environment. Another source of concern is the lack of utility software (e-mail, data bases, etc.) that we have grown to expect. A number of vendors have indicated commitment, but implementations are still the exception rather than the rule. Software development costs are not yet known.

Perhaps the greatest source of concern when talking about DCE is the lack of support from major organizations such as Microsoft and Novell. While both have indicated that they will provide limited support, their full acceptance and support of the technology would be extremely helpful in providing assurance to users who have major investments in infrastructure based on products from these two companies. It is unlikely that LAN-based organizations will be tempted to convert if their existing software portfolio will not run in DCE.

Client/server

Another approach that is being discussed when it comes to integration is client/server. There has been a great deal of press coverage on this topic. At various times, it has been hailed as the savior of processing, the only direction to follow, filled with hidden costs, the destroyer of mainframe technology, and even the key to all the secrets of the universe (well, perhaps we are stretching the point on the last item)! Filled with conflicting reports, client/server represents a huge technological pit that can easily swallow a tremendous amount of resources.

A client/server analogy

Our approach is somewhat simpler. Consider the following analogy:

You are a customer (client) walking into a shopping mall (network). The mall is filled with stores (servers). As an interested observer, I

follow your progress through the mall. I soon discover a number of interesting points. Your behavior as a customer cannot be predicted. When you walk into the mall, I have no idea which store(s) you will visit, and in what order. However, I find it much easier to predict your behavior when you walk into any particular store. While you obviously have a different interaction with the salesperson in a shoe store (e.g., "I would like a size 8 black dress shoe") from one in a book store ("Do you have a copy of that great book by Fortino and Golick?"), your method of interaction (messaging) remains consistent for each store type. In other words, you use more or less the same interaction for *every* shoe store and *every* book store. Also, I detect certain similarities regardless of the type of store you enter (e.g., "How may I help you?", "Will that be cash or charge?", "Have a good day."). This generally has to do with the beginning and the end of the interaction.

Client/server is a lot like our example. Consider your organization as providing a shopping mall of servers. Users, from their desktop, act as customers who visit different "shops" as their needs require.

Client/server motivation

Why should we want to do this? As a counterargument, it could be said that it is more desirable to put all items into a single "superstore" (mainframe) and use a single interface to access them. In general, this is true. The problem is one of scale. It is impossible to build a single store that can satisfy all needs of all customers. If this were not, there would be no need for shopping malls. Also, it is generally easier to create a new store and advertise its presence than to keep trying to add new services/products to our superstore.

So client/server represents an opportunity to work more effectively with smaller processors. Does this mean the death of the mainframe? Not at all. We discuss this question in the next section. But before we can start our discussion of client/server in detail, we need a definition.

Client/server defined

We define *client/server* as "A cooperative processing environment that provides a single-system image to the user." While there is no

single industry-accepted definition, we feel that ours is sufficiently generic to accommodate most others.

A cooperative processing environment implies that more than one process, generally on more than one hardware platform, is jointly involved in the execution of some task. A good analogy might be a baseball team. While any individual might be a superb athlete, it requires a team of individuals to play baseball. No individual can successfully compete against a group in this particular situation. The same is true of client/server.

It is important to note that clients and servers represent process, not platform. While the client process is normally implemented in hardware and software that exist on the user's desktop, and servers are implemented in other locations, there is nothing in the client/server model that requires this always to be true. In fact, it is useful from the perspective of design not to couple platform and process too tightly. Client and server represent concepts that are implemented in process. No one vendor can claim to have the "true" client/server.

The relationship between the processes is primary/secondary, even though the communication is generally implemented as peer-to-peer. In client/server, the client is an initiator of requests. As a general-purpose program, the client can launch any number of servers concurrently. Furthermore, it is difficult to predict the behavior of a client. While nowadays many clients are written to address a specific task, it is expected that future generations of clients will be much more general. Think of the client as a general-purpose container. Items placed in the container represent services that are available. As new services are offered, they are "advertised" to the client container. The user initiates any service in the same manner, generally by clicking on the icon that represents the server process.

The server (the secondary process in the relationship) satisfies specific requests. Generally written to provide a certain function, servers can accomplish a wide variety of tasks, including data access, data transformation, graphics generation, and communication.

It is not necessary for the client and server to understand how each performs its function. We can see a parallel here with object

technology. In essence, the client and the server are both objects that communicate across a message boundary.

Of course, all vendors have their own vision of client/server and are quick to point out that unless you follow their definition you are not doing "real" client/server. The definition is less important than attempting to leverage existing assets.

A prime requirement of a client/server environment is the presentation of a *single-system image* to the user. Generally implemented on the user's desktop, the single-system image provides a consistent interface through which to access all available services (i.e., servers). While the image might take many forms, the generally accepted norm today is that of a graphical user interface (GUI). The nature of the interface is such that it provides the user with a virtual view onto both local and remote resources. By providing an interface that is consistent, and that uses a single command structure and lexicon, it is hoped that users will become more productive and require less training on new applications. While some research has been done, this point is still open to debate.

As a final note, our definition ends with the phrase "to the user." The implication is that client/server has a user-centric focus. The prime goal of client/server is (at least at this stage of development) satisfaction of user needs and desires. In some ways, this is the most difficult change in thinking that traditional system developers must accomplish. In classical system design, the focus is generally on data or process. With client/server, we switch the focus to the user.

In the early 1980s, Sun Microsystems coined the phrase "In the future the network is the computer." To this we should now add, "To the user, the interface is the system." The use of client/server does not preclude building large, mission-critical applications using traditional analysis and design techniques. Rather, it enhances development by enabling smaller systems to be rapidly built and deployed. Smaller client/server systems are generally developed from the interface inward. In other words, first the screens are designed (prototyped), process and data are handled.

The many faces of client/server

Figure 6-3 is generally credited to the Gartner Group. It represents one of the early attempts at defining a client/server environment. The idea is that programs consist of three major elements, presentation, business logic (application), and data access (management). Using client/server technology, it is possible to fragment the application within or among any of these elements. Client/server, in this view, becomes a continuum of possible implementations, rather than a single framework.

Figure 6-3

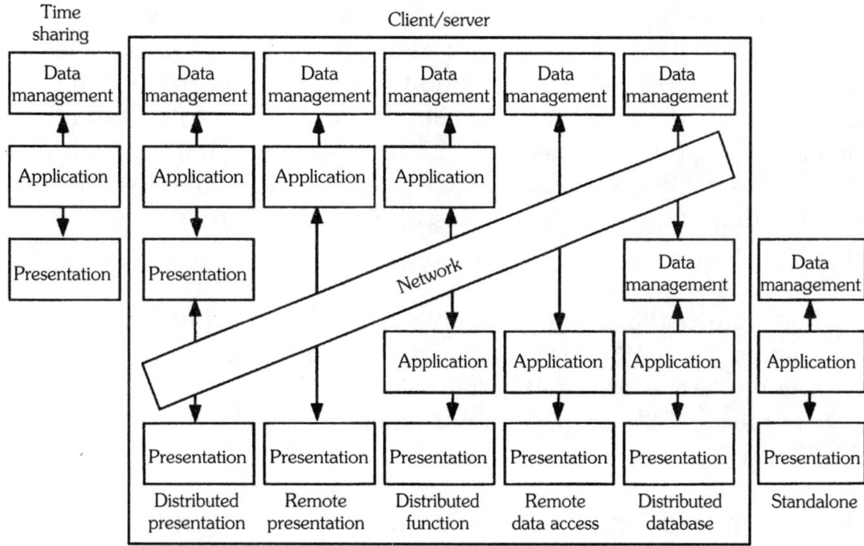

Source: Gartner Group.

The Gartner Client/Server frameworks. Learning Tree International

At either extreme, we have traditional monolithic applications. On the left-hand side of the graphic is "time sharing." This is a traditional terminal/host environment. At the other extreme we have the standalone, single-user machines generally represented by equipment on the user's desktop.

Over the last few years, a number of vendors have introduced products that fit into various categories of the Gartner model. In

addition, the model is useful, because it provides a starting place for building application solutions based on requirements. This gives the designer more options.

We now discuss the various client/server frameworks that are available, and show some possible implementations. Since each framework has its own set of strengths and weaknesses, we also examine the implications of each approach.

⁕ Distributed presentation: Host-based client/server

In many cases, centralized processing and data management makes good sense. In particular, for those applications that have high security requirements or high transaction loads against a central database, or where the cost of the user's desktop is an issue, this framework may be a possible solution. Using this approach, the application code and the data remain resident on a central server. Each application is provided with a virtual terminal that can provide a graphical user interface. At the user's desktop, a graphics terminal is provided that integrates the various applications onto a single physical screen.

The best (and perhaps only) example of this sort of framework is provided by the X-Windows environment that is available in most UNIX implementations. X-Windows was developed to enable UNIX users to have a GUI and to attach to multiple processors concurrently. (See Fig. 6-4.)

For environments with a significant investment in UNIX-based processing, this framework can be very useful. The end user is provided a single-system image regardless of the host accessed. Furthermore, multiple processors can be accessed at the same time, so it is possible to provide basic functions such as "cut and paste" with little or no difficulty.

The Open Software Foundation's Motif provides extensions to the basic X-Window specifications that afford additional functionality and ease of use. Coupled with the work being done by the Common Open Software Environment (COSE) to define a standard desktop for all UNIX systems, the Distributed Presentation framework can provide an excellent foundation for client/server development.

Figure 6-4

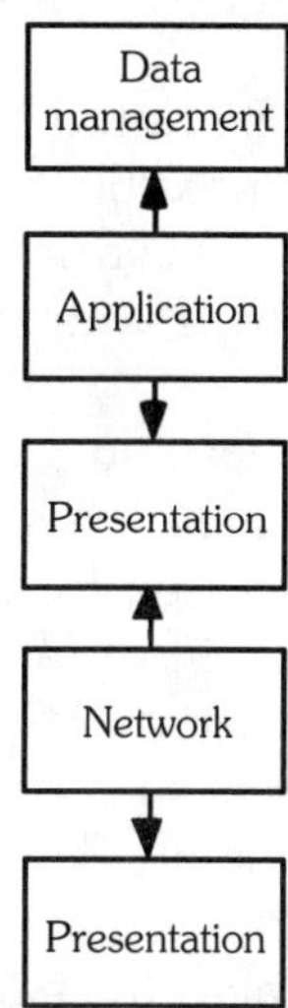

Distributed presentation framework generally implements X-Windows. Learning Tree International

A question has been raised as to whether this framework is "really" client/server, since significant portions of the application still reside on a central machine. However, once we get past the desire to have a single definition for client/server, we can see that this framework is a legitimate way to fragment applications, and therefore can be considered to be client/server.

A possible use for this framework might be in the publishing industry. A large publishing firm might have several computers that are used in various stages of text preparation and layout. Assuming that most of these systems were using UNIX, it would be possible for a journalist or editor to be involved in each step of production, from original draft to final typesetting, at the same terminal. This might be especially useful when researching information, since text from various sources could be easily moved between applications.

To a certain degree, the various graphical interfaces now available for the Internet World Wide Web (Mosaic, Netscape, WebExplorer) are also implementations of a Distributed Presentation framework. Portions of the interface exist on both the server and the client. This

standard interface enables Internet users to access multiple foreign hosts, and still provides a common interface.

It should also be noted that since X-Windows is a terminal-handling protocol, it is possible to implement the interface on microcomputers and workstations. Various terminal emulators are currently available for most major desktop operating systems.

Although this framework provides a potential cost savings over other approaches, there are some concerns. In general, this framework should only be used for new application development. Also, since it is primarily based in the UNIX world, it may be less suitable for other operating environments. Finally, since all of the processing and data access still occur at a central site, most of the problems of time sharing still remain. There is a single point of failure, inadequate central resources, and an inherent set of problems in trying to provide custom interfaces for each user. As such, this approach should only be considered for a relatively small percentage of client/server implementations.

Distributed Presentation can be useful for UNIX environments that would like to provide a GUI for end users, but may not be suitable for other implementations.

✳ Remote presentation: Migrating legacy applications to client/server

One of the significant issues facing many traditional MIS Groups is managing the transition from existing applications to client/server. The major applications that are currently running in many organizations were developed years ago, and were intended to operate in a terminal/host environment. Most user groups are now demanding that these applications be converted to client/server. In addition, as companies merge, they often find themselves with multiple operating environments that must somehow be integrated. Placing multiple terminals on the user's desktop is not an option.

The issues regarding legacy applications are significant. Much of the initial development and design time was spent ensuring the quality of these applications. Over the years, they have matured, stabilized, and now represent a significant investment to the organization. The cost to rewrite these applications for a client/server environment is substantial. In addition, since it is difficult to calculate the savings that might result from carrying out this migration, many MIS shops are reluctant to perform the conversion. However, the general trend to move to a distributed processing environment cannot be ignored. (See Fig. 6-5.)

Figure 6-5

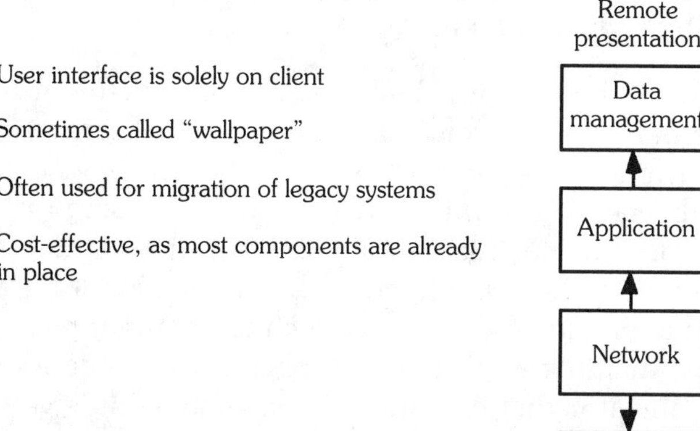

Remote presentation framework is a primary migration method. Learning Tree International

The Remote Presentation framework provides a possible solution. In essence, the application is not modified in any way. It is kept at the central site and still operates as if it were in a standard terminal/host environment. In other words, it still uses terminal screen handling as its primary form of interaction with the end user. From the user's perspective, however, this framework does provide a single-system image. This is accomplished by converting the terminal screens to a GUI within the user's desktop. A terminal emulation program provides the gateway between the legacy application and the user. Following the philosophy that "the interface is the system" from the end user's perspective, this framework should satisfy the requirements of many user groups to migrate to client/server.

There are many obvious advantages to this approach. The integrity, security, and investment in the legacy application is preserved. Since no changes are made to the original application, its quality is assured.

All of the work is done at the client. From the user's point of view, the system now behaves, and can be operated, with the same interface currently used for most applications. This approach can also be considered "low cost," since a relatively small number of individuals would have to be assigned to the conversion. The work can be facilitated by the use of a number of commercially available products that capture the incoming terminal screen and convert them to the graphical environment.

There appear to be two principal methods to implement Remote Presentation. The first involves placing the actual screen image in a window on the user's screen. In essence, the user sees a smaller version of the same screen that they are already using on their terminals. This approach is useful in organizations where the user community is already well trained in the application interface, but wants some additional functionality such as "cut and paste" or the ability to run multiple applications concurrently. Since it only requires a terminal emulator that can draw to the window, it is very quick, easy, and inexpensive to implement.

More challenging is to perform an actual conversion of the terminal screen image into a true GUI interface. As mentioned, a number of packages exist to assist in this migration. Often, however, reengineering the interface is also an excellent time to consider how the business process operates. It might be possible to integrate the information from a number of different terminal screens into a single user window. In addition, scripts could be created that would address additional services such as e-mail, fax, or printing. While the cost of performing this type of Remote Presentation is higher than simply writing the terminal screen to the window, it might also provide significant productivity improvements. In general, we believe that the legacy process, running on the host, is well done. What needs additional work is the user interface.

Another benefit of Remote Presentation is that it does allow for process reengineering after the fact. Since the interface is the only part of the system that is seen by the user, the other parts of the application can be modified, as required, with minimal impact on the user community. Therefore, when the legacy application does require significant modification, it can be rewritten using a more distributed model of interaction.

This framework has sometimes been called "wallpaper," "front ending," or "surround strategies." All of them refer to approximately the same thing. Unfortunately, the Gartner frameworks do not give enough details for the desire to begin. The placement of such functions as user-program routers, section concentrators, and other distributed services are not defined. Since this book is not about client/server design, we will avoid these questions.

The question inevitably arises as to which legacy applications make good candidates for using Remote Presentation. The trivial answer would be to say all of them. In fact, it is likely that operational data-entry systems could be the chief beneficiaries. These systems often have numerous esoteric codes and are difficult to learn. The use of a graphical front-end, especially if it is coupled with multimedia capabilities such as image, sound, or video, should provide a much more productive environment. Strangely, much of the remote presentation work is being done for analytical decision-support systems. Many packages are available that enable users to have a graphical interface into an existing database or file system. As we will see later, this is generally not advantageous.

While the standard advice about starting small should be applied, there is no reason why Remote Presentation frameworks cannot be quickly deployed. In many cases, this might be sufficient to satisfy the user's requirement for "client/server."

✻ Distributed function: The "real" client/server?

In many minds (including our own), client/server is about process-to-process communication. The client is viewed as a general-purpose requester of services, while the server is a specialized responder to these requests. In this framework, it is assumed that both the client and the server have separate parts of the application process, and exchange messages in order to accomplish a particular task. While the location of the client and the server are not specified by the framework, they are generally split between the user's desktop and some larger processor. Before discussing the framework in more detail, it is perhaps useful to spend some time discussing the motivation, advantages, and concerns about this approach to application design and development.

Distributed Function is a special case of cooperative processing. Cooperative processing is defined as two or more processes cooperating in the execution of some task. The relationship between these two processes is generally not defined, and is assumed to be peer-to-peer. In other words, either process can initiate or respond. In the client/server case, the relationship is more strictly defined. The client is the primary process, and the server is secondary. This makes sense, because the client will generally initiate operations at the request of the user. Furthermore, since the behavior of the client is unpredictable and is asynchronous in nature, it makes sense for it to take the primary role. The server process can either be initiated at the request of the client, or it can keep itself available waiting for requests. A good example of the latter case would be a DBMS or security server. (See Fig. 6-6.)

Since client/server is based on the assumption that the two processes are independent of one another, it is desirable that the message interface be as neutral as possible. We consider a neutral interface to be one that does not require a particular compiler, operating system, or network protocol. In other words, the message interface should

Figure 6-6

- "True" client/server
- Client and server are decoupled
- Split application across network
- Message passing between programs
- Requires sophisticated infrastructure

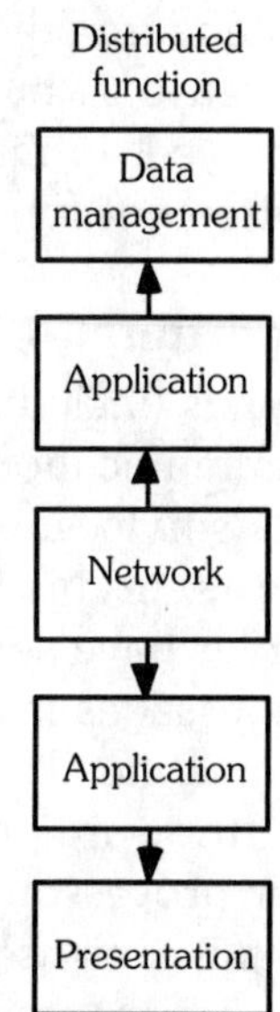

Distributed function framework maps well to object technology. Learning Tree International

enable any client to communicate with any server. Very few of today's messaging interfaces satisfy this requirement for neutrality. For example, most database access is still performed via SQL, and the standard RPC call used in many networks requires TCP/IP. A number of neutral syntaxes are available, including *Common Programming Interface for Communications* (CPI-C), *Transport Independent RPC* (TI-RPC), and *Message Oriented Middleware* (MOM). Perhaps the best example of a neutral message interface is the one being proposed by the OMG in their CORBA specifications.

Regardless of the interface chosen, developing an application that uses a message-based interface increases the complexity of the design/development process. The question is therefore why anyone should do it. There are a number of answers. From an economic point of view, it makes sense to leverage the existing investment in desktop computing by using more of the CPU cycles that are available. As increasingly powerful desktop devices become affordable, organizations will want to use them for tasks beyond e-mail, word processing, and spreadsheets. Having a general-purpose application that acts on the user's behalf also makes a great deal of sense. Such an environment would facilitate customization of the client to meet the individual requirements of each user.

Another reason to consider this framework is that it provides environments that have better reliability, availability, and serviceability. Since the application is now distributed, the failure of any one process should not halt the entire system. The various services offered by the OSF/DCE provide a good example. Most of the servers are replicated, so the failure of one server will generally not be noticed by the users.

This framework also facilitates the migration to object technology. Both the client and the server can now be considered to be objects. This should expedite the creation of new clients and servers. In fact, an argument could probably be made that client/server and object technology will eventually converge. At the current time, client/server is generally considered to be a distributed technology, while objects are generally not distributed. In the new technology now becoming available, there is less and less of a distinction between the two.

The Distributed Function framework can expedite the deployment of new applications. Since work on the client and the server can happen independently, the elapsed time to complete a project should decrease. Also, new functions can be easily added at both the client and server, making incremental changes easier to control.

To be fair, the tools for change control on the user's desktop in a distributed environment are still quite weak. Part of the problem is that it is difficult to control hardware that is located on the user's desk.

As mentioned, the reliability of client/server systems should be better than older monolithic applications. This is due in part to the fact that servers can now be highly specialized. No longer is it required to have very large programs that can accomplish multiple functions. Building smaller, highly focused applications should improve the reliability of the entire system. Note that this has not yet been proven, since empirical information about the maintenance of client/server systems is, to a large extent, still not available.

While these benefits appear attractive, there are a number of concerns about using this framework for application development. Earlier it was mentioned that Distributed Function is more complex to develop than traditional applications. This leads to higher costs not only in the building stage of the project, but also in the support and administration of the existing system. Consider the problems associated with managing a few local processors, and then extend the issues to cover a large, heterogeneous, distributed community. Even if the application can be developed, how will it be maintained? For example, if significant business logic is placed on the client side of the system, and a variety of user desktop devices must be supported (IBM, Macintosh, Sun, etc.), a change to the code will require that patches be developed for all of these environments. Maintenance might take a great deal longer. Security is another important issue. If application logic on the client can be modified locally, the integrity of the system as a whole may be compromised.

Additional staff will probably be required to support the client/server environment. These staff members have to be distributed along with the application. Central control of a distributed application makes no sense. Training will also be an issue, since the design, development, coding, and support of this environment will use tools and technologies that are different from those currently available. Users will also require retraining.

Performance is also an issue. It seems intuitive that an application running in a single location will operate more efficiently than if the process is distributed across a power. Not only is process-to-process communication slower—it is difficult to optimize the network for all possible applications. One area where performance should improve is response time. Since the client is the principal source of interaction with the user, the fact that it is available locally should provide instant responsiveness. When a request is made to a server, the client can inform the user that it will require a certain amount of time to complete, thereby enabling the user to run other applications or utilities.

Adding to all these problems are the conflicting and competing technologies and standards that are being offered by vendors. While interoperability is still a "strategic direction," today's reality is a fragmented, closed environment. Except in some special cases, selection of an operating system, network protocols, or compiler can force the developer to make choices.

Of all the frameworks discussed in this section, Distributed Function is the most challenging to implement. While the long-term gains derived from using this approach are substantial, organizations must be prepared to invest significant resources in both staff and infrastructure in order to build the proper foundation for implementation. The growing acceptance of object technology as a methodology for application design will certainly expedite the acceptance of this form of cooperative processing. However, since this model of interaction places the greatest demand on development/design teams, it is recommended that it be viewed as a strategic approach, and that the necessary commitment and support be made available.

* Remote data access: Client/server for the masses

The most popular form of client/server in use today is Remote Data Access. In this framework the application and presentation logic reside on the client, with a DBMS on the server to provide data access capabilities. The application logic and DBMS generally communicate via SQL. While not a requirement, rapid prototyping techniques are generally used to develop the application as quickly as possible.

There are many reasons for the appeal of this framework. Most of the components required to implement a Remote Data Access system can be easily purchased "off the shelf." A variety of products for most operating systems exist that facilitate the design of the screens, automatically generate SQL, and allow for the use of component technology so that parts of the application can be reused. The DBMS can reside on any platform, since the end user is sheltered from direct data access. Depending on the development package chosen, tools such as test generators, CASE, dummy responders/requesters, and terminal emulation may also be provided. (See Fig. 6-7.)

Figure 6-7

- Uses server as database engine
- Migration from LANs to client/server
- Useful to assess mini/mainframe data

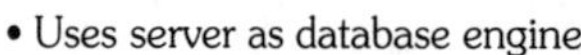

- Based around structured query language (SQL)

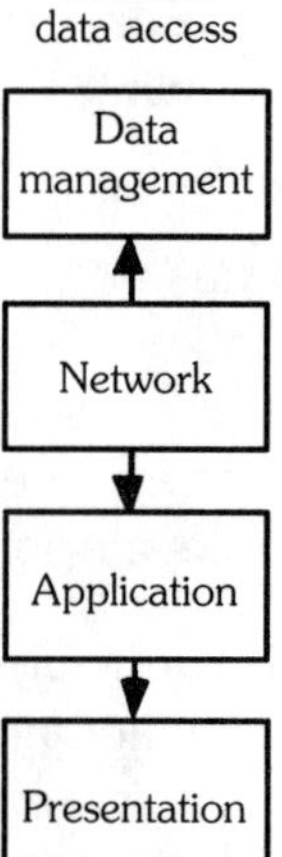

Remote data access framework is the most popular form of client/server. Learning Tree International

The cost of these development tools is also quite attractive when compared to mainframe-based development systems. For a relatively small investment, even small departments can begin to develop their own applications. The fact that these tools also use GUI interfaces increases their appeal, since developers will be working in a well-understood environment.

Given the current backlog in MIS shops for application development, and considering that a large percentage of these requests are for analytical, read-only data access, Remote Data Access seems to make a great deal of sense—so much so, that it is estimated that the majority of client/server applications in use today use this framework.

Organizations should be aware that there are a number of inherent limitations to using this framework. To begin with, the use of SQL as a message syntax is somewhat limiting. For example, it is difficult in SQL to execute the command "RUN SIMULATIONS" or "PERFORM GRAPHICAL TRANSFORMATION." These sorts of commands are better handled by the Distributed Function framework.

To counter these problems, many DBMS vendors now offer the ability to execute stored procedures on the server. These procedures have generally been coded in advance of execution. While this is certainly a possibility, it further limits options, since these procedures are generally specific to the particular vendor's DBMS and have little portability or scalability.

A word should be added here about the ongoing debate between the OLTP (On-Line Transaction Processing) community and the DBMS camp. While many of the DBMS vendors now offer functionality somewhat equivalent to a transaction processing monitor (TPM), the functionality of such systems is very narrow when compared. This is especially true if multiple resources must be managed concurrently. If a significant transaction load is predicted for an application, it is recommended that a TPM be considered.

Performance can also be an issue when using SQL to communicate between the client and the server. In many cases, the nature of the interaction, with respect to the use of the underlying network protocols, might not be well understood by developers or well implemented by the vendor. This might lead to poor response time or excessive bandwidth consumption on the network. It is difficult to identify the network overhead associated with this framework without performing traffic analysis. Additionally, it is recommended that the system be stress-tested prior to deployment. A system that provides satisfactory performance for a single station might not do the same for multiple users. As a general rule, a safety factor of three is recommended. In other words, stress test the application to three times the expected loading.

A more subtle danger in using this framework is the rapidly changing desktop environment. Investment in development and coding generally requires that all future generations of operating system

software be upward compatible. There is no guarantee of this compatibility in the desktop marketplace. In fact, it can be argued that the reverse is true. Vendors of desktop software generally do not receive warranty or maintenance revenue. In order to sustain their cash flow, new revisions of the product are required. One way to ensure new sales is by making the new product sufficiently incompatible with the older version and then dropping support. While this strategy might not be attractive to purchasers, it seems to have worked. Developers, especially those who are new to the industry, would be well advised to ensure that a upward path exists for their code. Otherwise they will be forced to rewrite the application when the next generation of operating system software becomes available.

Another area of concern is the emphasis that this framework places on having the client manage/coordinate the interaction among multiple databases. For example, if an update is required that will access two different databases on two different platforms (especially if they are from different vendors), the client process must implement the two-phase commit required to properly execute. In addition, the client will probably have to implement various audit and log files so that recovery is possible in the event of failure. This is challenging work, and will generally make the client portion of the application larger. Developers should be aware that when accessing multiple data bases concurrently, each DBMS is unaware of the existence of the other. All coordination must take place at the client.

This leads to another area of concern. Depending on the number of clients and their locations, support and change control increase by using this framework. Since it requires sophisticated client code, portability might also be an issue.

Nonetheless, this framework is recommended for many cases. In particular, for those environments with a high ad hoc query requirement, or when local departmental systems are being encouraged, this framework will provide the best results. This type of application satisfies many of today's requirements. As long as integrators keep in mind the limitations and concerns mentioned, there should be few problems. Attempting to scale this framework to an enterprise level is probably not acceptable.

In Chapter 4 we briefly discussed a number of ways that data access could be facilitated by the use of technologies such as Microsoft ODBC and IBM DRDA. The remote Data Access framework would be one of the principal users of this technology. Since the current trend is to provide heterogeneous database access through common SQL syntax, organizations would be well advised to consider the implications of choosing a client-driven, gateway-driven, or database-driven approach. This selection is considered to be strategic middleware, and it is difficult to switch to another system once the selection has been made.

It is anticipated that all organizations will require some amount of Remote Data Access to one degree or another. Our intuitive feel is that this solution will work well for smaller to mid-range applications, but is probably not suitable for larger organization-wide systems.

⁕ Distributed database: A vision of data

For years, the database industry has speculated about the possibility of providing seamless access to heterogeneous databases in a multivendor environment. In principle, the vision provides a virtual view onto any number of databases in a fashion that provides a single-system image to the user. While the dream is as yet unfulfilled, it is still alive and well.

The Distributed Database framework assumes that the presentation, application, and database logic reside on the client, and additional data access logic is spread across a series of distributed platforms. In addition, to a standalone machine, the client might be a multiuser host, or perhaps a UNIX host that is using the Distributed Presentation framework to interact with the users. Regardless of the platform chosen for the client, the magic of this framework is its ability to present the databases as a single virtual entity. (See Fig. 6-8.)

The advantage of this type of framework is that all data that is held by the organization can be accessed using a common interface. Furthermore, because the databases control access to information, the developer has fewer concerns about performing resource management and control. Unfortunately, many of the problems associated with concurrent update across distributed databases have

Figure 6-8

- Multiple databases that can be accessed concurrently
- Requires close cooperation among all participants
- Provides single image of database
- Generally implemented using proprietary technologies

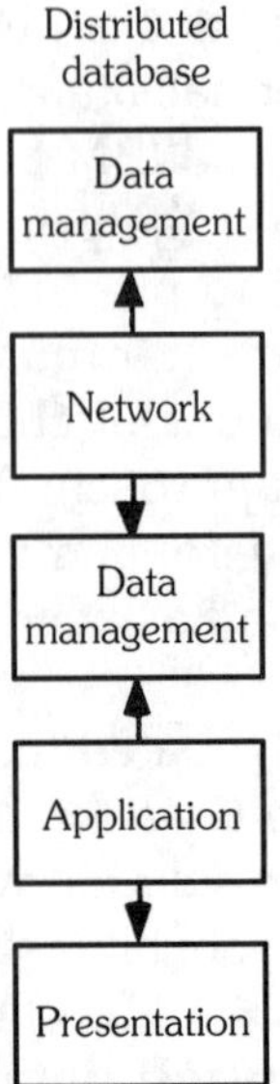

Distributed database framework represents a future seamless vision of data access. Learning Tree International

not yet been fully solved. This means that the primary use of this framework will be for data access. An excellent example of this would be access to a series of distributed database warehouses.

The problem of performing concurrent updates has been with the industry for many years. At the current time, the primary method of resolving this issue is via an algorithm known as the *two phase commit*. It is not our intent to discuss how this algorithm works; however, it should be noted that it will not necessarily recover from every possible system fault. The implication is that in certain instances, data integrity can be compromised even if the commit process is used.

While a number of individuals (most notably C. J. Date) have speculated as to the nature of a distributed heterogeneous database environment, today's reality still falls short of the vision. For those organizations that require such an environment, it is generally recommended that a single vendor be selected to provide all databases. Such a vendor would have to provide the databases across a wide range of operating platforms. This generally would rule out the selection of the lower-cost microcomputer-only systems.

This framework probably has the fewest actual implementations. In part this is due to the fact that a great deal of organizational data is still kept at a central location or under a single administration (such as a Database Administrator). Is unlikely that this will change until updates can be successfully performed in all cases.

On the other hand, using this framework for data access can be advantageous in some situations (such as the data warehouse). Once again, though, since the framework depends on a "heavy" client, our concerns about support and administration remain.

Sizing and the integrated enterprise

As the price of computing continues to drop, many organizations are exploring the possibility of replacing their centralized, large-scale, mainframe computers with smaller systems, the reason being that significant costs savings can be achieved by migration to these lower-cost platforms. At first glance, the argument seems compelling. The same twenty million dollars spent on a mainframe can purchase 10,000 desktop machines (assuming a purchase price of $2,000/machine). Assuming that mainframe applications can be migrated to these smaller machines, considerable savings can be achieved. Integration, client/server, and open systems would be the principal methodologies used to achieve this vision of a highly decentralized computing environment. This effort is generally called downsizing, rightsizing, or smartsizing.

A number of additional potential benefits are listed as reasons to consider resizing. Beyond cost reduction, there is the greater productivity that the new technology offers. Having machines customized to their user's or application's requirements should yield better price/performance. Placing the equipment and data closer to the end user should also improve performance. Cost allocation should improve, since local departments would then control their own hardware, software, and staff. By deploying technology directly in the user department, application development should be improved as well, especially when coupled with the new rapid prototyping tools that were mentioned in the client/server section of the chapter.

Business productivity should be another area of potential benefit. Since a well-integrated network enables individuals to access and disseminate information more efficiently, it should be possible to take advantage of flatter organization structures. This would also lead to more autonomous work groups, which should provide an organization that can be more responsive to changing market conditions. Furthermore, the widespread availability of low-cost mobile computing resources (laptops, notebooks, PDA, cellular technology) suggests a workforce that can take advantage of telecomputing, which provides an attractive alternative to single-site offices, and can also reduce operating costs.

Perhaps most importantly, resizing the organization avoids excessive reliance on a single processor. One of the reasons that MIS departments must expend considerable effort on disaster/contingency planning is that loss of the mainframe for any significant period of time would be catastrophic to the modern organization. The data/program assets must be preserved. By distributing this portfolio across a series of smaller processors, it should be possible to achieve a point where the loss of any single processor could be tolerated for some period of time. Additional cost savings might be achieved by reducing staff that are currently involved in ensuring that the central host is always operating.

As mentioned, the motivation to switch to the smaller platforms seems to make a great deal of sense. Unfortunately, the empirical information to date seems to indicate that the savings will probably not be achieved in the majority of cases. While detailed case studies are not yet available, certain trends seem to be evident. What they indicate is that there is a substantial set of hidden costs associated with resizing that are not recognized until after the new systems are deployed. These include reduced productivity, unreliable software, a dynamic marketplace, and inappropriate expectations. Most importantly, the support, administration. and training costs for a downsized environment seem to have been grossly underestimated. Some further explanation is required to identify why this has been happening.

There are a number of reasons why productivity can decrease in a downsized environment. Since the interface has been radically altered (from character-based to GUI), users require some time to learn how to operate the new system. Furthermore, since microcomputer software is not as mature as the large-scale systems, they are likely to require more frequent resets. The resets can have considerable impact if they have left a database in an undefined state, or a transaction only partially completed.

The complexity of the technology also makes it more fragile. Consider the number of components involved in linking the client to the server. The failure of any component will lead to a perceived system failure. Probability theory tells us that the probability of failure increases with the number of components in a serial system. These local outages can have considerable impact. If a LAN supporting one hundred users has an average repair time of four hours, then the potential productivity loss can exceed 400 hours!

The dynamic nature of the microcomputer software industry also engenders a new set of problems. Since most of the software is sold in a "shrinkwrap" state, there is no implied warranty or support available. The software vendors are at liberty to modify their products in any way, at any time, and in general are not compelled to provide upward compatibility. In fact, as new generations of system software become available, support for the old versions is quickly dropped as a way to "motivate" existing users to upgrade. Since microcomputer software/hardware vendors get a very small percentage of their revenues from maintenance activities, ongoing cash flow requires frequent new releases. The cost of implementing these new releases by the users is quite high. Beyond the purchase costs, there are additional costs associated with installation, compatibility problems, retraining, code conversion, data conversion, new software bugs, etc. It is difficult to assign a cost to these items, but their cumulative effect can be substantial.

Another interesting productivity issue that is beginning to surface is the amount of time that users spend "exploring" the GUI. The end user can now modify color, fonts, wallpaper, screen savers, sound bites, etc. The amount of time being devoted to this activity seems to grow with each new system software release. While an argument can be made that this ability of the end user to "personalize" the desktop will lead to higher efficiency, the short-term effect is a productivity hit. The widespread availability of low-cost computer games also seems to be taking its toll. This has now spread beyond the single-

user machine. LAN-ready games are now becoming available. Their impact on productivity and network performance has led many LAN administrators to ban them from the network.

Perhaps the most surprising aspect of downsizing is the impact it has on support and administration. The annual cost of supporting a user desktop machine has been estimated to be anywhere from $3,000/ $10,000. This represents the investment in additional support staff and frequent upgrades, plus the productivity hits previously mentioned. It would appear that capital acquisition costs for desktop hardware/ software are probably the smallest percentage of total costs over the unit's lifetime. A figure of 10/20% is probably not unrealistic.

Does this mean that downsizing is a bad idea? Only if the intent is to save money! On the other hand, it might be possible, over the longer term, to achieve a more productive organization by making the best use possible of all computing technologies. There is a place for each type of equipment, from the micro to the mainframe. Selecting the appropriate tasks for each unit can be a challenge. Furthermore, to be successful, the organization must balance the desire that local groups have to be autonomous with the organization's need to integrate. Figure 6-9 summarizes some of these issues.

The upper right-hand quadrant (Centralized/MIS driven) represents how many organizations have traditionally dealt with the introduction

Figure 6-9

	User driven	MIS driven
Centralized	• Products meet guidelines • Possible user "buy-in" • Cost containment	• Support cost reduced • Cost containment • Integration • Global optimization • Organizational focus
Decentralized	• User "buy-in" • Rapid technology deployment • Custom fit • Local optimization • Focus on individual	• The "new" MIS • Mix global/local optimization • Focus on departments

Authority for sizing issues. Learning Tree International

of new technology. In this view, end-user departments would approach MIS with a series of business problems. MIS would examine the problems (analyze), develop a solution (design), and then deploy the technology required to implement the solution (build). This view is sometimes know as an *MIS-push* philosophy, since technology was introduced, and controlled, by MIS. The principal advantage of this approach is that MIS can ensure that integration exists among all technology solutions.

The lower left-hand quadrant (Decentralized/User driven) characterizes the opposite approach. Here, the end-user department selects the technology that is thought to best suit the requirements. Applications are generally developed internally, without MIS involvement. MIS can be called in to deploy the technology chosen, or, at some later date, to integrate the system into the corporate backbone. This is generally accomplished at a relatively high cost, since the technology has been purchased without consideration of integration. This approach is sometimes referred to as *Client-pull*. The principal benefit of this approach is the high degree of commitment on the part of the end-user department to making the technology work. Having selected their own technological solution, workers in the end-user department are usually anxious to prove that it was the right choice. This leads to a high degree of motivation that is generally not seen when using the MIS-push approach. The effect of this motivation should not be underestimated. A dedicated end-user department can make or break the introduction of a new system.

The lower right-hand quadrant (Decentralized/MIS-driven) might represent the best possible compromise between these two approaches. Here we have a new form of the MIS department, which actively encourages local departments to choose and implement their own technology. This is done by active mentoring of local support staff, and by providing funding for investment in new technology. Integration is achieved by MIS retaining control over the corporate backbone.

The public phone system provides a good analogy. Any company is at liberty to select its own PBX. However, the standards for interfacing to the public phone system are controlled by the public carrier. In this way, companies can remain autonomous and still have

the opportunity to integrate their technology. Note that the phone company also provides value-added service, such as directories, conferencing, etc. In some cases, the public phone company can be hired to implement a local (i.e., private) phone system. This is generally done for a fee, and in this case the phone company is competing with other PBX vendors.

In the same way, MIS can offer local services to the end-user departments. Access to the backbone would be provided by a set of standards. Centralized services could be offered, such as directories, security, data warehouses, etc. Integration with other departments' data would be handled on a fee basis. The focus should be on providing the departments with the support required to implement their technologies. The intent is to build synergies, by leveraging technical knowledge in MIS with business knowledge in the local department.

This is not easy. In many organizations a great deal of animosity exists between MIS and the local groups. Trust might be difficult to achieve in the short term. In addition, many MIS staffers might feel threatened by this decentralization. One impact this approach is likely to have is a reduction in staff within the central MIS group. On the other hand, we can anticipate that there will be a rise in staff devoted to technology within the local groups. On the whole, the total number of individuals involved with technology will probably rise. This is due in part to both the greater sophistication of the technology, and the fact that certain support functions that were once handled centrally now have to be replicated at the local level. However, the increased productivity achieved by local commitment and rapid technology deployment should offset these higher costs. In the short term, it is expected that the migration from MIS-push to Client-pull will lead to a variety of political turmoil within the organizations. The change will not be easy!

One way or the other, things will change. In the same way that a single central processor cannot satisfy the total set of processing requirements that any organization will have, so too a central MIS department will have difficulty being responsive to the needs of each local department. By migrating staff into these departments, MIS can achieve a better understanding of these needs, and guarantee the degree of end-user commitment required to deploy successful systems.

Summary

Enterprise integration requires a strategic approach. The OSF/DCE technology represents an excellent technological foundation to achieve this integration. Unfortunately, the technology is still relatively new, and still lacks many of the tools and utilities required to make implementation successful. This should change over time as more software becomes available and early implementers share their experience. While it might be too soon to implement, it is not too early to begin exploring this technology and the impact that it is likely to have within the organization.

One of the principal models that will be used for new application development will be client/server. A variety of frameworks exist that can be used for particular situations, such as legacy migration, new system development, and ad hoc local systems. We favor a surround strategy for legacy migration, since it focuses on the user interface and enables the investment in the process to be preserved. Remote Data Access is a popular way of handling local system development. A convergence of object technology and client/server is likely, with standards such as the OMG's CORBA playing a principal role.

It is unlikely that downsizing leads to significant cost savings in the short term. Nonetheless, organizations should consider the decentralization of their MIS departments so that a higher level of user commitment can be achieved. A new model of MIS is required. While the short-term impact will likely have a negative effect on productivity, the long-term gains seem substantial. This can only be achieved by a strategic approach and commitment.

By the way, here is your review.

Concepts review

1. What does "DCE" stand for?

   D________________ C________________ E________________

2. What problem is DCE attempting to solve?

____

3. How is DCE solving the problem?

____

4. Who is behind DCE and why?

________, ________, ________, ________, ________

____

5. Name the five client/server frameworks and a major characteristic of each:

   1. ________________, ______________________
   2. ________________, ______________________
   3. ________________, ______________________
   4. ________________, ______________________
   5. ________________, ______________________

6. Name three sizing trends and give an example of each:

   1. ________________, ______________________
   2. ________________, ______________________
   3. ________________, ______________________

Network support and management

CHAPTER 7

DISTRIBUTED systems and multivendor networks present new and unique problems to the network manager. Traditional models of network management cannot be used. The classic role of the network manager as someone whose primary responsibility is to keep the network running is much too narrow a view to encompass the total set of duties required by enterprise systems.

In this chapter, we investigate some of the issues raised by modern multivendor networks and some of the proposed solutions. This includes a new model of the Network Control Center (NCC), new tools and protocols such as Simple Network Management Protocol (SNMP), security concerns raised by easily accessible networks, and some of the support issues such as decentralized and centralized staffing. While not a comprehensive examination, we hope that it provides the reader with the information required to begin a careful analysis of his organization's needs for network management and support.

In order to be successful, the network manager has to implement a vendor-neutral strategy that can provide the information required to support the accounting, performance monitoring, configuration, security, and fault management of distributed multivendor networks. This can only be accomplished by careful planning and a high-level commitment to elevate the NCC to a much higher level in the organization chart.

The network management problem

Back in the good old days of vendor-specific, monolithic, terminal/host computing, network management was a complex but not impossible task. The network manager was mandated to deliver application systems from the host computer to the user's fingertips. This required control of equipment from the I/O port on the mainframe to the terminal on the user's desktop. In essence, the network manager provided a support service to the MIS director. As such, the position was generally located within the MIS organization, and frequently as a subgroup in the Technical Support or Operations department. In many ways, the Network Control Center (NCC) was equivalent to the Operations group. The Operations manager was responsible for the efficient operation of the host computer, and the network manager had the same role with respect to the network. For many network managers, their job definition could be defined in a single sentence, "Keep the network up and running!" They approached this primarily as a technical task, and as such, most network managers took a real "hands-on" approach to their job with the emphasis on technical excellence.

To assist the network manager in fulfilling this mission, vendors provided the tools required to control their proprietary network systems. Each vendor had a set of tools unique to their product line. Tightly coupled to the underlying network protocol, these network control systems provided information on both the logical and physical well-being of the network. Training was supplied by the vendor, as well as by specialists who could be called in to help optimize the system as required. While not perfect, the fact that they were operating within a set of rigid boundaries defined by the proprietary nature of the protocol provided a fair degree of control over the environment that enabled the network manager to accomplish his mandate.

As we have seen throughout this book, the world is changing. Networks are no longer homogeneous. A variety of protocols, subnetworks, operating systems, and applications now guarantee that the network manager can no longer look to a single vendor to provide a solution for network management. In addition, as networks become more accessible, network security becomes a rapidly growing concern. To add to this increasingly complex situation, the network manager is being called upon to deploy applications such as e-mail, directories, file transfer, distributed file systems, and gateway services that provide interoperability among heterogeneous environments. Many of these applications, such as groupware and multimedia systems, are being deployed by user groups, who seem to believe that network bandwidth is free and unlimited, without consultation.

To compound this problem, the tools that are required to handle the support and management of this complex environment are either immature, inadequate, incompatible, or unavailable. Network management staff have generally not been trained to handle this level of diversification. As LAN Administrators build larger and larger

subnetworks, the lines of authority and responsibility between the NCC and local support staff become less clear, and this leads to political situations that further complicate administrative issues. Finally, at senior levels of the organization, there is little appreciation of these problems or the potential for disaster. In fact, within MIS, the NCC is still being viewed by the CIO in the same traditional role, so that funding required to handle these new tasks is generally not available.

We have often mentioned the concept of the "network as an organizational computer." If the backbone network represents the backplane of this new enterprise computer, then those who control the backbone, by inference, now control access to the entire set of information-processing assets available to the organization as a whole. The individual who has the responsibility to administer and manage these assets must have a senior role to play.

Earlier we stated that the traditional mandate of the network manager was to keep the network running. This mission statement must now change. We suggest a much more strategic view of network management. This new view might be best stated as, "How will network technology assist the organization in reaching its strategic objectives?" The answer to this question is the primary responsibility of the network manager and the NCC.

The network management model

A model is required to define the various roles and the duties that must be performed prior to determining the tools and staff required to implement an NCC. This model must be flexible enough to incorporate future services that the NCC will offer, and capable of being inserted at various levels of the organization chart. While a great deal of literature is available on the technical aspects of network management, the amount of material available on how to model the NCC is somewhat lacking.

The OSI network management model

As part of OSI, the ISO has defined the role of the network management. This is illustrated in Fig. 7-1. There are five major areas defined.

Figure 7-1

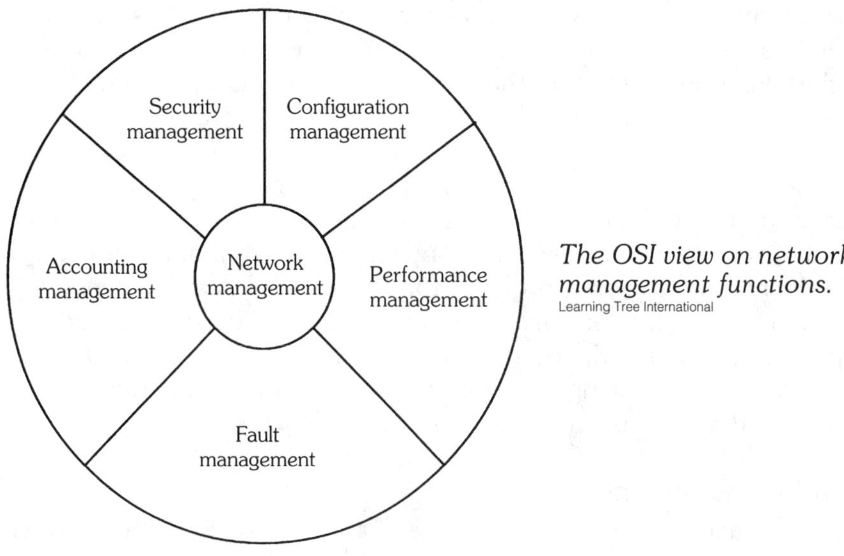

The OSI view on network management functions.
Learning Tree International

✳ Configuration management

Configuration management is that set of activities involved with the installation and modification of network elements. This includes both hardware and software, such as bridges, routers, network interface cards, drivers, file systems, etc. Assignment of network addresses is also included in this activity, and those activities involving change control are an important element. In a distributed environment, it becomes increasingly difficult to retain any control over hardware and software once it has been released to the end-user community. Policies and procedures are not enough. Some automated method of auditing changes must be made available. As a side aspect to change control, the task of software distribution must also be included as part of configuration management. How should new software be distributed, installed, and configured? There are numerous tradeoffs between electronic distribution and the use of hard media such as diskettes or CD-ROMs. We return to this subject a little later.

✳ Performance management

Performance management refers to activities related to monitoring the state of the network. These are required both to determine if the network is operating within specified metrics, and to provide valuable input for capacity planning. Part of the problem relates to the total

volume of data available to the network manager. There is so much raw data available that the task of filtering and analyzing it is a major undertaking. Furthermore, the act of collection places an additional load on the network.

In essence, performance management is a two-edged sword. To understand how the network is performing, we must collect data, but the act of collection impacts the network's performance! Resolving this dilemma has been a problem for network managers for many years. This problem has grown more complex as both the quantity and variety of network components have increased. Interpretation of the data is also challenging, since third-party software is required to integrate the various information sources.

⁜ Fault management

Fault management refers to activities related to isolating and correcting network errors. Of all network management activities, this area can be the most complex. Any network error can cause a cascade of fault reports. For example, a fault in a modem will probably cause reported errors from modem-, router-, and session-monitoring systems. Where should the network technician begin in isolating the problem? In complex systems, this fault-isolation activity requires sophisticated software systems that can interpret the incoming fault reports and isolate the original error. Fault recovery must also be managed. The total set of activities required to restore a system to a viable state includes fault reporting, response time, isolation time, diagnostic effort, the correction process, the restart process, recovery time, and user sign-off. At each step of the process, proper management procedures are required in order to ensure a successful recovery.

The fault management system must also collect information so that the network manager can estimate both the Mean Time Between Failure (MTBF) and the Mean Time To Repair (MTTR) of network components. This information is valuable, because it enables the network manager to take a more proactive role in the maintenance of network components. Having this information also enables the network manager to calculate the cost of network outage. LANs are particularly vulnerable to outage, since the potential productivity loss is the product of the number of users and the length of the outage.

Does the network manager have the time to collect all this information? Not without the assistance of sophisticated tools.

Assigning a value to the outage is difficult. Much depends on the nature of the work being performed on the LAN. However, as mission-critical applications migrate to LAN-based client/server systems, the cost of this outage is sure to rise. Fault management should also provide expert-system or automatic diagnostic capabilities, so problems can be corrected without the need for human operator intervention. While this capability is still beyond many of today's systems, it is found in sophisticated telecommunication systems, and is likely to migrate into the data communication world.

✳ Security management

The next area of network management that must be considered is security. The security system requires the capability to handle all aspects of network authentication, authorization, access control and validation. The security system must also be responsible for the integrity of data in transit between source and destination.

The problems associated with network security, particularly with LAN-based networks, are obvious. At any point in the communication system, information can be tapped, disrupted, or usurped. While encryption techniques can solve some of these problems, they cannot solve all of them.

One security problem deserves special mention. As application services are distributed, it is likely that a user (i.e., client) will have to contact many servers in order to accomplish a particular task. Sending authentication information, such as passwords, to each server increases the likelihood of a security breach. Later in this chapter we discuss one possible solution to this problem.

✳ Accounting management

Accounting rapidly becomes an important requirement for distributed networks. Many local departments and users presently consider corporate network bandwidth to be free and unlimited. New applications are added without any consideration being given to

capacity utilization. Network managers are faced with an increasingly heavy load of traffic, and have no way to determine the responsible parties or applications. Until such time as users can be made aware of the cost of transmission, we can anticipate that the load will continue to increase at an escalating pace. This is especially true as multimedia and groupware applications become more readily available, not to mention real-time video conferencing.

The network management system must have the capability to charge users for bandwidth utilization. When this is done, users become more sensitive to how they use the networks, and network managers are able to perform between capacity planning. As an additional benefit, the use of an accounting system provides the basis for the NCC to be viewed as a profit center, rather than a cost center. This is important if network managers expect to have the budgets required to deploy the more sophisticated, higher-speed networks that will be required in the future. Unfortunately, this capability is one of the last considered by network managers who are trained to think in terms of reliability, serviceability, and availability, rather than accountability.

A management model for the NCC

While the OSI network management view indicates areas that require attention, it tells us little about the management structure of the NCC. What organization chart should be used? What are the functional areas to be implemented? Clearly, we cannot limit our examination of network management to process.

This leads us to the general structure of the NCC. It is unfortunate that most of the emphasis in network management is currently focused on the tools to be used, rather than on the people. Network managers bemoan the fact that the tools available are not adequate to properly administer their domain, and they ignore the fact that this activity must be performed by their staff, not the tools.

In fact, the term "network management" has become synonymous with the technical aspects of network control, rather than the consideration of how best to manage the staff that performs these activities. As a group, network managers take more technical courses, and fewer people-management courses, then any of their technical peers. Part of

the reason has to do with their highly technical background and their focus on keeping the network "up and running." However, as we have mentioned earlier, the network manager must now be prepared to transcend this strictly technical definition and consider how the NCC will play an active role in assisting the organization to reach its strategic objectives. This implies that the network manager must be prepared to focus more energy on management endeavors, and less on the actual running of the network.

To accomplish this task, two things are required. First, the network manager requires more in-depth training in management. Second, a model of the NCC is required that will define the activities, and the staff, required to accomplish its mandate. While it is not possible for us to plan the career development for the network manager in this book, it is possible to present a modest proposal for an NCC model.

Figure 7-2 breaks down the NCC into three functional areas. These are defined as *Administration*, *Information*, and *Operations*. Each group requires a particular set of technical and interpersonal skills in order to succeed. The list of tasks assigned to each area is not intended to be complete, merely a starting point for the development of each function. Although we will examine each one, it should be obvious that the staff required for each area varies considerably. One of the principal missions of the network manager is to ensure that the proper staff and staff development program is available for members of each group.

✻ NCC administration group

The Administration function has the mandate to ensure that the required resources are available for the NCC to satisfy its mission. In addition, this function must also provide the bureaucracy required for the NCC to be able to operate efficiently and scale to the size required for the organization. Principal tasks of the Administration group are:

- Contract negotiations with vendors for services and products
- Development of budgets to ensure adequate funding
- Staff development programs
- Maintenance of inventory of all hardware and software managed by the NCC
- Liaison to senior management

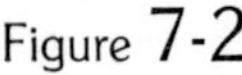

Figure 7-2

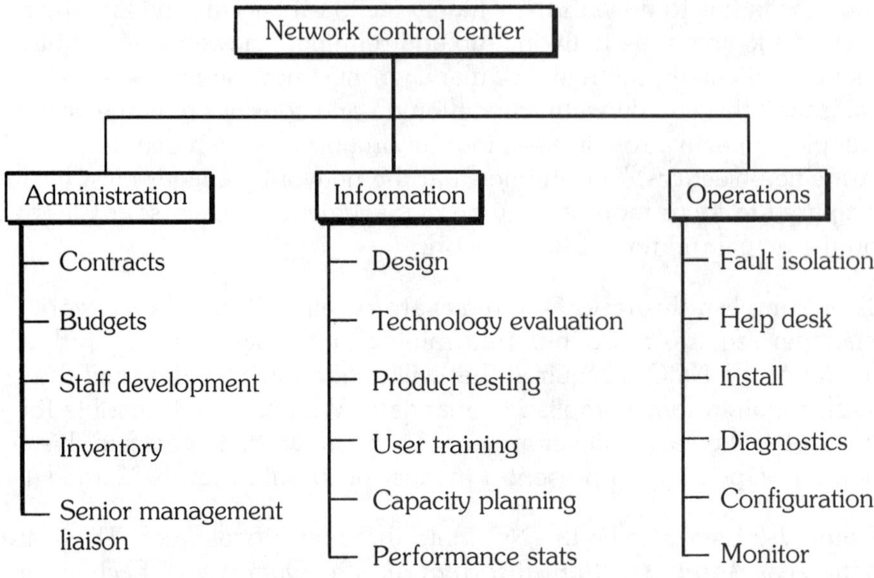

A management model of the NCC. Learning Tree International

Beyond these tasks we anticipate the Administration group to be instrumental in the development and distribution of policies and guidelines relating to the use of the network. While the principal focus is on the backbone network, this group might also be called upon to assist local administrators in developing their own administrative functions, as required, and to purchase network technology, depending on whether centralized or decentralized capital acquisition is being used.

Additionally, this group may be responsible for charge-back of services if an accounting system is in place. As new user accounts are created, the Administration group must be involved, even though they don't actually perform the technical task of opening these accounts. Also, as members of the Information group perform services for local departments, it is the responsibility of this group to ensure that the user is charged for services rendered.

To accomplish these tasks, the Administration group requires staff that favors an administrative background over technical skills. Such individuals might have backgrounds in a variety of areas, including methods and procedures, accounting, law, or management. While some exposure to data communications would be advantageous, the emphasis favors business skills. It might even be advantageous to draw the network manager from this group (though this might cause some friction with the other groups in the NCC). Interpersonal skills should include negotiation and supervision. Attention to detail would also be desirable. It is unlikely that these individuals can be drawn from the other groups in the NCC; they will have to be imported from other areas of the organization.

✻ NCC information group

The Information group has a diverse charter that includes planning, design, evaluation, training, and consulting. This group is responsible for project-oriented activities that emphasize the longer-term viability of the network.

Working primarily in teams, members of the Information group focus on long-term issues, rather than the day-to-day running of the network.

An important group function is technology assessment. As new technology becomes available, it must be examined to determine what role, if any, it might play in the network, and how it might benefit the organization as a whole. This function might be developed either at the request of external groups, or as part of the normal technical monitoring that the Information group performs; it enables the NCC to anticipate new requirements.

Capacity planning is another important function. In order to ensure that adequate bandwidth is available, the members of the Information group evaluate the impact of new applications on the network, and provide recommendations on how upgrades might be phased into the existing environment.

This group is also called upon to perform new product testing. This testing might be motivated either by technology assessment or product acceptance.

The group develops user training programs. This training focuses on local administrators, ensuring that they have the requisite skills to accomplish their tasks. Fee services can also be made available to provide additional technical assistance to local groups in the design of individual networks. This requirement cannot be overemphasized,

since the NCC must be a mentoring agency to local groups, providing specialized technical skills that might not be available at the local level.

Staffing of the Information group draws primarily upon the technical data communication staff. Since most of the tasks are project-oriented, it is important that the individuals be strong team players with good analysis and design skills. Interpersonal skills are required to deal with local groups so that successful analyses can be completed. Documentation skills are also important. Staff development should include user requirements analysis, in addition to the standard complement of data communication training.

✻ NCC operations group

The Operations group concerns itself with the day-to-day running of the network. Primarily task oriented, this group ensures that the network is operating efficiently, that faults are isolated and repaired, and that new equipment is properly configured.

The Help Desk is located within this group, and handles user questions and fault reporting. Most of the technology used to implement the network is the responsibility of this group. In many ways it is equivalent to the Operations group of MIS, which is responsible for the mainframe and production activities.

The group collects the statistics used by the Information group for performance monitoring and capacity planning. Proactive and reactive maintenance activities are planned and scheduled within the Operations group. An automated trouble ticket tracking system is highly recommended.

It is important to mention that the focus of the Operations group is the backbone, and not on each individual subnet. It is probably unwise to attempt to monitor local networks—the volume of data required would likely overload the backbone network. While this group might be called upon to assist in the diagnosis of particular local problems, this should be the exception, rather than the rule. However, local groups that do not wish to have their own local monitoring resources might contract the Operations group for this service. The funding provided would be used to hire additional resources to accomplish these activities. Wherever possible, though, there should be a clear division between the efforts of the central NCC and local groups.

Staffing for the Operations group should draw from a pool of highly technical individuals. Unlike the Information group, these individuals must be task and skills oriented, since the activities of this group cannot be classified as projects, and there will likely be a requirement to manage a number of open problems with constantly changing priorities. Staff interpersonal skills require the capability to deal with external customers/clients during high stress (i.e., fault) periods. A fair degree of autonomy is also required, since it is difficult to monitor the wide variety of activities that must be accomplished. This confusion can be resolved by dividing this group into a series of subgroups, each focused on a different activity, such as Help Desk, Installations, Faults, etc.

✳ Other NCC functions

Security is perhaps the most challenging issue in terms of its placement within the NCC. Does it belong in Administration, Operations, or Information? An argument can be made for each. In addition, it is possible to argue that Security should be a function separate from all these groups, since it will also be responsible for risk analysis, contingency planning, and disaster recovery. We are tempted to place it within the Operations area, but as the complexity of the network grows, a separate group might be required.

Another area that we have not mentioned is address and directory management. This important activity becomes a crucial one as the size of the network increases. The temptation is to place it within the Operations group as part of its configuration activity, but the network manager might feel more comfortable with this activity as an administrative function.

There are other activities that we have not mentioned that must be included within the NCC. However, it has not been our intention to provide a comprehensive discussion of the NCC activity (that will be the subject of another book!), but rather to raise the profile of this important area. Without a clear model of how the NCC will function, and its position within the organization chart, the network manager will find it challenging to provide the class of service that the organization demands. Furthermore, if the proper staff is not hired and developed, it is likely that some areas will receive more attention than others, causing a skewed approach to network management that

cannot satisfy the organization's requirements. We encourage network managers to focus on the management aspects of network management, rather than the network itself.

Network management standards

Regardless of the lack of available models for network management, we still find that the emphasis today is on the selection of tools and protocols in multivendor networks. How can vendors supply products that will enable monitoring, control, and security of a wide variety of devices that are connected together with a variety of transport protocols? Furthermore, how can we ensure that as new protocols and products become available, our existing systems will remain upward-compatible?

There appear to be three major approaches that can be taken.

In the first, we can use a variety of network management systems specifically tailored for the management of a particular class of device, or a particular vendor's product line. This approach is sometimes called *proprietary*, since we rarely find standards.

The second approach attempts to integrate messages from a variety of devices so as to provide a coherent view of the network, which can then be managed as a single entity. This approach is sometimes called *open*.

In the third approach, we attempt to mix the availability of open protocols for communication among devices with a proprietary integration system, or we combine proprietary messages with an integration platform that uses open system specifications. This approach is sometimes called *hybrid*.

The bad news is that none of these systems is likely to fulfill all the requirements of the network manager. Adding to the misery is the fact that there is no single platform that currently provides functionality in the five basic areas of network management mentioned in the last section. And, to ensure that the reader does not get too optimistic, the industry trend seems to be toward the fragmentation of these standards, with new consortiums and vendor initiatives coming out on a daily basis. What strategy should the network manager adopt?

Network management standards have been extensively covered by Black, Rose, Terplan, Stallings, and other authors. In addition, a wealth of information is available from vendors such as IBM, Sun Microsystems, HP, Cabletron, and most vendors of data communication equipment. We limit our discussion to the most popular and commonly available protocol, *Simple Network Management Protocol* (SNMP). There are a number of reasons we have chosen to focus on this set of specifications.

- SNMP represents the most commonly used open specifications for network management
- There are too many proprietary specifications to cover in the space available
- Most of today's internetworking components support the SNMP standard

Simple Network Management Protocol (SNMP)

Originally intended as an interim solution for router management on the Internet, SNMP has quickly become the de facto standard for the storage, description, and communication of network management data. As an interesting side note, it should be mentioned that SNMP was supposed to be replaced by CMOT (CMIP Over TCP/IP) eventually, since TCP/IP was going to be replaced by OSI. Part of the reason for the early acceptance of SNMP has to do with the ease of implementation on the part of manufacturers. In the best tradition of the Internet community, the SNMP standards were designed to be easy to understand and deploy. The emphasis was on the word "simple." This fact, coupled with the lack of any alternative for integrated network management in a multivendor environment, has led to the rapid deployment of SNMP in many organizations. Does this mean that SNMP-based systems are the answer to all of our problems? Hardly—we will see there are still some outstanding issues with SNMP that must be addressed.

✻ The SNMP management framework

SNMP defines a management framework for network control (see Fig. 7-3). There are two basic components to the framework. The first is the Network Management Console (NMC). The NMC is the device (generally a high-end workstation) that contains the programs that implement the network management services mentioned previously. SNMP does not define these functions, or describe how they should be implemented. That is left to the vendors of the NMC. The second major component of the framework is the agent. An agent is any device or process that can be managed from the console. Anything can be an agent: a bridge, hub, router, etc. As mentioned previously, a process can also be an agent. For example, it is possible to manage a user's operating system, a file server, or a database using SNMP. This is done by providing additional software which implements the agent function.

There are some additional components that should be mentioned. At each agent, there is a special record implemented called a *Management Information Base*, or MIB. The MIB contains all the

Figure 7-3

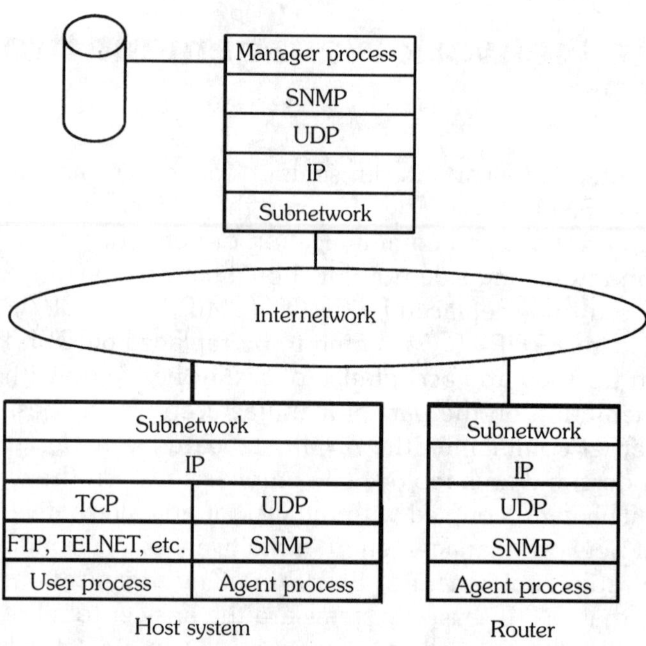

The SNMP model of network management.
Learning Tree International

information required to manage the agent. The MIB has a standard format, so that it is possible to manage the diverse agents in a heterogeneous environment. Extensions are permitted to the MIB to allow for vendor-specific information that might be required. Both the standard MIB format and the extensions are defined to the NMC so that it can query the agents and obtain information, which it can then integrate to provide a coherent view of the entire network. Furthermore, by modifying the values in the MIB, the NMC can change the agent's behavior so as to meet new requirements in the network.

✳ Functions provided by SNMP

The communication between the agent and the NMC is handled by SNMP. Four primitive functions are described in the first version of SNMP:

- ➢ GET—retrieve the value of a single data element in the MIB
- ➢ GET-NEXT—get the value of the following data element. This is especially valuable when reading variable length fields such as tables
- ➢ SET—modify the value of a data element in the MIB.
- ➢ TRAP—enables the agent to report an extraordinary event to the MIB

It should be noted that the latest revision to SNMP, known as SNMP v2, adds new functions for bulk data transfer, security, and interoperation with SNMP v1. This will greatly facilitate the functionality that can be offered. It is recommended that purchasers select SNMP v2-compliant products whenever possible.

✳ SNMP's Management Information Base (MIB)

The MIB is split into a number of "namespaces." This refers to groups of data elements that have some common function. For example, the "SYSTEM" namespace contains data elements such as serial number, contact name, or perhaps vendor name. Some of the namespaces are optional, while others are mandatory. This avoids having to include namespaces that would be blank for a particular type of agent. For example, there is no point including the Exterior Gateway Protocol (EGP) namespace in a hub, since it never interacts with an EGP device.

Since the MIB is a sophisticated record, a data descriptor language was required to define it. This language is called the *Structure of Management Information* (SMI). SMI is based on the OSI data descriptor language ASN.1 (Abstract Syntax Notation). In fact, SMI is a subset of ASN.1, which suggests that in the future there should be some level of compatibility between SNMP MIBs and those of the OSI network management scheme known as *Common Management Information Protocol* (CMIP).

Unfortunately, it may be some time before we start seeing commercial applications of CMIP-based systems for data communication networks. CMIP is complex and comprehensive, and it seems complete. However, it is also predicated on the belief that we are using OSI-based networks. Until this actually happens, it is unlikely that there will be much interest in CMIP.

* The bottom line on SNMP

To conclude, it appears that SNMP will be our open network management standard at least for the next few years. Given the general lack of acceptance of the OSI protocol suite, it is unlikely that we will see much interest in the development of CMIP-based management systems. This "Catch-22" will persist until there are enough OSI-based products and services to warrant implementing CMIP systems. If and when this happens, it is likely that we will be able to migrate the existing SNMP MIBs, since they have been defined with the same notational language as the CMIP MIBs.

Security and integrity issues

Consider the LAN. The closest analogy that we have in the telecommunications world is the party line. A party line is a shared telephone circuit that enables a number of separate phone numbers to share a common line. Cooperation is required among the participants, since if one person is using the phone, no one else can gain access. Each phone number has its own unique ring indicator, which makes it possible to signal different locations. It should be noted that party lines used to be considered a cheap alternative to private lines, and were generally used when it was not economical to provide separate circuits to each location.

The LAN is much like the party line. Only one station can use the line at any time, and each station has its own unique address, so that it can determine if an incoming packet should be accepted. While this form of shared medium does, in some cases, provide a lower-cost alternative to individual cabling, it does present some problems, particularly in the area of security.

Figure 7-4 illustrates how "Bob the Spy" can easily listen to (i.e., tap) any communication between Fred and Sally. This is especially true if Fred and Sally are using a wireless LAN to communicate. While it might be possible to encrypt the data, this rarely happens in practice. As a general rule, we ask the following question of LAN security: "Does anyone have the ability to monitor or disrupt LAN services in your organization and not be caught?" If the answer is yes, then a criminal can probably also do so. Furthermore, Bob might also choose to disrupt the connection by any number of means. While the immediate cost of this form of tapping or disruption might seem minor at first glance, the cumulative costs can add up quickly.

Calculating the cost of LAN outage

Calculating the cost of LAN outage has always been challenging, because it has usually been difficult to assign a value to the work being

Figure 7-4

The LAN as a party line. Passive monitoring is only one of the problems.
Learning Tree International

performed. In general, we assume that LANs used for mission-critical applications have a higher cost of outage than those that might be used for office-automation support functions. The costs of the outage can escalate if the LAN participates in a bridged environment (as discussed in Chapter 3), where it might be a connection (cascade) point between multiple LANs. In this case, failure of the LAN can degrade connectivity over a much larger user community.

Let us try a conservative estimate. Suppose we have a LAN with 100 users that has a four-hour outage. If the LAN is required for the users to perform their function, then the potential loss is four hundred person-hours, plus the value that their function provides to the organization. For those four hours they become useless.

Of course, the reality is that they are probably able to operate in some degraded mode for the interim period. Let us say that they are able to operate at 75% of their former efficiency. This implies that the organization will lose one hundred hours of work from this group of people. As just mentioned, to this cost must be added the loss of business that might result from not being able to handle incoming work, representing income lost to the organization. While it is impossible for us to assign a dollar value to this exercise, it should be evident that the cost of LAN outage can quickly become significant as more and more critical applications are ported.

Note also that it is not necessary for a mission-critical application actually to be deployed on the LAN. As many organizations replace their terminal networks, they use their LANs as the primary connection path to the mainframe and their resident mission-critical legacy applications. Loss of the LAN means that the users can no longer access these applications, and this does have the same impact on these users as if the mainframe had suffered an outage.

The computer virus

While the cost of LAN outage is a significant concern, other dangers threaten as well. In particular, the LAN represents an excellent entry point for theft, espionage, and terrorism. Since it is easy to access, and since all computers are attached to the LAN, it becomes possible to easily spread virus programs. While it is not our intention to discuss virus programs in detail, we can list some of their characteristics:

- They are generally embedded in other programs
- They can remain dormant and undetected for extended periods of time
- They exhibit the ability to replicate across environments
- They can be programmed to perform a variety of destructive tasks

Virus engineering is still in its infancy, and the vast majority of virus programs have been relatively benign. They erase disks, corrupt screens, and cause general mischief. This may change as the engineers of these viruses are offered large sums of money to attack particular systems in particular ways. The use of virus programming for military purposes has been speculated upon for some time. Hospitals, power plants, and other critical infrastructure might also be targets. It is not too difficult to conceive of a virus that would investigate data and then report to its originator as required.

Methods are required to protect organizations from these threats. It is unlikely that diskless workstations will be considered an alternative (even though they would be effective), because most users would be unwilling to give up the flexibility and convenience that their current workstations provide. Furthermore, it must be recognized that most departmental servers must be considered insecure if they are accessible to the majority of individuals within the department. It must be also remembered that virus scanning programs generally protect against known viruses only. It is difficult to spot code that is original.

Authentication techniques

One approach to preventing unauthorized access to the network is authentication. Generally implemented in the form of password control, authentication enables the network servers to identify those

who wish access, and the set of privileges that they are allowed. Unfortunately, passwords are of limited value for a number of reasons:

- To be memorized they must be simple (suggesting that they can easily be discovered)
- To be secure they must be complex (suggesting that they will be written down—and discovered)
- They can be discovered by tapping the LAN (suggesting that it doesn't matter whether they are complex or simple)

See Fig. 7-5.

Figure 7-5

Passwords do not provide security! Learning Tree International

As many organizations begin to use domain login capabilities, the loss of a password becomes more significant, since it enables access to a greater percentage of the enterprise. While certification servers (to be discussed in the next section) help, they don't get around the basic problem of passwords.

Security people like to classify authentication systems into three broad categories:

- Something you know (e.g., a password)
- Something you have (e.g., a passcard)
- Something you are (e.g., fingerprint, retinal scan, etc.)

Most systems are still based on passwords. Since the cost of implementation of the more advanced authentication systems is still high, it is likely that passwords will remain the preferred method for some time. This presents certain problems when users require access to many heterogeneous servers. Since the LAN can be easily tapped, sending passwords across the network must be avoided whenever possible. How can this be done?

Certification servers

The problem mentioned above does have a solution. It is called a certification server. The best-known implementation was developed at MIT as part of Project Athena. Its name is Kerberos (other implementations of certification servers exist as well). A certification server is based on the principle of a "well-trusted third party." In this approach, neither the client nor the server is responsible for security.

Password files are kept in a secure location (the well-trusted server; see Fig. 7-6). When a user desires access to a particular server, he sends a request to the certification server, including his user ID. The

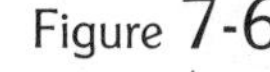
Figure 7-6

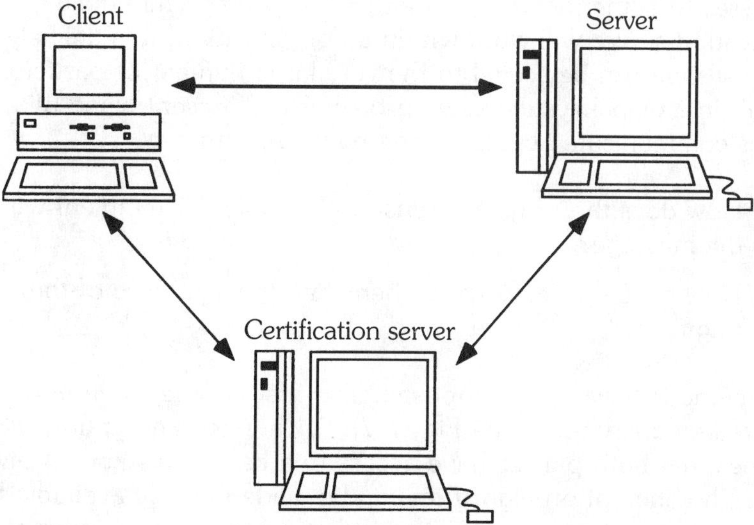

Certification servers are an excellent way to provide authentication in the well-integrated network. Learning Tree International

server uses the person's password as a key to encrypt a reply (the certificate or ticket). The reply is sent back to the user's desktop, where the user can decode the ticket using his password. Note how the password is never sent across the network.

A similar approach is used to access the server of choice. The ticket is sent to the server of choice. The server sends the ticket to the certification server for authentication. Since a time limit is placed on the ticket copying, the file only provides access for a finite time.

Detailed information about Kerberos is available from a number of sources, but the OSF security server is a complete implementation. It is recommended that the integrator consider an implementation of a security server in a heterogeneous environment. Agent software is available for most platforms, and the system is independent of the underlying network protocols, so installation should be possible in a wide variety of environments.

Public and private keys

As global systems become more widespread, the need for people and processes to authenticate one another increases. While the certification server will work within an organization, it is unlikely that such a system can be scaled to the very large number of participants present in a global system. Two problems are present when two parties communicate across a large public system:

- How does the originator ensure that only the recipient will see the message?
- How does the recipient authenticate the signature of the originator?

Two methods have been proposed. The first is called *symmetric* or *private-key* encryption (see Fig. 7-7). Private-key encryption assumes that both parties have access to a key that is known only to them. The encryption algorithm can be made publicly available to anyone who wants to use it. The best-known implementation of a private-key system is the *Data Encryption Standard* (DES).

Figure 7-7

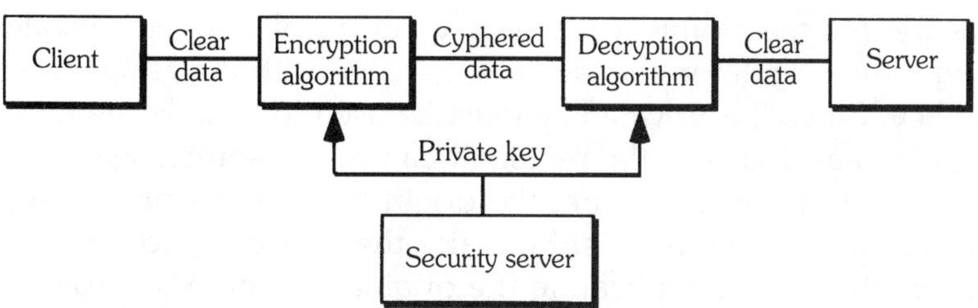

Private-key encryption provides very secure communication, which is offset by its lack of suitability for large public networks. Learning Tree International

Generally implemented in firmware, the DES standard has been in use for a number of years in many commercial environments that require high security.

The problem with private-key encryption is that it requires a prior agreement on the part of the participants. This is difficult when sending e-mail to someone who might not be known, or communicating with many different parties. It is impractical to maintain different passwords for each possible pair. This is where the second proposed encryption system is used.

Asymmetric or *public-key* encryption uses two different keys to encode and decode the file (see Fig. 7-8). Each system participant has two assigned keys:

- A private key known only to the individual
- A public key that is freely available and listed in an easily accessible directory. This key must be unique to each participant.

Figure 7-8

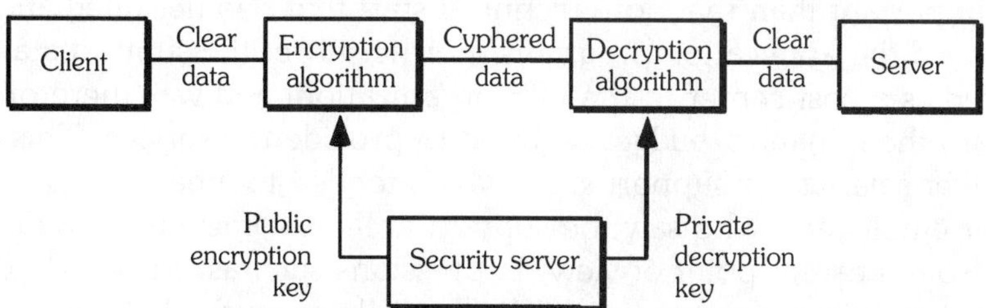

Public-key encryption provides both a public and private key to each network participant. Learning Tree International

To ensure that the recipient is the only individual who can decode the message, the originator encodes the message with the recipient's public key. Since the private key must be used to decode the message, this guarantees that only the recipient can read the message. To provide an electronic signature, the originator uses his private key to encode his name. The recipient can use the public key to decode the signature, thereby authenticating the originator. The best-known version of this approach is RSA (named after its designers).

Most of the criticism of asymmetric encryption concerns its safety. Some critics have said that given the public key, the encryption algorithm, and a sample of the encrypted text, it might be possible to derive the private key and break the system. At the present time, widespread implementation has not been done.

Support issues

Another area of our NCC model that requires special attention is the set of support services offered to the organization. In the past, support services could be centralized, since the environment was homogeneous (generally a mainframe computer and application programs). It made sense for a central group to answer questions and handle problems, since all users had access to the same set of screens and programs. Furthermore, this group was generally located within the MIS department. This gave the support staff easy access to programmers and developers when code had to be fixed, or when more expertise was required to solve a problem.

This had the unfortunate side effect of making the support staff appear less important than the more technical staff that had designed and deployed the application programs. Furthermore, the support staff was viewed as a cost center to the MIS organization, and was therefore kept to the minimum budget required to provide the service. This led to lower salaries for support staff, which tended to mean hiring more junior employees, who saw the support area as somewhat of a dead end from a career point of view. Expressions such as, "The help desk is our first line of defense" and "We'll put them on the help desk until they learn the ropes," did little to improve this perception. While many models were developed to improve the quality of support, they had little impact, since the organization as a whole was rarely interested in the commitment required to make the help desk work properly. The one exception to this otherwise gloomy situation was the information center, which provided help to users on leading-edge technologies such as personal computers and fourth-generation languages.

The well-integrated network requires sophisticated support services. There are many reasons for coming to this conclusion:

- The complexity of the environment requires staff that can deal with a wide variety of customized hardware and software environments.
- Support services are increasingly seen as the "moment of truth" between the NCC and its clients or customers. The quality of service offered directly reflects on the NCC.
- Revenue for warranty and maintenance can be directly traced to the quality of support offered.
- As mission-critical applications continue to migrate into client/server environments, the help desk will be required to ensure that a high level of system availability is maintained.
- A series of automated systems are required to support the help desk function. These include trouble-ticket systems, configuration databases, user profile databases, and expert systems to search for known problems and solutions.

Solving these problems requires a much higher level of commitment than has previously been offered to the group that provides these support services. Over time, they must be viewed as elite technical staff who combine a unique mix of technical and interpersonal skills.

In addition, the support group might offer a variety of new services. These can include

- a Bulletin Board System (BBS) for software updates, user forums, noncritical questions, and access to other NCC services
- remote control systems that enable the support staff to observe the problem on a user's desktop without having to travel to the user's physical location

- beta testing of new software and/or hardware
- quality control for new systems prior to their release
- market surveys of the existing user base
- software distribution

Unfortunately, we find ourselves once again without the space to cover this important issue to the depth it requires (sounds like another book project!). We can only leave you with our conviction that the well-integrated network must provide a comprehensive set of support services if it is to meet the RAS requirements that will be demanded. The help desk must make the transition from a cost center to a profit center. Failure to do so will mean a network that is fragile and unable to support the demands placed upon it by a growing list of mission-critical applications.

Summary

The well-integrated network requires a centralized management function. Issues such as performance, configuration, fault diagnosis, security, and accounting require automated tools if they are to be addressed. Some of these tools are available, others are still in the process of being developed. Beyond these functions and tools, the NCC requires a management model that accommodates the staff required to successfully support the well-integrated network. We outlined one such model, which divided the NCC into three groups—Administration, Information, and Operations.

The well-integrated network also requires that a protocol be selected to implement the Network Management System. We discussed why SNMP is presently a good choice. Proprietary systems should not be overlooked, since they often provide additional functionality beyond that offered by SNMP. If you are considering deployment of an SNMP-based system, try to use products that are compliant with the SNMP v2 specifications. While not yet as widely available as SNMP v1, this second release provides new functions and closes some security loopholes that existed in the first version.

Security is an ongoing issue. We favor the use of certification servers, since they provide better security than most server-based systems. Kerberos is the name of the algorithm that has been implemented in many systems, including the OSF's Security server. The use of a public-key encryption system will likely be a requirement for most enterprise-wide networks.

Please pay special attention to the help desk and support services. They are a focal point in the well-integrated network. The help desk is called upon to handle a wide variety of interoperability problems. The staff requires extensive training in a broad range of technologies. Remember that the help desk represents your organization's "moment of truth."

Well, we have almost reached the end. Our next chapter provides some speculations on what we believe is likely to occur over the next few years. We will probably be wrong, but we console ourselves with the knowledge that very few people make accurate predictions!

Here's your last concepts review.

Concepts review

1. What are the five major network-management concerns according to the ISO model?

    1. ______________________
    2. ______________________
    3. ______________________
    4. ______________________
    5. ______________________

2. What three service areas are managed by the network control center (NCC)?

   1. ______________________________

   2. ______________________________

   3. ______________________________

3. What are the two major network-management protocols and the protocol suites in which they belong?

   1. ______________, ______________________

   2. ______________, ______________________

4. What does "SNMP" mean?

   S __________ N __________ M __________ P __________

5. What does "MIB" mean, and what is it?

   M __________ I __________ B __________

   ____

6. Indicate the components of an SNMP network-management system and the network-management paradigm on the following network drawing:

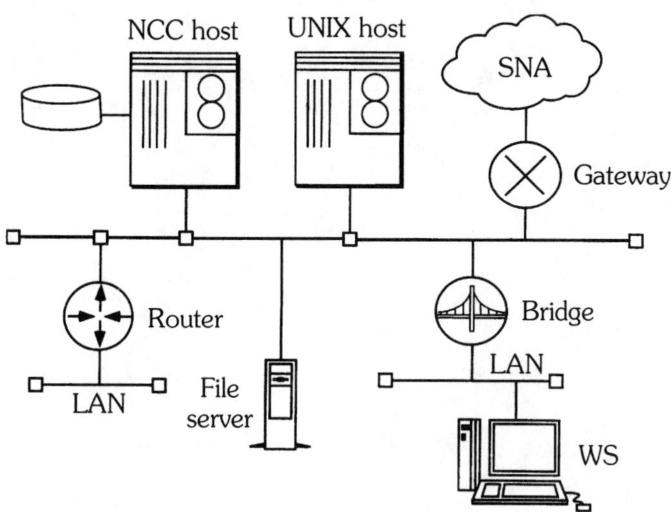
NCC host
UNIX host
SNA
Gateway
Router
Bridge
LAN
File
server
LAN
WS

Future trends in integration

CHAPTER 8

WHILE the computer industry may debate many issues, one thing is certain: dynamic, volatile change will be the norm for the foreseeable future. Each week brings new products, new specifications, new standards, and a host of upgrades to existing products. As new technology is introduced, many local user groups will be anxious to be early adopters. Consider the excitement over technologies such as multimedia, video conferencing, and groupware as prime examples. Nobody can predict what new technology will become "hot," and what impact it will have on the existing network infrastructure. How can the network manager perform capacity planning when new applications can be deployed by simply inserting a disk into a microcomputer?

An excellent, though perhaps trivial, example is the current craze of LAN-ready games. These games can easily be purchased at any local computer store and installed on LAN-connected machines without the knowledge of the LAN administrator. These games consume a great deal of LAN bandwidth. Frequently they impact other users who are trying to work! The LAN administrator may have to spend hours trying to track down the "problem" that is causing the LAN to behave in an unpredictable manner. As more games arrive, users will be eager to try them, and their office system will provide a readily accessible environment.

Given that there is no way to predict what new technology will become available, we have chosen instead to take a brief look at the current state of the industry. We indicate what we believe are the most important of the trends, so that the integrator can track developments over the next few years.

We have identified four major areas that require special attention:

- Partnerships and alliances
- Desktop operating systems
- Mobile computing
- Emerging technologies

We will look at each topic and identify the key players, major issues, important dates, and possible impact. The integrator would be well-

advised to keep track of the state of the art both through trade magazines and on-line services such as CompuServe and the Internet. Remember that new developments occur on a daily basis.

Some words of caution: don't believe everything you read! Our industry is filled with rumors and FUD (Fear, Uncertainty, Doubt). Even new technologies require 18/24 months to become widely available. Resist being a pioneer of new technology for as long as possible. This will have the twofold advantage of allowing the product to mature and the price to decrease. Early implementers always pay the most.

Partnerships and alliances

The industry is filled with a variety of consortia, alliances, industry groups, and partnerships that are all trying to set the standard for the next wave of technology. These groups may offer products, specifications, or industry direction. While we have mentioned some of these groups earlier, here is our current list of groups that we are watching.

Object Management Group

With the release of CORBA 2.0, OMG has now provided a clear direction as to how multivendor interoperability should be achieved in the area of object technology. This work has direct impact on the integrator, because it provides the best hope for truly interoperable multivendor environments. By decoupling object technology from the underlying operating system and network, the OMG hopes to offer an environment in which true process-to-process interaction can take place.

Given the number of organizations that participate in the OMG, and the broad base of support from major vendors such as IBM, HP, and DEC, we feel that OMG should be watched carefully. Integrators should question their suppliers as to what level of support for OMG specifications will be offered and when this support will be made available. In addition, a strategy for the adoption of OMG-compliant products should be developed by each organization.

Compliance with OMG specifications should provide a "level playing field" for the development of well-integrated applications. These applications will cover just about every type of processing that is done today. The well-integrated network will benefit by conforming to OMG specifications, due to purchasing flexibility, decoupling, and higher levels of interoperability. On the other hand, this is still an emerging technology. Products might not be available or fully compliant for a few years. So track this trend carefully, but look before you leap!

 OSF/DCE

While the Open Software Foundation has more or less satisfied its mandate of developing a distributed computing environment, we are just at the beginning of large-scale deployment. Vendors are beginning to sell developer's tool kits and implementations of the basic set of services. Notably lacking are applications such as DBMS, e-mail, groupware, etc. While some companies have promised to develop these applications, the marketplace is still quite sparse.

We expect this to change. There is a great deal of excitement about DCE. What OMG/CORBA will do for interoperable applications, OSF/DCE is likely to do for interoperable networks. DCE has the potential to realize a longstanding dream in data communications, namely a heterogeneous multivendor network that can operate in a seamless manner without the need for gateways or other forms of "spoofing." In fact, if it works as well as advertised, we may find ourselves out of a job!

This is not far from the truth. It is conceivable that the synergy of OMG/CORBA and OSF/DCE may make distributed applications as easy to use as those on a standalone microcomputer such as the Apple Macintosh. While it is still too early to say this with any degree of certainty, the possibility is intriguing. This is why we recommend that the integrator carefully track industry acceptance of OSF/DCE.

It is still probably too early to adopt this technology on a wide basis. As we have indicated, there is still a lack of utilities and basic applications. We are hopeful that this will change. When a critical

mass of applications is reached, we expect that adoption of OSF/DCE-based technology will be rapid.

It is not too early to begin pilot studies and evaluations of OSF/DCE. Suppliers should be questioned about their level of commitment and anticipated delivery dates. Test-bed systems can be set up to experiment with the technology. Also consider how the existing network infrastructure will be migrated to OSF/DCE. As we said, it's not too early to begin making these plans.

X/Open Group

X/Open may represent the last, best chance for UNIX-based operating systems to gain a wide degree of acceptance. While UNIX (and UNIX clones) has been with us for years, it has never achieved the degree of market penetration that many thought it would enjoy as an open system technology. Except for high-end workstations, UNIX has little penetration on the user desktop when compared with Microsoft DOS/Windows, IBM OS/2, Apple MacOS, etc. Large-scale operating systems such as IBM MVS & OS/400, DEC VMS, and others still represent a majority of commercial processing environments. While broadly used in the academic, engineering, and scientific communities, UNIX has never really been viewed as "mainstream" technology. In addition, we hear many dire predictions about how technologies such as Microsoft Windows NT may represent the death knell for UNIX.

We feel that, while the death of UNIX has been greatly exaggerated, there is still no unified approach on how it will be marketed to the industry. This has been largely due to the fragmentation of the marketplace, and the historical fact that the major vendors always preferred to sell their proprietary systems over their UNIX offering. The work of X/Open Group may change this perception.

X/Open is now the recognized unifying agency for UNIX. It offers a certification process that enables vendors to prove that their UNIX implementation is compliant with an agreed-upon definition. Furthermore, the number and caliber of the major supporters of X/Open would seem to indicate that people are getting serious about

UNIX. It is a preferred mid-range server platform because of its scalability and the wide range of products that exist. It offers purchasing flexibility. Given the fact that it has had a number of years to mature, most UNIX implementations should be stable. All the major vendors now offer a UNIX clone. We anticipate that all of these will become X/Open compliant. This makes X/Open Group an important organization to track.

The integrator should ask operating system suppliers about X/Open compliance. Software application vendors should indicate whether their products will be migrated to work in X/Open compliant systems. These efforts should be encouraged whenever possible. Internal application developers should examine X/Open compliant systems as an excellent server platform for a wide variety of tasks including DBMS, e-mail server, directories, security/certification, and communications. This can be accomplished without impacting the user's desktop environment, which will probably not migrate quickly to UNIX due to the high cost.

Of course, this still requires an industry-wide commitment to choose open operating system technology over the more popular proprietary systems. Will this occur? Frankly, we don't know. Nonetheless, it does seem that the UNIX world will make at least one last attempt to penetrate the mainstream in the emerging area of commercial distributed networks.

Motorola/Apple/IBM: PowerPC

Does this world need a new desktop processing platform? These companies feel that the answer is yes, and that its name is PowerPC. PowerPC represents an alternative to the Intel-based platforms that are now widely deployed. By the way, when we say Intel-based we are referring to both Intel and Intel-clones. According to the pundits, the PowerPC chips are faster, cheaper, and smaller then the comparable Intel offerings. This remains to be seen. However, given the companies that are involved, plus the fact that a specification has been developed that will allow any manufacturer to build a machine that will run the majority of PowerPC code, we feel that it is an important trend that should be tracked.

Early PowerPC machines have already been made available by both Apple and IBM. These machines are not compatible since they do not conform to the PowerPC association specifications. However, both companies are committed to building machines that do meet the specification. We anticipate that a number of operating systems will be migrated to the PowerPC, including Windows NT, OS/2, MacOS, and a number of UNIX clones. If this happens, the PowerPC will become a potentially important piece of hardware for the integrator.

Consider the value of having a single hardware platform that can run multiple operating systems. Local groups could still have the freedom to choose their preferred operating system, but support would be greatly improved by only having a single hardware platform to maintain. Furthermore, this platform could be purchased from a wide variety of vendors, which should result in a competitive marketplace with good price/performance metrics. Finally, if applications can be easily migrated to the PowerPC platform, users will require little training to take advantage of technology.

All this suggests that the integrator should track developments involving PowerPC technology. While the desktop marketplace is currently heavily oriented toward Intel-based machines, organizations are always in the process of upgrading their technology. A switch to PowerPC is possible if the price is right. In particular, application software developers should be questioned as to their intention to migrate their applications to PowerPC platforms. On the other hand, it is unlikely that we will see large-scale availability of low-cost PowerPC-based machines until the second quarter of 1996.

OSI and TCP/IP

At the current time, the industry appears to have accepted TCP/IP as the open protocol of choice for networking. This has largely been driven by the rapid growth of the Internet, and governments' general lack of support and commitment to OSI. We see nothing in the short term that is likely to change this attitude.

This does not mean that OSI is dead. Rather, it has been placed on a back burner for the foreseeable future. We still believe that OSI has a

major role to play in the implementation of secure, reliable network communications. However, it will take some time before a global network that offers this capability is required. In the meantime, if you need an open protocol to operate on the backbone, TCP/IP is our choice.

Nonetheless, the integrator might wish to track important OSI specifications such as X.400 and X.500. These will likely be deployed in advance of full OSI networks. There is a current requirement for secure e-mail and a worldwide directory structure. Given the growth of the Internet, the current DNS directory scheme will not be sufficient to meet these needs. It is likely that X.500 will be deployed on the Internet at some future time.

ATM Forum

The ATM Forum provides the guidelines and specifications for the wide scale adoption of ATM technology. Many believe that ATM will become the standard for linking local and wide area networks. It is scalable, provides a common local and wide area protocol, implements switched virtual circuits, and can support a wide variety of media types. ATM offerings are currently available.

The integrator will probably want to track these developments. As users outgrow the bandwidth of their existing LANs, the integrator will have to provide guidance for replacement technology. In Chapter 3, we outlined some of the alternatives. ATM heads our list, because it offers the best chance of interoperability at the subnetwork layer.

Vendors should be queried as to their support and commitment to ATM. Do they participate in the ATM Forum? Do they intend to offer products that will interoperate with other vendors' offerings? When will these products be ready? Once again, it is not too early to be running pilot studies. ATM may represent a strategic direction for organizations that are looking to build a well-integrated subnetwork.

As an interesting side note, there are some in the industry who are speculating that widespread acceptance of ATM may mark the end of routed networks and their protocols (such as TCP/IP). This is a

possibility. Given that ATM implements virtual switched networks, the need for routing is greatly diminished. However, it will take some time before applications are developed that can take advantage of ATM switching. Until then, it is likely that the routed networks will continue to grow in popularity. As organizations increase their commitment to routed networks, they are unlikely to give up the investment to switch to ATM. Although the long-term expectation may be a move away from routed networks, the foreseeable future belongs to routing.

The desktop operating system

Nowhere is the volatile nature of the industry more apparent than in the user desktop. In particular, we have seen the rise of what might be best described as "techno-religious" wars with respect to the operating system deployed on the desktop. The principal contenders are (in no particular order)

- Microsoft Windows 3.x and Windows for Workgroups
- Microsoft Windows NT
- IBM OS/2 Warp
- Apple MacOS
- a variety of UNIX clones, but Sun's Solaris in particular

Starting a discussion on desktop operating systems is a little like opening Pandora's box. Nothing good ever comes from the discussion. Entrenched positions already exist in each camp, and it is difficult to have a meaningful dialog. The arguments always seem to come down to a chest-beating exercise that sounds like "My technology is better than yours and I can prove it!" By the way, Andres uses Windows for Workgroups, and Jerry is an OS/2 Warp user. Even we can't agree!

We wished to avoid this entire discussion, but it is an area that the integrator needs to track. Luckily, we have found a way to meet both of these objectives. As you may recall, in Chapter 5 we discussed the

issue of client connectivity. We covered all the major desktop operating systems. One of the clear indications is that each client can access and be supported by virtually all servers. To a very large extent, interoperability has already been achieved with respect to these desktop environments. While programs cannot be ported between them, each is capable of being a full-fledged participant in today's heterogeneous networks. Furthermore, all major network operating systems support these environments, so selection of one does not preclude the use of another.

This leads us to the conclusion that the selection of the desktop operating system has now become largely a matter of personal preference. Since all of them will integrate, each becomes a viable alternative. In a very real sense, this puts the word "personal" back into the PC.

Of course, the integrator will not be happy with the prospect of having to support multiple desktop environments. We appreciate this and sympathize. However, short of creating universal standards, which have the unfortunate effect of negating local autonomy, we see no easy solution. However, we do suggest that the user desktop be supported using local resources, whenever possible. Perhaps the choice of operating system should be made at the departmental level.

One more thing. If you are interested in new developments in operating systems, follow the Taligent project. While it is still much too early to make predictions, the work that is being done appears exciting, and in many ways groundbreaking. Taligent is attempting to build an object-oriented operating system. If this can be done, operating systems of the future may be much leaner, and offer more custom functions.

Mobile computing

Over the next few years, many individuals will migrate away from large desktop machines toward laptop computing. This has largely been driven by lower costs and sophisticated docking stations that allow the user to work with a full-size screen and keyboard during the day, while still having the flexibility to take the machine home at night. We anticipate that as trends such as telecommuting become more popular, this move toward mobile users will accelerate.

The integrator will face unique challenges in the area of mobile computing. Applications that were built on the assumption that high-speed LANs would be available tend to fail at the much lower speeds of dial-up communications. This is especially true when large amounts of data must be sent or received. In addition, support of mobile users is almost impossible. There will never be a support person near the mobile user. How, then, can problems that occur at home or while traveling be corrected?

Security will be another problem. Mobile computing implies the need for many dial-up lines. How will these be secured? Furthermore, if a laptop machine is stolen or lost, and if it contains scripts that perform automatic login, how will unauthorized access be prevented?

We expect trends in laptop devices to accelerate at a rapid pace. In addition, there will be new categories of devices such as the *Personal Digital Assistant* (PDA). All of these will have to be integrated.

We have no solutions at the current time. We are raising the issue as one more area that the integrator should be considering as he builds his networks. Mobile computing is a primary direction of the future. Plan for it.

Emerging technologies

Here is a brief list of other technologies that we feel should be tracked:

- Groupware—this refers to a wide variety of applications that assist local groups in leveraging networking technology. It may include scheduling, project management, conferencing, application development, e-mail, and any other applications that would assist a group of people. Lotus Notes is one example of groupware technology. As groups go mobile, groupware may become one of the core technologies that enable the group to remain cohesive while being geographically distributed. We anticipate many new products in this area.

- Multimedia—multimedia refers to applications and data that operate on audio, video, image, or any other form of digital data representation. It is already being widely deployed as the cost of entry continues to decline. Multimedia will have a major impact on the integrator, since it places a heavy demand on both the local and backbone networks. Users will push for rapid deployment. The integrator must be prepared.
- Video mail—perhaps the next "really big" application. Why type when you can show it in words and pictures? The appeal of video mail is obvious to many end users. Touch a button, record your message, press send. No messy keyboards. If a picture is worth a thousand words, then what is the worth of 60 seconds of video at 15 frames/second? About 10 megabytes of storage space will be required to store a single 60-second message. Multiply this by the number of e-mail messages you get a day. Add all the messages that you save for future reference. Get the picture? It's not with us yet, but video mail will require substantial upgrades to networks, servers, and user desktops.
- Virtual reality, expert systems, artificial intelligence, . . .—All of these technologies will begin to appear over the next 3–5-year period. Not enough is known for us to make any prediction, or even indicate a trend that should be followed. We feel that this is very much a "blue sky" issue at the present time that should not require a great deal of attention on the part of the integrator.

Summary

You now have our ideas on some of the trends that you should be tracking. Of course, there are other trends that will be important to each organization, but to list them all would have required another book. We have tried to highlight those areas that we think are of primary importance.

As always, we encourage the integrator to track those technologies that will lead to better interoperability and internetworking. Chief among these is OMG/CORBA and OSF/DCE. While both

technologies are still emerging, they seem to represent an excellent mechanism to achieve our vision of the well-integrated network.

In closing, we would like to say that we have enjoyed the opportunity to share our views and experiences in building multivendor networks. We hope that you have found the material to be of value, and that it will assist you in planning your own integrated environment. There are still numerous details that you will have to research, but we trust that we have started you on your way. The well-integrated network can be built. By balancing local autonomy and integration, the needs of both local departments and the organization can be achieved. Good luck in your integration projects, and thank you for taking the time to read this book.

A

Online services, Internet, and CD-ROM vendor resources

Many vendors make available information in electronic form. Four key vendor sources of information are:

- On Line Services forums: The most popular is CompuServe.
- Internet: Anonymous FTP and World Wide Web servers on the Internet. These are becoming increasingly popular.
- CD-ROM: Many vendors make available copious amounts of technical and product information in CD-ROM format. They are available to qualified customers.
- A bulletin board: Some vendors maintain, in addition to the services above, a private bulletin board for patches, fixes, and a place to answer questions from customers.

We highly recommend that a multivendor network manager obtain an account on both CompuServe and the Internet. The Internet account should be a SLIP or PPP account, if it is a dial-up account, or do it through your company's network. But it must be more than just a TELNET access or e-mail-only account. It has to have FTP access at minimum, or better yet, full WWW access. It is very useful to access the Internet via the graphics interface of the Web via Mosaic or an equivalent interface. It is also important to have an e-mail account both on CompuServe and the Internet.

The following list is sorted alphabetically by vendor. We tried to get as much information as possible about each vendor's offering by press time. Many new services are springing up all the time, so use this list as a starting point in your search.

Association for Computing Machinery

WWW server: http://info.acm.org

The ACM is the oldest and most influential of all computer organizations. Most of their work is academic, but they provide a valuable resource for research and the cutting edge of the industry. Join up! They deserve your support.

Apple Computer, Inc.

WWW server: http://www.apple.com

CompuServe: GO APPLE

Stick to the Mac Communications Forum for information about networks and the Mac. It is worthwhile going there for technical solutions to your Mac connectivity problems.

Banyan

WWW server: http://www.banyan.com

CompuServe Forum: GO BANYAN

The name of the forum you want is BANFORUM. It contains a wealth of information, tech notes, and product news—enough to satisfy any Banyan fan or network integrator who needs to deal with Banyan servers.

Bay Networks

WWW server: http://www.baynetworks.com

CompuServe Forum: GO BAYNETWORKS

Useful web site and forum for information on new products, updates, and general information on Bay Networks' internetworking products. The Synoptics and the Wellfleet companies merged to form Bay Networks.

Cisco Systems

WWW server: http://www.cisco.com

Useful web site for info on new products, updates, and general info on Cisco's internetworking products.

Digital Equipment Corporation

WWW server: http://www.digital.com

CompuServe Forum: GO DEC

Digital maintains their computer catalog on line at either location (CompuServe or the Web server). They also have a service for integrators, which can be reached through CompuServe forum GO DECNIDEV. There are several DEC forums dealing with the issues of networking.

Digital makes available a wonderful CD-ROM called DEC Direct Interactive. The CD is their entire catalog and has a great deal of information on it. Call 1-800-344-4825 to get your copy. It is a subscription, so you will get one every month. Or order by e-mail at converge@world.std.com.

Disaster Recovery Journal

CompuServe forum: GO DRJNRL

The Disaster Recovery Journal (DRJ) is a journal dedicated to business continuity and news gathering. DRJ has been a primary force in the Disaster Recovery arena for over eight years, and is published quarterly by Systems Support, Inc. Subscriptions are free to all qualified personnel in the U.S. and Canada involved in managing, preparing, or supervising contingency planning. The DRJ also hosts two annual conferences. One is in the spring in San Diego, California, and one is in the fall in Atlanta, Georgia.

IBM

WWW server: http://www.ibm.com

CompuServe Forum: GO IBM

Gopher Sites: www01.ny.us.ibm.net

This site contains entry into many IBM public domain files that cover various areas, including new technology, client/server, product announcements, and case studies. The files are available for a free download.

A yearly subscription to the OS/2 Developer's Connection brings about six CD-ROMs per quarter, containing product demos, LAN connection tools, developer's tools, and other useful files. The low price makes it attractive to anyone interested in OS/2 development. They also include a migration system to facilitate migrating programs from MS-Windows to OS/2.

Hewlett-Packard

WWW server: http://www.hp.com

CompuServe Forum: GO HP

Check out the HP Systems Product Forum under the main menu. There is a great deal of information on OpenView and other networking products.

Internet Society

WWW server: http://www.isoc.com

FTP server: ftp.isoc.org

There is a great deal of information at these servers about the Internet, Internet providers, RFCs, the process of creating a protocol, copies of the protocols, and many more important facts about TCP/IP as well. One interesting URL contains information on Internet providers, and it is well worth the visit. The Network Service Providers Around the World (NSPAW) file, courtesy of Barry Raveendran Greene, can be found at URL: http://www.isoc.org/~bgreene/nsp-index.html

Institute of Electrical and Electronics Engineers (IEEE)

WWW server: http://www.ieee.com

The IEEE maintains a very active computer group that publishes many magazines of interest. Membership also gains you access to a vast array of specifications and standards of interest.

Lotus

WWW server: http://www.lotus.com

CompuServe: GO LOTUS

You are welcome to peruse the entire forum, but we recommend that you stick to network products—Notes and cc:Mail. Be sure to choose Lotus Comm Forum under the main menu.

Microsoft

WWW server: http://www.microsoft.com

Microsoft maintains a collection of product information and other resources on this Web site. The server features Windows NT Server Evaluation kits, which can be downloaded, as well as many white papers. It includes information on the use of Windows NT as an Internet server. Network integrators can use this Web site to search the Microsoft Software Library and Microsoft TecNet. You can get Developer Network News, as well as Windows News on-line. The Web site provides access to ftp.microsoft.com as well.

CompuServe Forum: GO MSNETWORKS

This is one of their forums, which deals with networking, out of a larger collection of forums known as The Microsoft Connection. There is a great deal of information here, including how to hook up a Windows NT workstation to a network, and information on all of their server products.

There are several CD-ROMs available from Microsoft. The Windows NT Resource Guide comes with a CD-ROM loaded with technical information on networking Windows NT. It is well worth getting. Also, Microsoft makes available a CD called Network Developer's CD (the word network has nothing to do with data networking, but with people networking between developers). You get a few free copies before you have to sign up and pay for the subscription. The

networking information is not worth the price of a subscription. Get a few sample copies and see for yourself.

National Aeronautics and Space Administration (NASA)

URL: http://epims1.gsfc.nasa.gov/engineering/engineering .html

NASA maintains a Virtual Library staffed by volunteer "maintainers." It contains a comprehensive list of links to dozens of subjects, including engineering topics ranging from aerospace and control to nuclear and power engineering. There are dozens of categories and lists, such as information resources, manufacturers/vendors, and educational and research institutions.

Novell

WWW server: http://www.novell.com

CompuServe Forum: GO NETWIRE

Netwire is a comprehensive forum that has been around for many years. This has traditionally been a NetWare technician's resource of last resort. If you could not find the answer to your question here, you simply gave up. The Novell staff always promptly answers questions. If you have anything to do with Novell products, getting a CompuServe account is a must.

Novell makes a lot of information available on CD. You can get a free subscription to their Market Messenger CD-ROM. It contains a complete description of their entire product line, and a lot of technical information on both NetWare and UNIX. Then there is the NetWare Support Encyclopedia, which is full of good technical information. You get a free copy when you become a CNE (Certified NetWare Engineer). After that, it costs several thousand dollars for a yearly subscription. You can subscribe even though you don't have a CNE. Lastly, most NetWare manuals these days come on CD, so you can get an inexpensive copy to carry around or access from your CD-ROM player. It is easier than the bookshelf of red-bound books you could be carrying around!

National Computer Security Association

CompuServe forum: GO NCSA

The National Computer Security Association is an organization which provides educational materials, training, testing, and consulting services to improve computer and information security, reliability, and ethics. Training is delivered through public and in-house seminars, and NCSA's annual security conference provides a meeting ground for members and nonmembers to share experiences and learn about current technology and solutions. NCSA manages this CompuServe forum dedicated to computer security and ethics (GO NCSAFORUM or NCSA). NCSA can also be reached via Email at 75300.2557@compuserve.com.

Oracle

WWW server: http://www.oracle.com

CompuServe Forum: GO ORACLE

This is run by the Oracle User's group, called the IAOG.

Retix

WWW server: http://www.retix.com

Sun

WWW server: http://www.sun.com

CompuServe Forum: GO SUNSOFT

Don't confuse this forum with GO SUN, or you will wind up buying sunglasses. Lots of demos of all their products.

Sun makes Solaris demonstrations available on CD-ROM for Windows workstations. The demo also runs on Solaris 2.2 SparcStations. There is some limited information there. If you are considering buying Sun SparcStations, this may be of use to you. Their e-mail is address is cdtimes@sun.com.

3COM

WWW server: http://www.3com.com

CompuServe Forum: GO ASKTHREECOM

Xerox

WWW server: http://ww.xerox.com

CompuServe: GO XEROX

A great deal of information on printers and office products. Some limited information on software products that work over a network.

Other resources

A wonderful set of CD-ROMs with standards and other utilities can be obtained from InfoMagic. They have the complete set of Internet, CCITT, IEEE, ANSI, and ISO standards on CD. They even have the complete set of RFCs in Hypertext. Their phone number is 1-602-526-9565, e-mail: info@infomagic.com. Get their catalog!

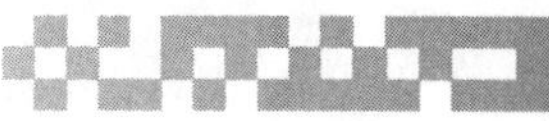

B

Glossary of terms

The following is from RFC 1242, and is useful for those who need a standard set of definitions for performance metrics and related terms.

Network Working Group
S. Bradner, Editor
Request for Comments: 1242
Harvard University
July 1991

Benchmarking terminology for network interconnection devices

Status of this memo

This memo provides information for the Internet community. It does not specify an Internet standard. Distribution of this memo is unlimited.

Abstract

This memo discusses and defines a number of terms that are used in describing performance benchmarking tests and the results of such tests. The terms defined in this memo will be used in additional

memos to define specific benchmarking tests and the suggested format to be used in reporting the results of each of the tests. This memo is a product of the Benchmarking Methodology Working Group (BMWG) of the Internet Engineering Task Force (IETF).

1. Introduction

Vendors often engage in "specsmanship" in an attempt to give their products a better position in the marketplace. This usually involves much "smoke & mirrors" used to confuse the user. This memo and follow-up memos attempt to define a specific set of terminology and tests that vendors can use to measure and report the performance characteristics of network devices. This will provide the user comparable data from different vendors with which to evaluate these devices.

2. Definition format

Term to be defined. (e.g., Latency)

Definition: The specific definition for the term.

Discussion: A brief discussion about the term, its application and any restrictions on measurement procedures.

Measurement units: The units used to report measurements of this term, if applicable.

Issues: List of issues or conditions that affect this term.

See Also: List of other terms that are relevant to the discussion of this term.

3. Term definitions

✳ 3.1 Back-to-back

Definition: Fixed length frames presented at a rate such that there is the minimum legal separation for a given medium between frames over a short to medium period of time, starting from an idle state.

Discussion: A growing number of devices on a network can produce bursts of back-to-back frames. Remote disk servers using protocols like NFS, remote disk backup systems like rdump, and remote tape access systems can be configured such that a single request can result in a block of data being returned of as much as 64K octets. Over networks like ethernet with a relatively small MTU this results in many fragments to be transmitted. Since fragment reassembly will only be attempted if all fragments have been received, the loss of even one fragment because of the failure of some intermediate network device to process enough continuous frames can cause an endless loop as the sender repetitively attempts to send its large data block.

With the increasing size of the Internet, routing updates can span many frames, with modern routers able to transmit very quickly. Missing frames of routing information can produce false indications of unreachability. Tests of this parameter are intended to determine the extent of data buffering in the device.

Measurement units: Number of N-octet frames in burst.

Issues:

See Also:

✻ 3.2 Bridge

Definition: A system which forwards data frames based on information in the data link layer.

Discussion:

Measurement units: n/a

Issues:

See Also: Bridge/router (3.3); Router (3.15)

✻ 3.3 Bridge/router

Definition: A bridge/router is a network device that can selectively function as a router and/or a bridge based on the protocol of a specific frame.

Discussion:

Measurement units: n/a

Issues:

See Also: Bridge (3.2); Router (3.15)

⁜ 3.4 Constant load

Definition: Fixed length frames at a fixed interval of time.

Discussion: Although it is rare, to say the least, to encounter a steady state load on a network device in the real world, measurement of steady state performance may be useful in evaluating competing devices. The frame size is specified and constant. All device parameters are constant. When there is a checksum in the frame, it must be verified.

Measurement units: n/a

Issues: unidirectional vs. bidirectional

See Also:

⁜ 3.5 Data link frame size

Definition: The number of octets in the frame from the first octet following the preamble to the end of the FCS, if present, or to the last octet of the data if there is no FCS.

Discussion: There is much confusion in reporting the frame sizes used in testing network devices or network measurement. Some authors include the checksum, some do not. This is a specific definition for use in this and subsequent memos.

Measurement units: octets

Issues:

See Also:

✻ 3.6 Frame loss rate

Definition: Percentage of frames that should have been forwarded by a network device under steady state (constant) load that were not forwarded due to lack of resources.

Discussion: This measurement can be used in reporting the performance of a network device in an overloaded state. This can be a useful indication of how a device would perform under pathological network conditions such as broadcast storms.

Measurement units: Percentage of N-octet offered frames that are dropped. To be reported as a graph of offered load vs frame loss.

Issues:

See Also: Overhead behavior (3.11); Policy based filtering (3.13); MTU mismatch behavior (3.10)

✻ 3.7 Inter frame gap

Definition: The delay from the end of a data link frame as defined in section 3.5, to the start of the preamble of the next data link frame.

Discussion: There is much confusion in reporting the between frame time used in testing network devices. This is a specific definition for use in this and subsequent memos.

Measurement units: Time with fine enough units to distinguish between 2 events.

Issues: Link data rate.

See Also:

✻ 3.8 Latency

Definition: For store and forward devices: The time interval starting when the last bit of the input frame reaches the input port and ending when the first bit of the output frame is seen on the output port.

For bit forwarding devices: The time interval starting when the end of the first bit of the input frame reaches the input port and ending

when the start of the first bit of the output frame is seen on the output port.

Discussion: Variability of latency can be a problem. Some protocols are timing dependent (e.g., LAT and IPX). Future applications are likely to be sensitive to network latency. Increased device delay can reduce the useful diameter of net. It is desired to eliminate the effect of the data rate on the latency measurement. This measurement should only reflect the actual within-device latency. Measurements should be taken for a spectrum of frame sizes without changing the device setup.

Ideally, the measurements for all devices would be from the first actual bit of the frame after the preamble. Theoretically a vendor could design a device that normally would be considered a store and forward device, a bridge for example, that begins transmitting a frame before it is fully received. This type of device is known as a "cut through" device. The assumption is that the device would somehow invalidate the partially transmitted frame if in receiving the remainder of the input frame, something came up that the frame or this specific forwarding of it was in error. For example, a bad checksum. In this case, the device would still be considered a store and forward device and the latency would still be from last bit in to first bit out, even though the value would be negative. The intent is to treat the device as a unit without regard to the internal structure.

Measurement units: Time with fine enough units to distinguish between 2 events.

Issues:

See Also: Link speed mismatch (3.9); Constant load (3.4): Back-to-back (3.1); Policy based filtering (3.13); Single frame behavior (3.16)

✻ 3.9 Link Speed Mismatch

Definition: Speed mismatch between input and output data rates.

Discussion: This does not refer to frame rate per se, it refers to the actual data rate of the data path. For example, an Ethernet on one side and a 56KB serial link on the other. This has also been referred

to as the "fire hose effect." Networks that make use of serial links between local high speed networks will usually have link speed mismatch at each end of the serial links.

Measurement units: Ratio of input and output data rates.

Issues:

See Also: Constant load (3.4); Back-to-back (3.1)

✻ 3.10 MTU-mismatch behavior

Definition: The network MTU (Maximum Transmission Unit) of the output network is smaller than the MTU of the input network, this results in fragmentation.

Discussion: The performance of network devices can be significantly affected by having to fragment frames.

Measurement units: Description of behavior.

Issues:

See Also:

✻ 3.11 Overhead behavior

Definition: Processing done other than that for normal data frames.

Discussion: Network devices perform many functions in addition to forwarding frames. These tasks range from internal hardware testing to the processing of routing information and responding to network management requests. It is useful to know what the effect of these sorts of tasks is on the device performance. An example would be if a router were to suspend forwarding or accepting frames during the processing of large routing update for a complex protocol like OSPF. It would be good to know of this sort of behavior.

Measurement units: Any quantitative understanding of this behavior is by the determination of its effect on other measurements.

Issues: Bridging and routing protocols; control processing; icmp; ip options processing; fragmentation; error processing; event logging/statistics collection; arp

See Also: policy based filtering (3.13)

⁕ 3.12 Overloaded behavior

Definition: When demand exceeds available system resources.

Discussion: Devices in an overloaded state will lose frames. The device might lose frames that contain routing or configuration information. An overloaded state is assumed when there is any frame loss.

Measurement units: Description of behavior of device in any overloaded states for both input and output overload conditions.

Issues: How well does the device recover from overloaded state? How does source quench production affect device? What does device do when its resources are exhausted? What is response to system management in overloaded state?

See Also:

⁕ 3.13 Policy based filtering

Definition: Filtering is the process of discarding received frames by administrative decision where normal operation would be to forward them.

Discussion: Many network devices have the ability to be configured to discard frames based on a number of criteria. These criteria can range from simple source or destination addresses to examining specific fields in the data frame itself. Configuring many network devices to perform filtering operations impacts the throughput of the device.

Measurement units: n/a

Issues: Flexibility of filter options; number of filter conditions

See Also:

✻ 3.14 Restart behavior

Definition: Reinitialization of system causing data loss.

Discussion: During a period of time after a power up or reset, network devices do not accept and forward frames. The duration of this period of unavailability can be useful in evaluating devices. In addition, some network devices require some form of reset when specific setup variables are modified. If the reset period were long it might discourage network managers from modifying these variables on production networks.

Measurement units: Description of device behavior under various restart conditions.

Issues: Types: power on; reload software image; flush port; reset buffers; restart current code image without reconfiguration

Under what conditions is a restart required? Does the device know when restart needed (i.e., hung state timeout)? Does the device recognize condition of too frequent auto-restart? Does the device run diagnostics on all or some resets? How may restart be initiated? Physical intervention remote via terminal line or login over network

See Also:

✻ 3.15 Router

Definition: A system which forwards data frames based on information in the network layer.

Discussion: This implies running the network level protocol routing algorithm and performing whatever actions that the protocol requires. For example, decrementing the TTL field in the TCP/IP header.

Measurement units: n/a

Issues:

See Also: Bridge (3.2); Bridge/router (3.3)

✻ 3.16 Single frame behavior

Definition: One frame received on the input to a device.

Discussion: A data stream consisting of a single frame can require a network device to do a lot of processing. Figuring routes, performing ARPs, checking permissions, etc., in general, setting up cache entries. Devices will often take much more time to process a single frame presented in isolation than they would if the same frame were part of a steady stream. There is a worry that some devices would even discard a single frame as part of the cache setup procedure under the assumption that the frame is only the first of many.

Measurement units: Description of the behavior of the device.

Issues:

See Also: Policy based filtering (3.13)

✻ 3.17 Throughput

Definition: The maximum rate at which none of the offered frames are dropped by the device.

Discussion: The throughput figure allows vendors to report a single value which has proven to have use in the marketplace. Since even the loss of one frame in a data stream can cause significant delays while waiting for the higher level protocols to time out, it is useful to know the actual maximum data rate that the device can support. Measurements should be taken over an assortment of frame sizes. Separate measurements for routed and bridged data in those devices that can support both. If there is a checksum in the received frame, full checksum processing must be done.

Measurement units: N-octet input frames per second, input bits per second

Issues: Single path vs. aggregate load; unidirectional vs. bidirectional; checksum processing required on some protocols

See Also: Frame loss rate (3.6); Constant load (3.4); Back-to-back (3.1)

4. Acknowledgements

This memo is a product of the IETF BMWG working group:

Chet Birger, Coral Networks; Scott Bradner, Harvard University (chair); Steve Butterfield, independent consultant; Frank Chui, TRW; Phill Gross, CNRI; Steve Knowles, FTP Software, Inc.; Mat Lew, TRW; Gary Malkin, FTP Software, Inc.; K.K. Ramakrishnan, Digital Equipment Corp.; Mick Scully, Ungerman Bass; William M. Seifert, Wellfleet Communications Corp.; John Shriver, Proteon, Inc.; Dick Sterry, Microcom; Geof Stone, Network Systems Corp.; Geoff Thompson, SynOptics; Mary Youssef, IBM.

Security Considerations

Security issues are not discussed in this memo.

Author's Address

Scott Bradner
Harvard University
William James Hall
1232 33 Kirkland Street
Cambridge, MA 02138
Phone: (617) 495-3864

EMail: SOB@HARVARD.HARVARD.EDU

Or, send comments to: bmwg@harvisr.harvard.edu.

Here are some additional terms to help keep track of the alphabet soup.

AFS: Andrew File System
API: Application Programming Interface
APPC: Advanced Program-to-Program Communication
APPN: Advanced Peer-to-Peer Networking

ASN.1: Abstract Syntax Notation One
ATM: Asynchronous Transfer Mode
BER: Basic Encoding Rules
B-ISDN: Broadband ISDN (see ISDN)
BLOB: Binary Large OBject
BSD: Berkeley Software Distribution
CASE: Computer Aided Software Engineering
CDE: Common Development Environment
CICS: Customer Information & Control System
CIL: Components Integration Laboratories
CLI: Call Level Interface
CMIP: Common Management Information Protocol
CMIS: Common Management Information Services
CMOT: CMIP over TCP/IP
COM: Common Object Module
CORBA: Common Object Request Broker Architecture
COSE: Common Open Software Environment
CPI-C: Common Programming Interface for Communications
CUA: Common User Access
DAP: Directory Access Protocol
DBA: Database Administrator
DCE: Distributed Computing Environment
DDL: Data Definition Language
DES: Data Encryption Standard
DFS: Distributed File System
DME: Distributed Management Environment
DNS: Domain Name System
DOE: Distributed Objects Everywhere
DPMI: DOS Protected Mode Interface
DRDA: Distributed Relational Database Architecture
DSA: Directory Service Agent
DSOM: Distributed System Object Model
DSP: Directory System Protocol
DUA: Directory User Agent
EDI: Electronic Data Interchange
FAP: Formats and Protocols
FAT: File Allocation Table
FDDI: Fiber Distributed Data Interface
FTAM: File Transfer Access and Management
FTP: File Transfer Protocol

GDS: Global Directory Service
GUI: Graphical User Interface
HPFS: High Performance File System
ICMP: Internet Control Message Protocol
IDAPI: Integrated Database Application Programming Interface
IDL: Interface Definition Language
IETF: Internet Engineering Taskforce
IP: Internet Protocol
IPX: Internet Packet eXchange
ISDN: Integrated Services Digital Network
LFS: Local File System
LLC: Logical Link Control
MAC: Media Access Control
MAPI: Messaging API
MIB: Management Information Base
MTBF: Mean Time Between Failure
MTTR: Mean Time to Repair
NCP: Netware Core Protocols
NDIS: Network Driver Interface Specification
NFS: Network File System
NLM: Netware Loadable Modules
NMS: Network Management System
NOS: Network Operating System
NTFS: Windows NT File System
ODBC: Open DataBase Connectivity
OLE: Object Linking and Embedding
OLTP: On-Line Transaction Processing
OMG: Object Management Group
OML: Object Manipulation Language
ONC: Open Network Computing
OOUI: Object Oriented User Interface
ORB: Object Request Broker
OSF: Open Software Foundation
OSI: Open Systems Interconnect
POSIX: Portable Operating System Interface
POTS: Plain Old Telephone System
RFC: Request for Comment
RFP: Request for Proposal
RIP: Routing Information Protocol
RMON: Remote network MONitoring

RPC: Remote Procedure Call
SFT: Software Fault Tolerant
SLIP: Serial Line Internet Protocol
SMB: Server Message Block
SMDS: Switched Multi-megabit Data Services
SMI: Structure of Management Information
SMP: Symmetric MultiProcessing
SMTP: Simple Mail Transfer Protocol
SNA: System Network Architecture
SNMP: Simple Network Management Protocol
SOM: System Object Model
SPX: Sequenced Packet eXchange
TFTP: Trivial File Transfer Protocol
TP: Transaction Processing
USL: Unix System Lab
VIM: Vendor Independent Messaging

C

Answers to concepts reviews

Chapter 1

1. What are some benefits of a well-integrated network? *Purchasing flexibility, transparency, information flow*
2. What are some current problems in the definition of terminology in multivendor networks? *Definition of standards, competing models, too many TLAs*
3. What characterizes the vendor's view of multivendor networks? *Products, profits*

   What characterizes the customer's view of multivendor networks? *Problems, solutions*
4. What are the two components of integration? *Interoperability and internetworking*
5. Define the following terms:
   Internetworking: *Getting data across the net*
   Interoperability: *Network applications working together across the net*
   Subnet: *Smallest network addressable unit*
   Internet: *Collection of subnets*
   Protocol: *Specific rules for data communications*
   Model: *Conceptual view of network protocols*

6. Name some subnet protocols. *SDLC, 802.3-Ethernet, 802.5 Token Ring*
7. Name the seven layers of the OSI Model. *Physical, data link, network, transport, session, presentation, application*
8. Name some end-to-end service protocols. *TCPOs IP, ISO IP, X.25, IPX*
9. Name some components of internetworking. *Repeaters, bridges, routers, gateways*
10. Name the major components or areas of interoperability. *Remote terminal access, remote procedure execution, file transfer, distributed file systems, PC network operating system, e-mail, database access*

Chapter 2

1. In today's multivendor networks, what is a key activity of an integrator? *Choosing a blueprint, and blueprint components, for integration*
2. Name three major network protocol suites that a network integrator must deal with in today's networks, and the major promoters and developers. *TCP/IP Users (Internet), OSI ISO (government), SNA (IBM)*
3. Name two popular LAN protocols that are used in today's PC networks, and name the major promoters and developers. *802.3-Ethernet (DEC, Intel, Xerox), 802.5 Token Ring (IBM)*
4. Name the major components of a successful integration blueprint. *Applications, APIs, common set of services, common interfaces, transport services, subnet services*
5. Name three successful protocol integration strategies. *Standard protocol backbone, single-protocol solution, multiprotocol stacks at servers and workstations*

Chapter 3

1. What is the major problem facing the integrator in the area of internetworking? *Big mix of subnetworks, must deal with many different protocols*
2. What are the major integration components for internetworking? *Repeater, bridge, router, gateway*
3. What addresses do bridges use to forward frames? *Destination physical (frame) address*
4. What addresses do routers use to forward packets? *Destination network (NET ID) address in packet*
5. Name two types of routing protocols used by bridges. *802.3 (TST), 802.5 (source routed)*
6. Bridges that are based on geographic coverage are called what? *Local, remote*
7. What component must a protocol suite have for routers to be a viable integration technology with that suite? *A network layer*
8. Why are routers, rather than bridges, generally the preferred integration component? *Scalability, higher performance, better use of LAN and WAN bandwidth*
9. Name some examples of gateways. *SNA, X.25, e-mail*
10. When are gateways an applicable integration solution? *When there is a need to translate protocols*

Chapter 4

Figure C-1

- Can you identify the one category in which these protocols/programs belong?

	Transport/ network protocol	Remote program execution	File transfer	Database access	Remote terminal access	DFS	PC NOS	E-mail	Network management	Directory services
TCP	■									
NetWare							■			
FTP			■							
NFS						■				
X.500										■
X.25	■									
TELNET					■					
ROSE		■								
DNS										■
TP4	■									
APPC	■									
DB2				■						
TFTP			■							
NCP						■				
SMB						■				
NetBIOS	■									
Advanced Server							■			
X.400								■		
Sockets		■								
UDP	■									
IPX	■									
VT					■					
SNMP									■	
CMIP									■	
SMTP								■		
POP3								■		
PROFS								■		

Discover the acronyms and what they mean. Learning Tree International

Chapter 5

1. What are some of the most popular client workstation platforms, by operating system type? *DOS, DOS with Windows, Windows NT, OS/2, Mac, UNIX*
2. What server and host types do these workstations typically connect to? *Novell NetWare Server, Microsoft Advanced Server, LAN Manager Server, IBM LAN Server, UNIX hosts, DEC PATHWORKS server, DEC VAX, IBM mainframes, AS/400*

Figure C-2

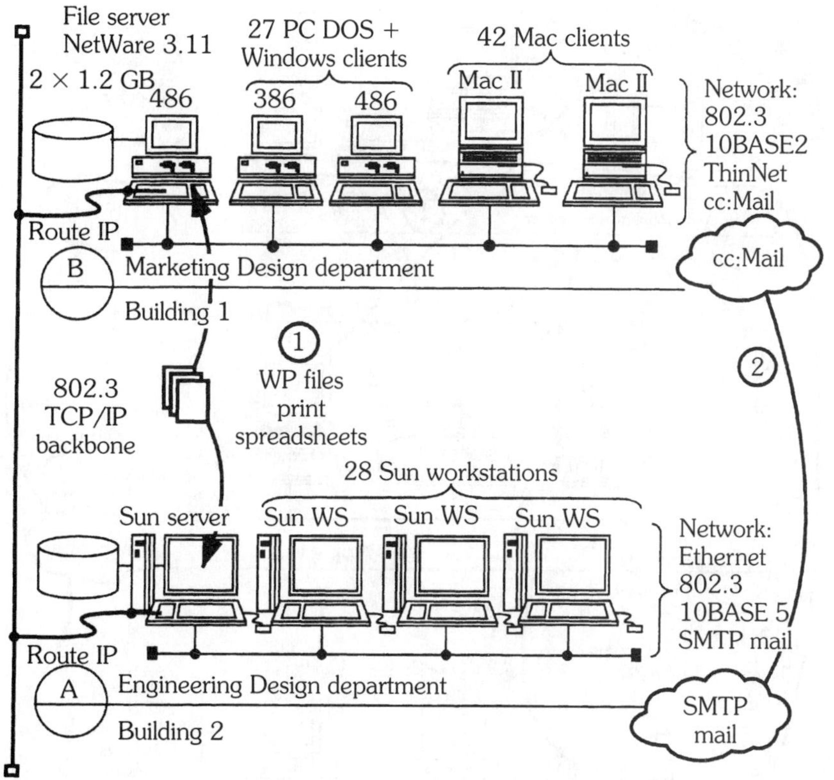

One possible solution to the requirements of Workshop 1. Learning Tree International

Proposed interoperable solution:
cc:Mail gateway SMTP/cc:Mail
NFS on NetWare server

Deposit all files on NetWare server or FTP
NetWare print GW between NetWare UNIX

Proposed changes to internetworking:
Connect NetWare server to UNIX Ethernet via separate NIC, load TCP/IP on NetWare server

Proposed interoperable solution:
FTP for file transfer from Macs
TELNET for Macs and PCs
MS MailDSMTP gateway from Microsoft

Figure C-3

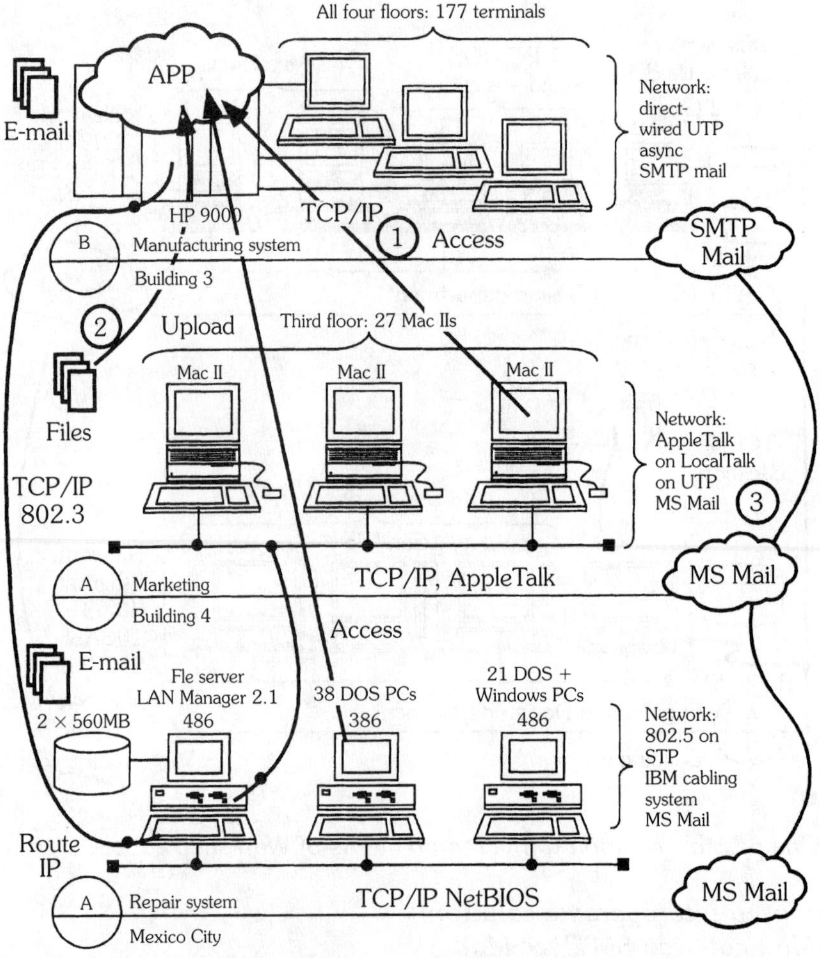

One possible solution to the requirements of Workshop 2. Learning Tree International

Run TCP/IP on PCs and Macs, NDIS, or Packet Drivers
TCP/IP on the HP host

Proposed changes to internetworking:
Macs on Ethernet, connect to LAN MAN server
Ethernet adapter on HP, connect to LAN MAN server
Let LAN Manager server an IP router, better yet, change the LAN Manager server to Win NT and Advanced Server

3. When dealing with a workstation requirement to employ multiple protocols simultaneously, what are the most popular interoperable solutions and the vendors or organizations that promote them? *NDIS, ODI, Packet drivers: Microsoft, Novell, Public domain*
4. What are some solutions to the problem of proliferation of user login and passwords as the degree of connectivity of a workstation increases? *UNIX trusted hosts, LAN Manager domain login, Banyan ENS*

Chapter 6

1. What does DCE stand for? *Distributed Computing Environment, a set of vendor-independent interoperable set of APIs for distributed computing.*
2. What problem is DCE attempting to solve? *There are too many competing standards for distributed computing.*
3. How is DCE solving the problem? *One organization (OSF) has source code for pieces and sells to all vendors.*
4. Who is behind DCE and why? *IBM, HP, DEC, Sun, Transarc*
5. Name the five client/server frameworks, and a major characteristic of each.
   Presentation on client X Windows
   Wallpaper: GUI on WS, old application on MF
   Split application: New
   File server: Data on file-server files
   Distributed database: SQL on backend, database server
6. Name three sizing trends and give an example of each.
   Downsizing: Moving from MF to PC networks
   Upsizing: Increasing size and capacity of file servers
   Rightsizing: Designing the correct platform for new applications

Chapter 7

1. What are the five major network-management concerns according to the ISO model? *Fault management, security, accounting, configuration, performance*
2. What three service areas are managed by the network control center (NCC)? *Administration, information, operations*
3. What are the two major network-management protocols and the protocol suites in which they belong? *SNMP (TCP/IP), CMIP (OSI)*
4. What does SNMP mean? *Simple Network Management Protocol*
5. What does MIB mean, and what is it? *Management Information Base, a collection of SNMP-managed objects on which we collect information*
6. Indicate the components of an SNMP network-management system and the network-management paradigm on the following network drawing (Fig. C-4).

Figure C-4

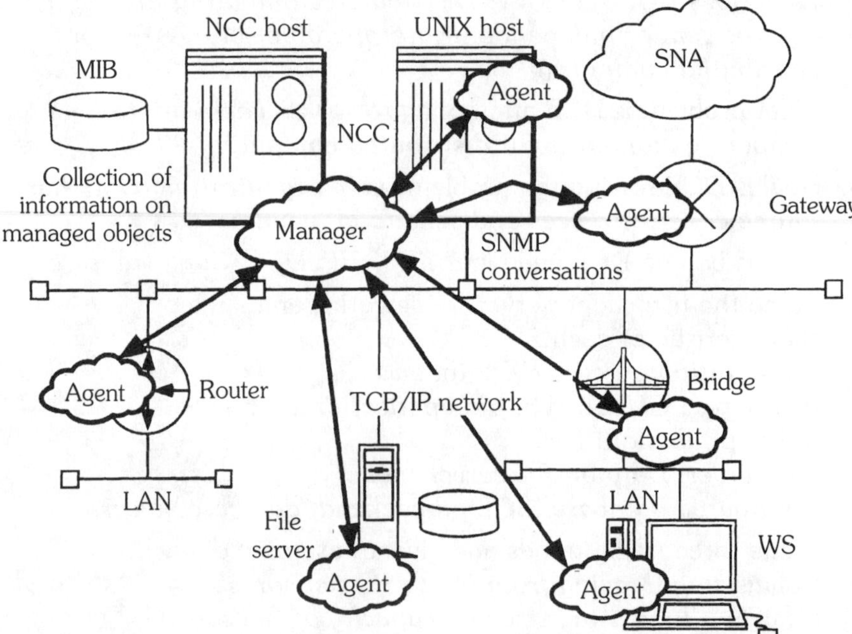

An SNMP network management system and its components. Learning Tree International

D

Annotated bibliography

Basic books

Comer, Douglas, *Internetworking with TCP/IP*, Vol. I, 2nd ed., Prentice-Hall, Englewood Cliffs, NJ, 1991.
Douglas Comer provides an overview of TCP/IP. The main emphasis of this book is on transport and network services offered by the protocol. Addressing, routing, and internet protocols are covered. The section on application services is weak, but provides a reasonable introduction. This book provides a good starting point, but the reader will have to use Comer's reading list to delve deeper into the subject.

Orfali, Robert and Dan Harkey, *The Client/Server Survival Guide with OS/2*, Van Nostrand Reinhold, New York, NY, 1994.
Jerry's favorite reference text. A comprehensive look at all the technologies required to build integrated networks. The book covers everything from operating systems to object technology. All the major standards are examined. Orfali and Harkey bring their own unique view to the subject. For those who do not require the OS/2 related topics, they have written another book called *The Essential Client/Server Survival Guide*, which contains much of the same material.

Rose, Marshall T., *The Open Book, A Practical Perspective on OSI*, Prentice-Hall, Englewood Cliffs, NJ, 1990.

Rose introduces OSI protocols. The book examines each layer of the OSI Model and details the relevant OSI protocols. Of particular interest is the amount of material on the application layers (Session, Presentation, and Application). The book also contains a major section on making the transition to OSI. For those contemplating a move to OSI (or for background information), the book is recommended. If your OSI plans are more far-reaching, you may want to wait for the second edition.

Rose, Marshall T., *The Simple Book: An Introduction to Management of TCP/IP-Based Internets*, Prentice-Hall, Englewood Cliffs, NJ, 1991.

Marshall Rose gives an excellent introduction to TCP/IP-based network management. The book is well organized, with emphasis on both the MIB and the SNMP primitives. The MIB information will also interest those contemplating CMIP-managed networks, due to the similarity between the protocols. The first few chapters review TCP/IP basics. Rose also gives good historical information on the development of SNMP.

Tanenbaum, Andrew S., *Computer Networks*, 2d ed., Prentice-Hall, Englewood Cliffs, NJ, 1988.

Considered a reference text by many, Andrew Tanenbaum's book is used by many universities. Using the OSI Model, Tanenbaum examines most of the major (and many of the minor) data communication protocols in use today. While some of the information is not current, the book is still an excellent source of information. One strong point is that Tanenbaum does not assume that the reader has a degree in electrical engineering. The index and reading list are both comprehensive.

General multivendor network references

Perlman, Radia, *Interconnections, Bridges, and Routers*, Addison-Wesley, Reading, MA, 1992.

One of the best books for a detailed examination of internetworking technologies. This book gives all the information required to make informed choices between bridge and router technology. Recommended for network designers.

Madron, Thomas W., *Enterprisewide Computing: How to Implement and Manage LANs*, John Wiley and Sons, New York, 1991.
A book that covers the micro-mainframe link heavily, but has very good insights for the enterprisewide computing environment. The focus is IBM mainframe connectivity, OSI, and TCP/IP. It treats PC LANs as connectivity tools for the big iron.

IBM SNA networking

Ranade, Jay and George C. Sackett, *Introduction to SNA Networking: A Guide to Using VTAM and SNA*, McGraw-Hill, New York, 1989.
The first three chapters are a great review of the hardware and software components of the IBM SNA environment. The rest of the book is for those who are looking for in-depth VTAM and NCP help.

DEC DECnet

Malamud, Carl, *DEC Networks and Architectures*, McGraw-Hill, New York, 1989.
A comprehensive look at DECnet, including Phase V. Somewhat hard to follow.

UNIX and TCP/IP

Comer, Douglas, *Internetworking with TCP/IP*, Vol. I, 2nd ed., Prentice-Hall, Englewood Cliffs, NJ, 1991.
A comprehensive introduction to TCP/IP. See Basic Books.

Kochan, Stephen G. and Patrick H. Wood, *UNIX Networking*, Hayden Books, Carmel, IN, 1989.
A collection of articles on UNIX applications and application interfaces dealing with networking, including LAN Manager on UNIX and NFS. Not for the faint of heart.

Rose, Marshall T., *The Simple Book: An Introduction to Management of TCP/IP-Based Internets*, Prentice-Hall, Englewood Cliffs, NJ, 1991.

Good review of SNMP. Gives tongue-in-cheek understanding of the struggle between TCP/IP and OSI camps. Some parts are very technical.

X.25 packet networks

Schlar, Sherman K., *Inside X.25: A Manager's Guide*, McGraw-Hill, New York, 1990.
An excellent review of the X.25 and related protocols. Also gives a very good review of the industry and hardware vendors, plus an overview of how to purchase X.25 services.

Local area networks

Martin, James and Kathleen K. Chapman, *Local Area Networks*, Prentice-Hall, Englewood Cliffs, NJ, 1989.
LAN technology and the IEEE 802 protocols are explained well. Also includes an explanation of how SNA works. A good overall reference.

OSI protocols

Rose, Marshall T., *The Open Book, A Practical Perspective on OSI*, Prentice-Hall, Englewood Cliffs, NJ, 1990.

Magazines and trade newspapers

Data Communications, McGraw-Hill, New York.
Monthly technical magazine with well-written, in-depth articles. Subscription cost: $38/year. Well worth looking at past issues from the last two years to review latest technology of multivendor networks.

Network World, Network World, Inc., Framingham, MA
Free weekly newspaper to help keep up with the latest developments in the multivendor networking world.

Index

A

Illustrations are indicated in **boldface.**

D

E

G

H

I

N

S

X

Z

About the authors

Andres Fortino is an electrical engineer, computer consultant, author, and seminar leader. Fortino teaches courses on networking topics at both The Learning Tree International and the University of California at Berkeley. He holds a Ph.D. in electrical engineering from CUNY and holds several certifications, including Master Electrician, Certified NetWare Engineer (CNE), and Professional Engineer (PE). He is the author of Several other books, including *Handbook of Computer Network Protection* (1994) and *The PC Network Administrator's Tool Kit* (1995).

Jerry Golick is a network consultant, author, seminar leader, and columnist. Golick teaches high technology courses for The Learning Tree International and has consulted for such organizations as the Governor's Office of the State of Florida and Bell Canada. He is a contributing columnist for Computing Canada.

CD-ROM WARRANTY

This software is protected by both United States copyright law and international copyright treaty provision. You must treat this software just like a book. By saying "just like a book," McGraw-Hill means, for example, that this software may be used by any number of people and may be freely moved from one computer location to another, so long as there is no possibility of its being used at one location or on one computer while it also is being used at another. Just as a book cannot be read by two different people in two different places at the same time, neither can the software be used by two different people in two different places at the same time (unless, of course, McGraw-Hill's copyright is being violated).

LIMITED WARRANTY

McGraw-Hill takes great care to provide you with top-quality software, thoroughly checked to prevent virus infections. McGraw-Hill warrants the physical diskette(s) contained herein to be free of defects in materials and workmanship for a period of sixty days from the purchase date. If McGraw-Hill receives written notification within the warranty period of defects in materials or workmanship, and such notification is determined by McGraw-Hill to be correct, McGraw-Hill will replace the defective diskette(s). Send requests to:

McGraw-Hill, Inc.
Customer Services
P.O. Box 545
Blacklick, OH 43004-0545

The entire and exclusive liability and remedy for breach of this Limited Warranty shall be limited to replacement of defective diskette(s) and shall not include or extend to any claim for or right to cover any other damages, including but not limited to, loss of profit, data, or use of the software, or special, incidental, or consequential damages or other similar claims, even if McGraw-Hill has been specifically advised of the possibility of such damages. In no event will McGraw-Hill's liability for any damages to you or any other person ever exceed the lower of suggested list price or actual price paid for the license to use the software, regardless of any form of the claim.

McGRAW-HILL, INC. SPECIFICALLY DISCLAIMS ALL OTHER WARRANTIES, EXPRESS OR IMPLIED, INCLUDING, BUT NOT LIMITED TO, ANY IMPLIED WARRANTY OF MERCHANTABILITY OR FITNESS FOR A PARTICULAR PURPOSE.

Specifically, McGraw-Hill makes no representation or warranty that the software is fit for any particular purpose and any implied warranty of merchantability is limited to the sixty-day duration of the Limited Warranty covering the physical CD-ROM only (and not the software) and is otherwise expressly and specifically disclaimed.

This limited warranty gives you specific legal rights; you may have others which may vary from state to state. Some states do not allow the exclusion of incidental or consequential damages, or the limitation on how long an implied warranty lasts, so some of the above may not apply to you.